Photocatalytic Nanomaterials for Environmental Purification

Photocatalytic Nanomaterials for Environmental Purification

Guest Editor

Xiaowang Lu

MDPI

Basel • Beijing • Wuhan • Barcelona • Belgrade • Novi Sad • Cluj • Manchester

Guest Editor
Xiaowang Lu
School of Material Science and Engineering
Yancheng Institute of Technology
Yancheng
China

Editorial Office
MDPI AG
Grosspeteranlage 5
4052 Basel, Switzerland

This is a reprint of the Special Issue, published open access by the journal *Catalysts* (ISSN 2073-4344), freely accessible at: https://www.mdpi.com/journal/catalysts/special_issues/N5DQ0M4R0X.

For citation purposes, cite each article independently as indicated on the article page online and as indicated below:

Lastname, A.A.; Lastname, B.B. Article Title. *Journal Name* **Year**, *Volume Number*, Page Range.

ISBN 978-3-7258-6674-8 (Hbk)
ISBN 978-3-7258-6675-5 (PDF)
https://doi.org/10.3390/books978-3-7258-6675-5

Contents

About the Editor

Xiaowang Lu

Xiaowang Lu currently serves as a Senior Experimentalist at the School of Materials Science and Engineering, Yancheng Institute of Technology. He has long been dedicated to design, synthesis, and application research of photocatalytic nanomaterials in the field of environmental purification. His main research directions include the construction of semiconductor heterostructures, development of visible-light-responsive catalytic materials, investigation into pollutant degradation mechanisms, and catalytic conversion technologies for new energy applications. In recent years, he has led and participated in multiple national- and provincial-level research projects, actively promoting the practical application of photocatalytic materials in areas such as wastewater treatment, air purification, and energy conversion. Xiaowang Lu has published over forty academic papers in internationally renowned journals, and he also serves as a reviewer for several journals in the fields of environmental science and materials. Through surface/interface engineering and structural optimization of materials, his research team has successfully developed various efficient and stable composite photocatalytic systems, providing promising material-based solutions for environmental pollution control.

catalysts

MDPI

Editorial

Photocatalytic Nanomaterials for Environmental Purification

Xiaowang Lu

School of Material Science and Engineering, Yancheng Institute of Technology, Yancheng 224051, China; luxiaowang@ycit.edu.cn

The accelerating pace of industrialization and continuous population growth have made environmental pollution, particularly of water and air, a severe challenge to global ecosystems and human health. Conventional water treatment and air purification methods often suffer from limitations such as limited efficiency, high energy consumption, or the generation of secondary pollution. Consequently, the development of efficient, sustainable, and environmentally friendly purification technologies is urgently needed [1–3].

Among various strategies developed to address environmental pollution, photocatalytic technology has emerged as a robust and promising solution for water and air purification. This technology utilizes light energy to drive redox reactions that degrade harmful pollutants into harmless byproducts, thereby offering a green, efficient, and sustainable pathway for environmental remediation [4–6]. In the field of environmental purification applications utilizing solar energy, nanostructured photocatalysts have attracted increasing research attention. Various photocatalytic nanomaterials demonstrate superior photocatalytic activity compared to their bulk counterparts, owing to their enhanced interfacial charge separation efficiency and larger surface area (which provides more active sites) [7,8].

This Special Issue, Photocatalytic Nanomaterials for Environmental Purification, brings together a collection of original research articles that highlight recent advances in the synthesis, characterization, and application of nanoscale photocatalysts for environmental purification. The contributions reflect the diversity and depth of ongoing efforts to address real-world pollution challenges through innovative material design and process optimization.

In Contribution 1, the study evaluates UVC photolysis with type C ultraviolet radiation (UVC) and UVC/TiO_2 photocatalysis of a mixture of four pharmaceuticals-atenolol (ATL), acetaminophen (ACM), clofibric acid (CLA), and antipyrine (ANT)-commonly found in treated urban wastewater. A comprehensive kinetic model was developed to describe their degradation, taking into account the generation of reactive oxygen species (ROS): hydroxyl, superoxide ion radicals, and singlet oxygen, along with their reactions with both the pharmaceuticals and dissolved organic matter. The model reproduces the inhibitory effect of natural organic matter in secondary effluent and, in most cases, treated effluent, accurately predicts the concentration profiles of the pharmaceuticals. This validated model provides a useful tool for understanding the degradation mechanisms of pharmaceutical mixtures and for supporting the design of effective water strategies based on photochemical processes.

Zhang Mei and Wang Xu (Contribution 2) present a green and eco-friendly egg white (EW)/TiO_2 hydrogel synthesized using waste eggs nearing expiration, designed as a multifunctional bio-adsorbent for wastewater treatment. The material exhibits a high adsorption capacity of 333.172 $mg \cdot g^{-1}$ for methylene blue (MB) following the Langmuir model. Under visible light irradiation, the 5 wt% TiO_2-doped hydrogel achieves 99% photocatalytic degradation of MB within 120 min. Additionally, it demonstrates significant antibacterial activity, achieving a 90.4% inactivation efficiency against Escherichia coli through synergistic effects of natural lysozymes in EW and reactive oxygen species generated by TiO_2. The hydrogel also shows low cytotoxicity, highlighting its potential as a sustainable, cost-effective, and reusable material for simultaneous organic dye removal and microbial disinfection in industrial wastewater.

In Contribution 3, Zhu et al. conducted a first-principles study to address the critical limitation of monolayer MoS_2—rapid electron–hole recombination—by designing four van der Waals heterojunctions: MoS_2/WSe_2, $MoS_2/MoSe_2$, MoS_2/AlN, and MoS_2/ZnO. All the heterojunction systems exhibit stable Type-II band alignments, which generate internal electric fields that effectively suppress charge recombination and reduce the bandgap by 0.2–0.5 eV compared to pristine MoS_2. Notably, chalcogenide-based heterojunctions (MoS_2/WSe_2 and $MoS_2/MoSe_2$) significantly enhance visible light absorption, with $MoS_2/MoSe_2$ demonstrating the highest photoresponsivity (absorption coefficient > 10^5 cm^{-1}). In contrast, non-chalcogenide systems (MoS_2/AlN and MoS_2/ZnO) exhibit comparatively weaker optical performance. These findings establish $MoS_2/MoSe_2$ as the optimal photocatalyst for optoelectronic applications, demonstrating that strategic pairing of chalcogenide elements can maximize charge separation efficiency and solar energy utilization.

Bobirică et al. (Contribution 4) focused on developing a novel photocatalytic system for the removal of estrogenic pollutants (estradiol valerate/norgestrel) from wastewater. A plug-flow reactor operating under UV-A light was employed, with TiO_2, ZnO, and TiO_2/ZnO composite photocatalysts immobilized on glass beads. These beads were strung on stainless-steel wires and arranged modularly within the reactor. The photocatalysts were synthesized via sol–gel, coated onto the glass supports thermally, and calcined at 500 °C for 2 h. Results showed that degradation efficiency strongly depends on catalyst loading, which can be conveniently tuned by adding or removing photocatalytic modules. The TiO_2/ZnO composite exhibited optimal performance, achieving near-complete mineralization within 120 min using only two modules.

In Contribution 5, Mehta et al. synthesized pure CdO nanoparticles, magnetic Fe_3O_4 nanoparticles, and Fe_3O_4-CdO nanocomposites via a solution combustion method using cetyltrimethylammonium bromide (CTAB) as a template. Under optimized conditions, the photocatalytic performance of the Fe_3O_4-CdO nanocomposites was evaluated for the degradation of methylene blue dye. Under visible light irradiation, the Fe_3O_4-CdO nanostructures exhibited efficient photocatalytic activity, achieving 92% degradation of methylene blue, and demonstrated excellent stability over multiple reuse cycles

In Contribution 6, Dong et al. fabricated Pt-M/WO_3 (M = Cu, Co, Ni) thin films through a process involving homogeneous precursor sol preparation, spin-coating, toluene etching, and calcination. Their microstructural, chemical, and electrochemical properties were systematically compared with those of pure WO_3. The Pt-M/WO_3 films exhibited significantly enhanced photocatalytic degradation of methylene blue (MB) under visible light. Based on experimental evidence, the authors proposed detailed charge transfer pathways and an improved photocatalytic mechanism. The large work function difference between Pt-M alloys and WO_3 induces interfacial band bending, while transition metal doping further adjusts the electronic structure of Pt-M/WO_3, effectively promoting charge carrier separation.

In Contribution 7, the study presented a facile one-pot refluxing strategy to synthesize a 2D/2D $CdIn_2S_4/In_2S_3$ (CISI) heterojunction photocatalyst for efficient visible light-driven Cr(VI) reduction. The intimate interfacial contact between $CdIn_2S_4$ and In_2S_3 nanosheets significantly enhanced charge carrier separation and transfer, effectively suppressing recombination. The optimized 0.5 CISI composite demonstrated superior Cr(VI) photoreduction performance, outperforming its individual components. Mechanistic analysis confirmed

https://doi.org/10.3390/catal16010013

the heterojunction's role in accelerating electron–hole separation and promoting interfacial redox reactions. This work illustrated a scalable approach to designing high-efficiency photocatalysts for toxic heavy metal remediation, highlighting the critical impact of 2D/2D heterostructure engineering on environmental photocatalysis.

In Contribution 8, this study reported the microwave solvothermal synthesis of two-dimensional CeO_2/Bi_2O_3 layered composites for visible light-driven photocatalytic oxidative desulfurization (PODS). Optimized at a 1:2 Ce/Bi molar ratio, the composite achieved the highest desulfurization efficiency for dibenzothiophene (DBT) in fuel. Active species trapping experiments identified hydroxyl radicals (•OH) as the dominant reactive species driving DBT oxidation. A mechanistic pathway was proposed in which the CeO_2/Bi_2O_3 interface facilitated electron transfer under visible light, enabling efficient conversion of DBT to sulfones for deep desulfurization. This work provided a sustainable, hydrogen-free strategy for fuel purification with potential industrial applications.

In Contribution 9, Zhang et al. fabricated isostructural Sillén–Aurivillius oxyhalides, $Bi_7Fe_2Ti_2O_{17}X$ (X = Cl, Br, I; denoted as BFTOX), for the first time for CO_2 photoreduction and degradation of organic pollutants. Density functional theory (DFT) calculations revealed that the valence band maximum (VBM) of BFTOC and BFTOB was dominated by the delocalized 2p orbitals of oxygen atoms, contributing to their narrow bandgap (E_g) and potentially enhancing stability against self-decomposition. The photocatalytic activities of BFTOX materials were significantly influenced by the halogen type (Cl, Br, I), with BFTOCl exhibiting the highest performance (3.74 $\mu mol \cdot g^{-1} \cdot h^{-1}$). This enhancement was attributed to efficient charge carrier separation, a small optical bandgap, and broadened light absorption. This study provided a useful reference for developing efficient and stable photocatalysts based on Sillén–Aurivillius layered oxyhalide materials.

In Contribution 10, Zhou et al. successfully constructed $BiPO_4$/Ov-BiOBr heterojunction composites by a two-step solvothermal method, in which bismuth phosphate ($BiPO_4$) with various molar ratios was in situ grown on the surface of oxygen-vacancy-rich bismuth oxybromide (Ov-BiOBr). By establishing a novel type-I high–low junction structure between semiconductor $BiPO_4$ and Ov-BiOBr, the system effectively retained more strongly oxidizing holes or reducing electrons, thereby enhancing the redox performance of the photocatalyst. Among these, the composite catalyst containing 10 mol% $BiPO_4$ exhibited the highest degradation efficiency for tetracycline (TC) while also demonstrating optimal photocatalytic activity for hydrogen peroxide (H_2O_2) production. The enhanced performance is attributed to suitable band alignment between $BiPO_4$ and Ov-BiOBr, an efficient electron transfer channel formed via bismuth bridges, and effective charge separation.

The studies presented in this Special Issue reflect the vibrant progress in photocatalytic nanomaterials for environmental purification. Through innovative material design, interface engineering, and process optimization, researchers are continuously enhancing the efficiency, stability, and applicability of photocatalysts. Future efforts should focus on scaling up synthesis methods, improving solar energy utilization, and advancing mechanistic understanding to facilitate the real-world deployment of photocatalytic technologies. Interdisciplinary collaboration will be key to addressing complex environmental challenges and achieving sustainable development goals.

Funding: This research received no external funding.

Conflicts of Interest: The author declares no conflicts of interest.

List of Contributions:

1. Jiménez-López, M.A.; Rubio-Gonzaga, M.C.; Beltrán, F.J. Removal from Water of Some Pharmaceuticals by Photolysis and Photocatalysis: Kinetic Models. *Catalysts* **2025**, *15*, 471. https://doi.org/10.3390/catal15050471.

2. Zhang, M.; Wang, X. Green and Eco-Friendly Egg White–TiO_2 Hydrogel with Enhanced Antimicrobial, Adsorptive, and Photocatalytic Properties. *Catalysts* **2024**, *14*, 899. https://doi.org/10.3390/catal14120899.
3. Liu, J.; Jin, Y.; Lei, B.; Zhao, X.; Huang, Y.; Zhang, L.; Zhu, Y. Studies on Electronic Structure and Optical Properties of $MoS_{2/X}$ (X = WSe_2, $MoSe_2$, AlN, and ZnO) Heterojunction by First Principles. *Catalysts* **2024**, *14*, 678. https://doi.org/10.3390/catal14100678.
4. Bobirică, L.; Orbeci, C.; Pîrvu, C.; Constantinescu, A.; Bîru, E.I.; Ionică, G.I.; Matei, E.; Berbecaru, A.C.; Bobirică, C. Modular Photocatalytic Reactor for the Removal of Estrogens from Aqueous Solutions. *Catalysts* **2024**, *14*, 661. https://doi.org/10.3390/catal14100661.
5. Albeladi, A.; Khan, Z.; Al-Thabaiti, S.A.; Patel, R.; Malik, M.A.; Mehta, S. Fe_3O_4-CdO Nanocomposite for Organic Dye Photocatalytic Degradation: Synthesis and Characterization. *Catalysts* **2024**, *14*, 71. https://doi.org/10.3390/catal14010071.
6. Pan, J.; Zhang, L.; Wang, C.; Gao, K.; Dong, P. Electron Regulation in Pt-M (M = Cu, Co, and Ni) Decorated WO_3 Thin Films for Photocatalytic Degradation Performance. *Catalysts* **2023**, *13*, 1044. https://doi.org/10.3390/catal13071044.
7. Hu, J.; Wu, J.; Zhang, S.; Chen, W.; Xiao, W.; Hou, H.; Lu, X.; Liu, C.; Zhang, Q. One-Pot Fabrication of 2D/2D $CdIn_2S_4/In_2S_3$ Heterojunction for Boosting Photocatalytic Cr(VI) Reduction. *Catalysts* **2023**, *13*, 826. https://doi.org/10.3390/catal13050826.
8. Lu, X.; Chen, W.; Hou, H.; Qian, J.; Zhang, Q. Preparation of Two-Dimensional Layered CeO_2/Bi_2O_3 Composites for Efficient Photocatalytic Desulfurization. *Catalysts* **2023**, *13*, 821. https://doi.org/10.3390/catal13050821.
9. Chen, J.; Gu, Y.; Xu, S.; Zhang, Y.; Zhang, Z.; Shi, L.; Mu, Z.; Zhou, C.; Zhang, J.; Zhang, Q. Band Gap Engineering in Quadruple-Layered Sillén–Aurivillius Perovskite Oxychlorides $Bi_7Fe_2Ti_2O_{17}X$ (X = Cl, Br, I) for Enhanced Photocatalytic Performance. *Catalysts* **2023**, *13*, 751. https://doi.org/10.3390/catal13040751.
10. Tang, M.; Li, X.; Deng, F.; Han, L.; Xie, Y.; Huang, J.; Chen, Z.; Feng, Z.; Zhou, Y. $BiPO_4$/Ov-BiOBr High-Low Junctions for Efficient Visible Light Photocatalytic Performance for Tetracycline Degradation and H_2O_2 Production. *Catalysts* **2023**, *13*, 634. https://doi.org/10.3390/catal13030634.

References

1. Chen, F.; Zhang, Y.; Huang, H. Layered photocatalytic nanomaterials for environmental applications. *Chin. Chem. Lett.* **2023**, *34*, 107523. [CrossRef]
2. Singh, K.K.; Singh, A.; Rai, S. A study on nanomaterials for water purification. *Mater. Today Proc.* **2022**, *51*, 1157–1163. [CrossRef]
3. Sivaraman, C.; Vijayalakshmi, S.; Leonard, E.; Sagadevan, S.; Jambulingam, R. Current Developments in the Effective Removal of Environmental Pollutants through Photocatalytic Degradation Using Nanomaterials. *Catalysts* **2022**, *12*, 544. [CrossRef]
4. Lv, H.; Liu, Y.; Bai, Y.; Shi, H.; Zhou, W.; Chen, Y.; Liu, Y.; Yu, D.-G. Recent Combinations of Electrospinning with Photocatalytic Technology for Treating Polluted Water. *Catalysts* **2023**, *13*, 758. [CrossRef]
5. Xie, A.; Jin, M.; Zhu, J.; Zhou, Q.; Fu, L.; Wu, W. Photocatalytic Technologies for Transformation and Degradation of Microplastics in the Environment: Current Achievements and Future Prospects. *Catalysts* **2023**, *13*, 846. [CrossRef]
6. Musial, J.; Mlynarczyk, D.T.; Stanisz, B.J. Photocatalytic degradation of sulfamethoxazole using TiO_2-based materials–Perspectives for the development of a sustainable water treatment technology. *Sci. Total Environ.* **2023**, *856*, 159122. [CrossRef] [PubMed]
7. Khan, K.A.; Shah, A.; Nisar, J.; Haleem, A.; Shah, I. Photocatalytic Degradation of Food and Juices Dyes via Photocatalytic Nanomaterials Synthesized through Green Synthetic Route: A Systematic Review. *Molecules* **2023**, *28*, 4600. [CrossRef] [PubMed]
8. Mishra, K.; Devi, N.; Siwal, S.S. Hybrid semiconductor photocatalyst nanomaterials for energy and environmental applications: Fundamentals, designing, and prospects. *Adv. Sustain. Syst.* **2023**, *7*, 2300095. [CrossRef]

https://doi.org/10.3390/catal16010013

catalysts

MDPI

Article

Removal from Water of Some Pharmaceuticals by Photolysis and Photocatalysis: Kinetic Models

Miguel A. Jiménez-López, María C. Rubio-Gonzaga and Fernando J. Beltrán *

Departamento de Ingeniería Química y Química Física, Instituto Universitario de Investigación del Agua, Cambio Climático y Sostenibilidad (IACYS), Universidad de Extremadura, Avenida de Elvas s/n, 06006 Badajoz, Spain; majimlopez@unex.es (M.A.J.-L.); mrubiowz@alumnos.unex.es (M.C.R.-G.)

* Correspondence: fbeltran@unex.es; Tel.: +34-924-289387

Abstract: Pharmaceutical residues are emerging contaminants of growing concern due to their persistence and poor removal efficiency in conventional wastewater treatment plants. This study evaluates UVC photolysis with type C ultraviolet radiation (UVC) and UVC/TiO_2 photocatalysis of a mixture of four pharmaceuticals—atenolol (ATL), acetaminophen (ACM), clofibric acid (CLA), and antipyrine (ANT)—commonly found in treated urban wastewater. A comprehensive kinetic model was developed to describe their degradation, taking into account the generation of reactive oxygen species (ROS): hydroxyl ($HO^{\bullet}$), superoxide ion ($O_2^{\bullet-}$) radicals, and singlet oxygen ($^1O_2$), along with their reactions with both the pharmaceuticals and dissolved organic matter. Direct quantum yields were determined as 8.05×10^{-3} mol·Einstein^{-1} for ATL, 1.93×10^{-3} for ACM, 3.12×10^{-1} for CLA, and 5.12×10^{-2} for ANT. In addition, rate constants of the reactions between singlet oxygen and pharmaceuticals were 9.93, 1.3×10^6, 1.18×10^2, and 1.14×10^4 $M^{-1} \cdot s^{-1}$ for ATL, ACM, CLA, and ANT, respectively. Scavenger experiments confirmed the key role of the ROS involved. The model reproduces the inhibitory effect of natural organic matter in secondary effluent and, in most cases, treated, accurately predicts the concentration profiles of the pharmaceuticals. Under photocatalytic conditions (0.10 g·L^{-1} TiO_2), all compounds were completely degraded in less than 15 min. This validated model provides a useful tool for understanding the degradation mechanisms of pharmaceutical mixtures and for supporting the design of effective water strategies based on photochemical processes.

Keywords: pharmaceutical contaminants; wastewater secondary effluent; radical oxygen species; UVC photolysis; UVC-TiO_2 photocatalysis; kinetic modelling

1. Introduction

The detection of pharmaceutical residues in aquatic environments has emerged as a matter of growing environmental concern [1–4]. Conventional wastewater treatment processes are often inadequate for removing these contaminants, allowing pharmaceutical substances such as atenolol (ATL), acetaminophen (ACM), clofibric acid (CLA), and antipyrine (ANT) (see Table 1) to remain in aquatic environments. Even though the concentrations of these compounds in the environment are significantly lower than therapeutic doses, it is recognized that some substances can interfere with vital processes in non-target and sensitive organisms, including vulnerable human populations like children and pregnant women [5]. These compounds, widely used in medical treatments, pose serious ecological and public health risks when released into natural water systems.

Table 1. Main physicochemical characteristics of the pharmaceuticals used in this study.

Pharmaceutical	Category	Chemical Formula	Molecular Weight (g·mol^{-1})	Solubility (g·L^{-1})	pKa	Structure
ATL (Atenolol)	β-blocker	$C_{14}H_{22}N_2O_3$	266.34	33.3	9.54	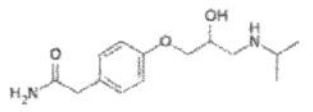
ACM (Acetminophen)	Analgesic Antipyretic	$C_8H_9NO_2$	151.16	14.0	9.50	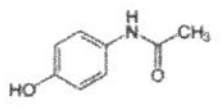
CLA (Clofibric acid)	Lipid regulator	$C_{10}H_{11}ClO_3$	214.65	0.58	2.9	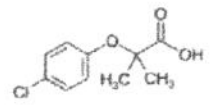
ANT (Antipyrine)	Analgesic	$C_{11}H_{12}N_2O$	188.23	56.0	1.4	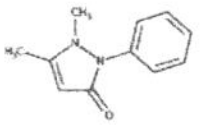

Atenolol, a beta-blocker prescribed for hypertension management, is frequently detected in surface waters and wastewater effluents. Concentrations ranging from 7 to 6897 ng·L^{-1} have been reported, with studies identifying levels up to 2300 ng·L^{-1} in urban water systems [6] or as high as 2346 ng·L^{-1} and 1707 ng·L^{-1} in influents and effluents, respectively [7]. ATL has also been widely detected in hospital sewage and wastewater in concentrations ranging from about 780 ng·L^{-1} to 6600 ng·L^{-1} [8]. Persistence of ATL is attributed to its resistance to conventional chlorination processes, while its ability to bioaccumulate in aquatic organisms raises additional ecological concerns [9].

Acetaminophen, a widely used analgesic and antipyretic, is another significant contaminant. Studies have reported variable concentrations across different types of water sources. For example, ACM levels of up to 500 ng·L^{-1} in Lagos and Ologe Lagoons [10] have been identified, and in treated wastewater effluents, concentrations from 8.1 ng·L^{-1} to 11,730 ng·L^{-1} have been measured [11]. Additionally, extreme values of 1.33×10^{10} ng·L^{-1} during periods of high usage in Aksaray, Turkey, have been reported [12], whereas an average concentration of 44,000 ng·L^{-1} in wastewater from San Marcos, Texas, has been observed [6]. Pozdnyakov et al. reported that irradiation at 254 nm leads to the disappearance of ACM with a $\phi_{254} = 1.4 \times 10^{-3}$ mol·Einstein^{-1} quantum yield [13]. The limited efficiency of traditional wastewater treatments to effectively remove ACM contributes to its accumulation in natural water systems, with potential adverse ecotoxicological impacts.

Antipyrine, used for its analgesic properties, has been studied less extensively but remains a persistent contaminant in aquatic environments. Roveri et al. detected ATN in river systems, highlighting its environmental persistence and potential for bioaccumulation [14]. In China, the concentration of ANT in the influent of a drinking water treatment plant was found to range from 1.34 to 2.22 ng·L^{-1} [15]. These characteristics underscore the importance of monitoring its impact on aquatic life and ecosystems.

Clofibric acid, a metabolite of lipid-regulating drugs, is another compound of concern due to its persistence in water. Concentrations ranging from 200 to 500 ng·L^{-1} have been reported in wastewater effluents [16]. Its resistance to conventional treatment methods and tendency to accumulate in aquatic organisms highlight the need to develop more effective removal strategies.

The available literature does not include studies reporting the UVC photolysis or photocatalytic removal of a mixture comprising the four selected contaminants using TiO_2 P25, nor the development of a kinetic model describing their simultaneous degradation. However, several works have investigated the individual direct photolysis and photocatalytic removal of these compounds using TiO_2 or TiO_2-based materials. Table S1 summarizes the experimental conditions and removal efficiencies reported for each contaminant.

To address the challenges posed by these contaminants, advanced treatment methods have been explored. Traditional processes, such as activated sludge systems, often fail to achieve adequate removal [17]. Advanced oxidation processes (AOPs), including ozonation and photocatalysis, have demonstrated significant potential in degrading pharmaceuticals. For instance, Lu et al. found that laccase-mediated oxidative coupling effectively removed ACM from water [18]. Floating treatment wetlands have also been shown to reduce organic contaminants, including ACM [19]. Bioremediation, leveraging microorganisms, represents another promising approach. Qutob et al. investigated the degradation of ACM and hydroxychloroquine, achieving substantial removal in controlled settings [20]. Similarly, adsorption techniques using natural materials like modified Moringa oleifera seed husks have shown potential for cost-effective removal [21]. Numerous AOPs have been explored for the removal of various pharmaceutical active compounds (PhACs) from water [22–26]. However, many studies have focused on the degradation of individual pharmaceuticals dissolved in water, often overlooking the influence of PhAC mixtures on treatment efficiency [27]. Photolysis, induced by ultraviolet (UV) radiation, plays a central role in photo-irradiated AOPs and can also be employed as a standalone process for wastewater treatment or sterilization. Despite its potential, the effectiveness of UV photolysis in treating water and effluents is often constrained by low photon absorption or limited quantum yields. However, when contaminants efficiently absorb UV radiation, photolysis can become a highly efficient treatment method [28].

Kinetic models have been developed to predict the degradation of these compounds under various treatment conditions. For instance, the degradation of pharmaceuticals in aqueous solutions often follows pseudo-first-order kinetics, as demonstrated in studies involving advanced oxidation processes (AOPs) [29]. The reaction rate constants for the degradation of specific pharmaceuticals have been reported, with ATL exhibiting a reaction rate constant with hydroxyl radicals ($HO^{\bullet}$) of approximately 1.1×10^9 $M^{-1} \cdot s^{-1}$ [30]. Similarly, ACM has been shown to react with $HO^{\bullet}$ at a rate constant of 1.4×10^9 $M^{-1} \cdot s^{-1}$ [31]. ANT and CLA also exhibit significant reactivity, with reported rate constants of 1.2×10^9 $M^{-1} \cdot s^{-1}$ and 7.5×10^8 $M^{-1} \cdot s^{-1}$, respectively [32,33]. The reaction rate constant of ATL with singlet oxygen ($^1O_2$) was determined as 7.0×10^5 $M^{-1} \cdot s^{-1}$ in D_2O, 8.0×10^6 $M^{-1} \cdot s^{-1}$ in acetonitrile, and 8.4×10^5 $M^{-1} \cdot s^{-1}$ in ethanol, highlighting the influence of solvent polarity and hydrogen donating ability on the reaction kinetics [34]. According to pulse radiolysis measurements and competitive kinetic approaches, the bimolecular reaction rate constants for $HO^{\bullet}$ and $^1O_2$ with CLA were found to be $8.47(\pm 0.33) \times 10^9$ $M^{-1} \cdot s^{-1}$ and $6.6(\pm 0.37) \times 10^6$ $M^{-1} \cdot s^{-1}$, respectively [35]. However, in another study on the kinetics and mechanism of CLA photodegradation under simulated sunlight irradiation, using the competition kinetic method, the bimolecular reaction rate constants with $HO^{\bullet}$ and $^1O_2$ were determined as $3.93(\pm 0.20) \times 10^8$ and $2.38(\pm 0.12) \times 10^6$ $M^{-1} \cdot s^{-1}$, respectively [36]. Literature provides limited information on the reaction rate constants of these contaminants with the superoxide ion radical, though values for other organics are between 10^5 and 10^8 $M^{-1}s^{-1}$ [37].

Novelty and main objectives of this study are as follows:

(a) Investigate the photolytic and photocatalytic degradation kinetics of a mixture comprising four pharmaceuticals, proposing a generic kinetic model to address the complexity of simultaneously removing multiple compounds, considering the formation of reactive oxygen species (ROS),

(b) Determination of the actual quantum yield at 254 nm of each pharmaceutical and identification of the specific ROS generated during their photolysis.

(c) Determination of the reaction rate constant of each pharmaceutical with $^1O_2$ to elucidate the role of this species in the photolysis of photosensitizing agents.

(d) Check the influence of TiO_2 concentration incorporated into the kinetic model in order to simulate the photocatalytic degradation of the four pharmaceuticals as a mixture, as well as the evolution of the ROS involved throughout the process.

2. Results and Discussion

2.1. UVC Photolysis

2.1.1. Effects of Scavengers

Several studies have shown that certain antibiotics (e.g., fluoroquinolones) and other organic compounds exhibit photosensitizing properties [38–40]. When exposed to light in aqueous solutions, these compounds can produce ROS, primarily $^1O_2$, $O_2^{\bullet-}$, and $HO^\bullet$, through mechanisms involving excitation to a triplet state. This process not only facilitates their photolysis but also accelerates the degradation of other substances present in the solution, a phenomenon known as indirect photolysis.

For this reason, in the present study, direct photolysis experiments are individually conducted for each of the selected pharmaceuticals in phosphate buffered ultrapure water at pH 7, both in the presence and absence of t-butanol (TBA), TIRON, and sodium azide (NaN_3), to determine whether $HO^\bullet$, $O_2^{\bullet-}$ and $^1O_2$ are formed and influence the degradation of each pharmaceutical when treated as a mixture. Figure 1 presents the results from the aforementioned experiments, illustrating the influence of these ROS on the photolytic degradation of the pharmaceuticals.

Figure 1 shows that the degradation rate varies depending on the scavenger used. The curve without scavenger (green line) represents the results of scavenger-free photolysis of the contaminant, where direct photolysis due to light absorption is likely combined with indirect photolysis mediated by ROS, such as $HO^\bullet$, $O_2^{\bullet-}$, and $^1O_2$, which are likely generated from the excited state of the contaminant. Subsequently, scavengers were used to determine which reactive species dominate in each case.

In the case of ATL, the addition of TBA (blue line), a selective scavenger of $HO^\bullet$, significantly reduces the degradation rate compared to the green curve. This indicates that $HO^\bullet$ plays an essential role in the degradation of this contaminant. The presence of NaN_3 (orange line), which eliminates both $HO^\bullet$ and $^1O_2$, produces a result almost identical to that obtained with TBA+TIRON (black line), which removes $HO^\bullet$ and $O_2^{\bullet-}$. The similarity between the orange and black lines suggests that both ROS could also be responsible for part of the ATL inhibition rate. However, as shown later (see Section 2.1.3), the low-rate constant of the $^1O_2$-ATL reaction seems to confirm that, in addition to $HO^\bullet$, the second most relevant ROS influencing ATL removal is $O_2^{\bullet-}$.

In the case of ACM, the addition of TBA reduces the degradation rate, confirming that $HO^\bullet$ is involved in the process, though the inhibition is less pronounced than in ATL. The use of NaN_3 produces a similar inhibitory effect to that of TBA, confirming the important involvement of $HO^\bullet$, with no contribution from $^1O_2$. However, the fact that the presence of TIRON leads to greater inhibition also indicates that $O_2^{\bullet-}$ participates to some extent.

The addition of TBA practically has little influence on the CLA reaction rate, indicating low participation of $HO^\bullet$. Inhibition is more pronounced with NaN_3, suggesting a possible contribution from $^1O_2$, although this ROS does not appear to be the predominant species. The combination of TBA+TIRON causes the greatest inhibition, with the curve above that corresponding to NaN_3. This behavior indicates that $O_2^{\bullet-}$ plays a more significant role in CLA degradation than $^1O_2$, since, when $HO^\bullet$ and $O_2^{\bullet-}$ radicals are removed, the degradation is reduced more than when $HO^\bullet$ and $^1O_2$ are eliminated. In the case of ANT, the addition of TBA slightly inhibits the degradation rate, confirming that $HO^\bullet$ is not significantly involved. The curve corresponding to TBA+TIRON is practically identical to that obtained with NaN_3, as is also seen for the case of ATL. For a similar reason (low-rate constant of $^1O_2$-ANT reaction, see Section 2.1.3). $O_2^{\bullet-}$ could be a more relevant species in the degradation of ANT. Based on these results, it is concluded that the dominant species in these processes are, to a greater or lesser extent, $HO^{\bullet}$ and $O_2^{\bullet-}$. This does not imply that $^1O_2$ fails to contribute to indirect photolysis, but rather that it may react more slowly with pharmaceuticals or be produced in a lower proportion [41,42]. In any case, considering the photolysis of the mixture of these pharmaceuticals, these three ROS will be formed, and their reactions need to be considered when modelling the kinetics (see Section 2.2.3).

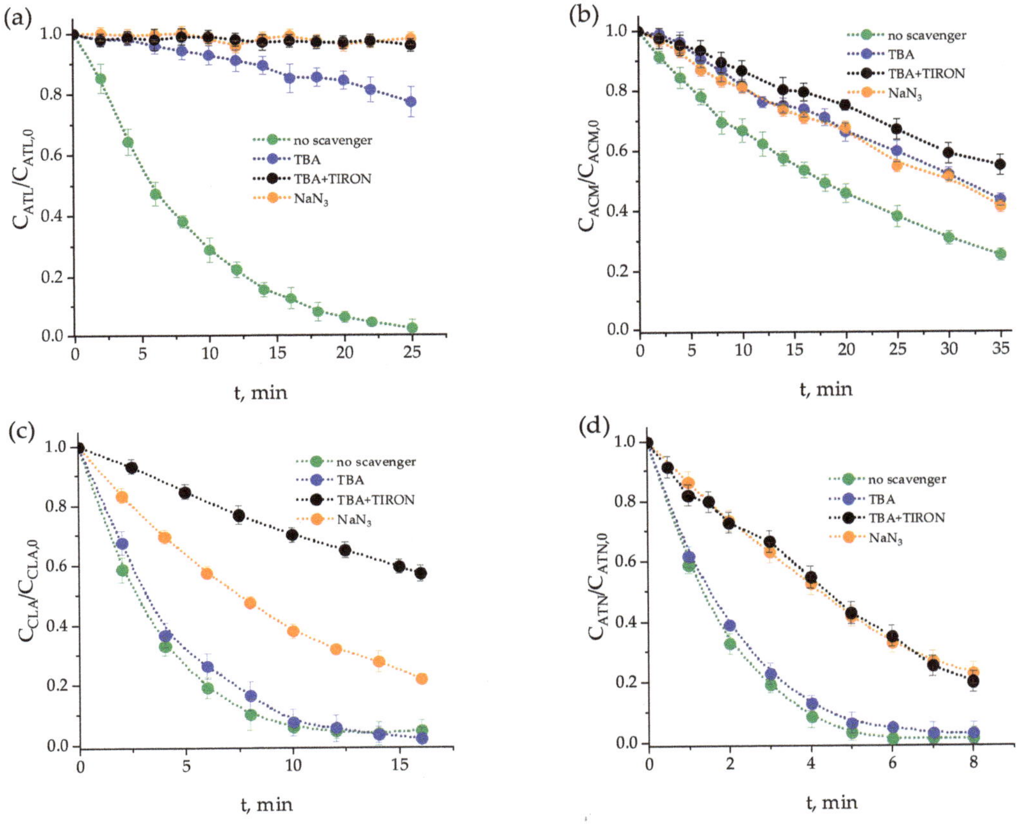

Figure 1. Impact of scavengers on the individual photolytic degradation of pharmaceuticals: (**a**) ATL, (**b**) ACM, (**c**) CLA, (**d**) ANT. Conditions: Initial concentration: 0.5 mg·L^{-1}, $C_{TBA,0}$ = 0.01 M, $C_{TIRON,0}$ = 0.001 M, $C_{NaN3,0}$ = 0.01 M, pH_0 = 7. $I_0 = 3.99 \times 10^{-6}$ Einstein L^{-1}·s^{-1}.

2.1.2. Comparison with Photolysis Results from Literature

Table S1 lists conditions and removal results of the few literature reports on the UVC photolysis of the pharmaceuticals studied that used similar photochemical equipment (UVC lamps emitting at 254 nm with similar intensity of radiation). For instance, Katsumata et al. [43] worked on the photolysis of ATL and ACM, among other pharmaceuticals. These authors found removals of 62% and 54%, respectively, of ATL and ACM after 1 h of treatment, while in the present study, complete removal of these compounds was achieved in less than 30 min. This difference can be mainly attributed to the initial concentration of the compounds, 5 mg·L^{-1} against 0.5 mg·L^{-1} used in this work. Also, Li et al. [44] have reported 92% removal of CLA after 1 h photolysis using an initial concentration of 2.14 mg·L^{-1}. However, in the present study, almost complete removal was achieved within just 15 min, which can also be attributed to the difference in CLA initial concentration used. Finally, Tan et al. [45] reported 40% ANT removal after 1 h, which negatively compares with the complete removal achieved in this work in just 5 min reaction. Once again, the initial concentration applied can be the reason for this difference.

2.1.3. Apparent and Actual Direct Quantum Yields

To determine the quantum yield of pharmaceuticals, the equation describing the mass balance of a compound undergoing photolysis is used [40], as follows:

$$\frac{dC_i}{dt} = -\phi_i e_i^a = -\phi_i\, I_0 F_i [1 - exp(-2.303\, LAbs_{254})] \quad (1)$$

where ϕ_i represents the quantum yield of component i, e_i^a corresponds to the photon absorption rate, I_0 is the volumetric incident radiation, L is the effective path length of radiation (see Section 3.2. for further information), Abs254 is the absorbance of solution at 254 nm, and F_i the fraction of incident radiation species i absorbs:

$$F_i = \frac{\varepsilon_i C_i}{Abs_{254}} \quad (2)$$

where ε_i is the molar extinction or absorptivity coefficient of i at 254 nm. It should be noted that Abs254 varies depending on the species present in the reaction medium, scavengers included, especially NaN_3 and TIRON, which absorb a significant fraction of 254 nm radiation (see Figure S1). However, when considering the transmittance of these compounds, it becomes clear that, although they absorb substantial amounts of radiation, they also allow a significant fraction to pass through (see Table 2). This ensures that the available radiation remains sufficient for the contaminants to undergo direct photolysis, while the scavenging function of both NaN_3 and TIRON is still preserved.

Table 2. Molar absorptivity coefficient, concentration of each pharmaceutical and scavenger used, absorbance at 254 nm, and transmittance at this concentration at pH 7.

Pharmaceutical	$\varepsilon_{254\ nm}$ $M^{-1}\cdot cm^{-1}$	Concentration M	$Abs_{254\ nm}$ a.u.	Transmittance %
ATL	441.2	1.87×10^{-6}	8.25×10^{-4}	99.8
ACM	10,645	3.31×10^{-6}	3.52×10^{-2}	92.1
CLA	251.2	2.06×10^{-6}	5.17×10^{-4}	99.9
ANT	8890.6	2.65×10^{-6}	2.36×10^{-2}	94.7
TBA	0.244	0.01	2.44×10^{-3}	99.4
NaN_3	5.35	0.01	0.053	88.5
TIRON	444.1	0.001	0.404	39.5
WWEf [a]	--	--	0.16	69.2

[a] WWEf: Wastewater effluent.

For the integration of Equation (1), a constant value can be assumed for Abs254, as it does not vary by more than 10% during the photolysis reaction (see Figure S2). Then, analytical integration of Equation (1) yields the following:

$$ln\frac{C_i}{C_{i0}} = -\phi_i\, I_0 \frac{\varepsilon_i}{Abs_{254}}[1 - exp(-2.303\, L\, Abs_{254})]\, t \quad (3)$$

By applying Equation (3) to the experimental results in the presence and absence of the scavenger (TBA+TIRON), both direct and apparent quantum yields of each pharmaceutical, respectively, can be determined from the slopes of the straight lines representing

Equation (3), as shown in Figure S3. Table 3 presents the quantum yield values obtained for each case. The apparent quantum yields were determined from scavenger-free photolysis experiments of each pharmaceutical.

Table 3. Apparent (ϕ_{app}) and direct (ϕ_d) quantum yields of pharmaceuticals studied.

Pharmaceutical	ϕ_{app} (mol·Einstein^{-1})	ϕ_d (mol·Einstein^{-1})
	No Scavenger	TBA+TIRON
ATL	0.338	8.05×10^{-3}
ACM	4.34×10^{-3}	1.93×10^{-3}
CLA	1.05	3.12×10^{-1}
ANT	6.44×10^{-2}	5.12×10^{-2}

The results of Table 3 show that ATL and ACM are the compounds with the lowest direct quantum yields. This explains why their removal requires the longest time, and their photodegradation rates were affected when scavengers were added. On the other hand, as expected, CLA and ANT exhibit higher quantum yields, which aligns with the fact that these were the compounds that degraded most rapidly. However, it can be observed that, although ANT is removed in under 10 min, its quantum yield is lower than that of CLA, which is completely degraded in approximately 30 min in the presence of TBA+TIRON. This apparent discrepancy is explained by the fact that ANT has a molar absorptivity coefficient of 8890.6 $M^{-1}·cm^{-1}$, while CLA has a coefficient of only 251.2 $M^{-1}·cm^{-1}$.

2.1.4. Determination of the Reaction Rate Constant with Singlet Oxygen

The influence of singlet oxygen is generally considered less significant than that of $HO^{\bullet}$ and $O_2^{\bullet-}$. However, due to the limited availability of data in the literature, the second-order reaction rate constant of singlet oxygen with the pharmaceuticals (k_{1O2}) was also determined in this work. For this purpose, the competitive kinetic method reported by Gao et al. [46] was used to determine these rate constants, employing Rose Bengal as a singlet oxygen generator and using furfural (F) as a reference compound, in accordance with Equation (4).

$$ln\frac{C_P}{C_{P0}} = \frac{k_{1O_2,P}}{k_{1O_2,F}} ln\frac{C_F}{C_{F0}} \quad (4)$$

where C_{P0} and C_{F0} represent the initial concentration of the pharmaceutical and furfural, respectively, and C_P and C_F are their corresponding concentrations at time t. Figure S4 confirms Equation (4), and Table 4 presents the results.

Table 4. Calculated rate constants of the reaction of singlet oxygen and pharmaceuticals were studied.

Pharmaceutical	k_{1O2} ($M^{-1}·s^{-1}$)
ATL	9.93
ACM	1.30×10^6
CLA	1.18×10^2
ANT	1.14×10^4

According to the rate constant data of Table 4, ATL exhibits the lowest reactivity towards singlet oxygen, showing an almost negligible reaction rate constant. This supports the earlier conclusion regarding the limited role of ROS in the indirect photolysis of ATL. A similar observation applies to singlet oxygen during CLA and ANT photolysis. ACM, by contrast, shows the highest value among the four compounds. However, according to results from Figure 1, formation of singlet oxygen from ACM photolysis does not seem important, and overall, it appears to be considerably less reactive than the hydroxyl radical and the superoxide ion radical as widely described in the literature [41,42]. Nonetheless, the influence of singlet oxygen must be considered when treating the UVC photolysis of an aqueous mixture of the four pharmaceuticals, as it could be mainly generated from CLA and/or ANT.

2.2. UVC Photolysis in Effluent of Urban Wastewater

The presence of dissolved organic matter (DOM) in water may have a significant impact on photochemical processes, as it could also generate ROS considered in this work ($O_2^{\bullet-}$, $^1O_2$, $HO^{\bullet}$). The production of these species varies depending on the source and composition of the DOM. For instance, humic and fulvic substances found in natural waters produce superoxide ion radicals ($O_2^{\bullet-}$) and hydrogen peroxide (H_2O_2), with reaction rates influenced by the molecular composition of the DOM [47]. Moreover, the aqueous matrix influences reaction kinetics, as natural organic matter (NOM) can modify both the generation and decay of ROS. Indicators such as specific ultraviolet absorbance (SUVA) are useful for predicting these kinetics [48]. These processes may involve direct photolysis, where a molecule absorbs light and undergoes transformation, or indirect photochemistry, mediated by photosensitizers that generate transient species. The presence of NOM can alter these mechanisms, influencing reaction kinetics and the formation of intermediates [49]. Based on these considerations, a study was conducted to determine the influence of organic matter in the wastewater used.

2.2.1. Determination of ROS Involved in the Photolysis of Secondary Effluent

Photolysis experiments were subsequently conducted to investigate the role of secondary effluent (WWEf) in generating ROS during photolytic processes. The focus was on the potential of this water matrix to absorb radiation and its ability to form ROS. To explore these mechanisms, direct photolysis experiments of WWEf were performed, both in the absence and presence of scavengers [40].

The results are presented in Figure 2, which shows the evolution of absorbance at 254 nm and total organic carbon (TOC) over time. An initial absorbance of 0.160 was recorded, indicating the presence of compounds that can absorb UVC radiation. In all cases, the absorbance attributable to the scavengers was subtracted to ensure that the absorbance loss was only due to the secondary effluent. While this value is relatively low, it is significant enough to compete for the radiation required for the photolytic degradation of contaminants. In contrast, TOC remained constant throughout the experiments, suggesting that the photolytic processes were insufficient to achieve significant mineralization of organic matter within the timeframe of the tests.

The results revealed no significant differences between the photolysis curves obtained under different scavenger conditions. This suggests that the water matrix used does not generate substantial amounts of ROS capable of influencing the degradation kinetics of the pharmaceuticals. Moreover, the absence of pronounced differences in the degradation behavior suggests that the effluent contains minimal photosensitizing compounds. If ROS were formed, their concentration and reactivity would likely be too low to significantly impact the degradation of any pharmaceutical.

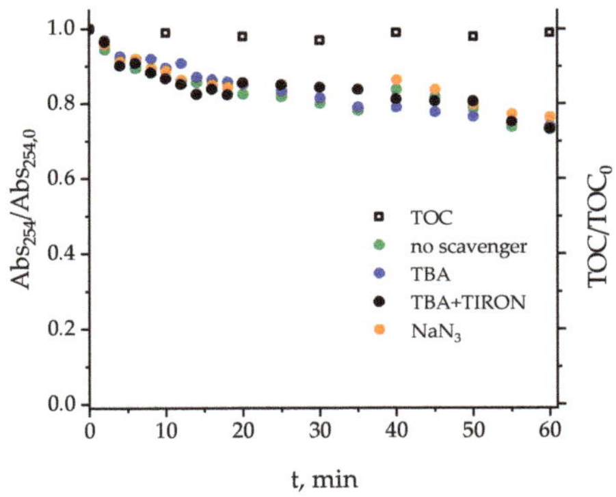

Figure 2. Impact of scavengers on the photolysis of pharmaceutical-free WWEf. Conditions: $Abs_{WWEf,0}$= 0.160 a.u., $Abs_{WWEf+TBA,0}$ = 0.163 a.u., $Abs_{WWEf+NaN3,0}$ = 0.213 a.u., $Abs_{WWEf+TIRON,0}$ = 0.604 a.u., TOC_0 = 7.3 mg·L^{-1}, $C_{TBA,0}$ = 0.01 M, $C_{TIRON,0}$ = 0.001 M, $C_{NaN3,0}$ = 0.01 M, pH_0 = 7.

The use of absorbance at 254 nm proved to be a valuable parameter for assessing the composition of the effluent. This measurement correlates with the presence of aromatic and unsaturated compounds, which are known to absorb radiation at this wavelength. The stability of TOC alongside changes in absorbance further supports the notion that photolysis predominantly induces structural modifications rather than complete mineralization of the organic matter present in the water. However, DOM would compete for ROS generated in the photolysis of the water matrix in the presence of pharmaceuticals studied, thereby hindering their indirect photolysis. Since the absorbance results of Figure 2 are similar regardless of the presence and absence of scavengers, there appears to be no significant role of ROS. Then, it can be inferred that only direct photolysis of the organic matter is occurring. Therefore, the quantum yield of the secondary effluent was also determined. As in the previous section, the variation in absorbance over time would be as follows:

$$-\frac{dAbs_{254}}{dt} = \phi_{WWEf} I_0 [1 - exp(-2.303\ L\ Abs_{254}\ )] \quad (5)$$

In this case, the term F_i is assumed to be unity, as it refers to the naturally occurring matter in water capable of absorbing radiation. This matter will be treated as a single entity, and thus the product $\Sigma\varepsilon_j C_j$ will correspond to the absorbance value of the water at each time point when measured in a 1 cm path length cell. The analytical solution of Equation (5) is as follows:

$$(Abs_{254,0} - Abs_{254}) + \frac{1}{2.303L} \ln\left(\frac{1 - exp(-2.303\ L\ Abs_{254}\ )}{1 - exp(-2.303\ L\ Abs_{254,0}\ )}\right) = \phi_{WWEf} I_0 t \quad (6)$$

A quantum yield value of 5.53 a.u.·L^{-1}·Einstein^{-1} was obtained from the experimental data using Equation (6) (see Figure S2).

2.2.2. Effect of the Aqueous Matrix on the Photolysis of Pharmaceuticals

Photolysis experiments were performed with a mixture of four pharmaceuticals in the secondary effluent to evaluate the effect of the aqueous matrix, particularly the presence of natural organic matter (NOM). The main characteristics of WWEf are described in Table S4 of Supplementary information.

The degradation kinetics of the pharmaceuticals were significantly slower in secondary effluent compared to ultrapure water, as shown in Figure 3 compared to Figure 1. This decrease in photolytic efficiency is attributable to the presence of NOM, which, as expected, influences both radiation absorption and the dynamics of reactive species formation. NOM absorbs radiation at 254 nm, reducing the fraction available for direct photolysis of the pharmaceuticals. This competitive light absorption limits the energy reaching the pharmaceuticals, thereby reducing their photodegradation rate. Although photolysis does achieve a certain degree of contaminant removal in secondary effluent (as shown in Figure 3), the extent of mineralization is not significant. The total organic carbon (TOC) remains practically constant throughout the process, with a maximum mineralization of only 4%, as illustrated in Figure S5.

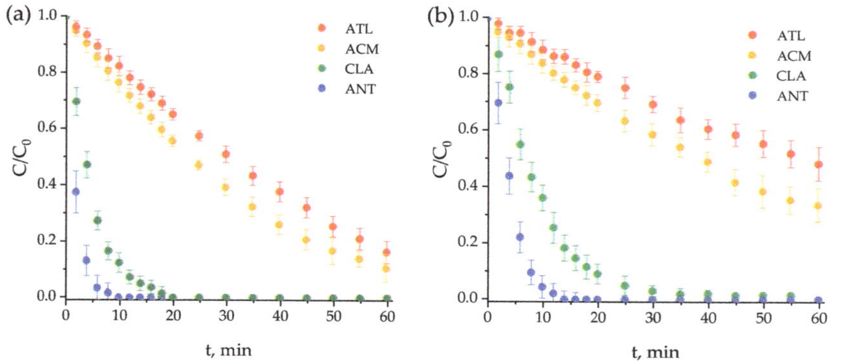

Figure 3. Experimental variation of the normalized concentrations of pharmaceuticals over time during the photolysis of the mixture in (**a**) ultrapure water and (**b**) secondary effluent. Conditions: $C_{Pharm,0} = 0.5\ mg \cdot L^{-1}$, $pH_0 = 7$, $I_0 = 3.99 \times 10^{-6}$ Einstein $L^{-1} \cdot s^{-1}$.

At first, NOM competes for UV radiation and ROS pharmaceuticals generated. However, the fact that scavengers practically have no influence on 254 nm absorbance suggests that NOM of wastewater is preferentially altered through direct photolysis. Thus, the decrease in pharmaceutical degradation rates in wastewater is mainly attributed to the photolysis of NOM, in contrast to what occurs in ultrapure water.

2.2.3. Kinetic Model of Photolysis

Ultrapure Water

The mechanism of direct photolysis involves both direct photon absorption and subsequent reaction of ROS with pharmaceuticals, including hydrogen peroxide generated during the process. In this mechanism, hydroxyl radicals are assumed to be generated from photolysis of ATL and ACM and singlet oxygen from that of CLA and ANT, as experimental results suggest (see Section 2.1.1). It is also assumed that the formation of superoxide ion radical results from the reactions involving hydroxyl radical and singlet oxygen with pharmaceuticals [39,40]:

$$ACM\ and\ ATL \xrightarrow{h\nu,\phi_M} P + \alpha HO^{\bullet} \quad (7)$$

$$CLA\ and\ ANT \xrightarrow{h\nu,\phi_M} \beta^1O_2 + P \quad (8)$$

$$HO^{\bullet} + M \xrightarrow{k_{HOi}} HO_2^{\bullet} + P \quad (9)$$

$$^1O_2 + M \xrightarrow{k_{Sing}} O_2^{\bullet-} + P \quad (10)$$

where M and P represent any pharmaceutical initially present in water and any product formed, respectively, and α and β are the stoichiometric ratios for $HO^\bullet$ and $^1O_2$ formation, respectively.

The hydroperoxide ion radical, $HO_2^{\cdot}$, is in equilibrium with the superoxide ion radical (Equation (11)), but this equilibrium is completely shifted towards the formation of the latter since pH was 7 and pK = 4.8. Hence, any hydroperoxide ion radical instantaneously becomes a superoxide ion radical [50]:

$$HO_2^{\bullet} \overset{pK = 4.8}{\rightleftarrows} O_2^{\bullet-} + H^+ \tag{11}$$

As a result, in reaction (9), the superoxide ion radical is effectively formed instead of the hydroperoxide radical.

The recombination of free radicals occurs via the following reaction [51]:

$$2HO^{\bullet} \xrightarrow{k_F} H_2O_2 \tag{12}$$

$$2O_2^{\bullet-} + 2H^+ \xrightarrow{k_F} H_2O_2 + O_2 \tag{13}$$

and reactions with the participation of hydrogen peroxide, including its direct photolysis, are as follows [52,53]:

$$H_2O_2 \overset{pK = 11.8}{\rightleftarrows} HO_2^- + H^+ \tag{14}$$

$$H_2O_2 \xrightarrow{h\nu,\phi=0.5} 2HO^{\bullet} \tag{15}$$

$$HO^{\bullet} + H_2O_2 \xrightarrow{k_{H1}} HO_2^{\bullet} + H_2O \tag{16}$$

$$HO^{\bullet} + HO_2^- \xrightarrow{k_{H2}} HO_2^{\bullet} + HO^- \tag{17}$$

and termination reactions [54].

$$HO^{\bullet} + P_i \xrightarrow{k_{HO-P}} End\ products \tag{18}$$

where

$$\sum k_{HO^{\bullet}-P} C_p = k_T \tag{19}$$

and

$$O_2^{\cdot-} + P_i \xrightarrow{k_{Sup}} End\ products \tag{20}$$

$$\sum k_{Sup} C_p = k_{TT} \tag{21}$$

It should be noted that there are other reactions involving hydroxyl radicals and primary intermediates that could also undergo photolysis, that have not been considered due to insufficient data. Reactions of superoxide ion radical with pharmaceuticals have also been considered as termination reactions:

$$O_2^{\bullet-} + M \xrightarrow{k_{O_2^{\bullet-}\ M}} End\ products \tag{22}$$

According to experimental results, the photolysis kinetic model for pharmaceuticals is based on mol balance equations of participating species, including both molecular species (pharmaceuticals and hydrogen peroxide) and ROS. All these equations follow a general form corresponding to the design equation for a perfectly mixed batch reactor for liquid phase reactions [55]:

$$-\frac{dC_i}{dt} = \sum r_i \tag{23}$$

The reaction rate term, r_i, for any participating species, i, takes the form presented below:

(a) For any pharmaceutical, the reaction rate comprises four contributions: direct photolysis and reactions with $HO^{\bullet}$, $O_2^{\bullet -}$, and $^1O_2$:

$$r_i = r_{UVi} + r_{HO^{\bullet}i} + r_{O_{2i}^{\bullet -}} + r_{1O_{2i}} \quad (24)$$

where [56]:

$$r_{UVi} = \phi_i e_i^a = \phi_i\, I_0 \frac{\varepsilon_i C_i}{\sum_j \varepsilon_j C_j}\left[1 - exp\left(-2.303\, L \sum \varepsilon_j\, C_j\right)\right] \quad (25)$$

$$r_{HO^{\bullet}i} = k_{HO^{\bullet}i} C_{HO^{\bullet}} C_i \quad (26)$$

$$r_{O_{2i}^{\bullet -}} = k_{O_{2i}^{\bullet -}} C_{O_2^{\bullet -}} C_i \quad (27)$$

$$r_{O_{2i}^{\bullet -}} = k_{1O_{2i}} C_{1O_2} C_i \quad (28)$$

here, ϕ_i represents the quantum yield of species i, and $e_i{}^a$ its volumetric photon absorption rate, $k_{HO\cdot i}$, $k_{O2\cdot -ii}$ and k_{1O2i} denote the rate constants for reactions with $HO^{\bullet}$, $O_2^{\bullet -}$, and $^1O_2$, respectively, and $C_{HO\cdot}$, $C_{O2\cdot -}$, and C_{1O2} represent the concentrations of these reactive oxygen species.

(b) For hydrogen peroxide, the net reaction rate is given by the following:

$$r_H = k_F C_{HO^{\bullet}}^2 + k_F C_{O_{2i}^{\bullet -}}^2 - r_{UV_H} - k_H C_{HO^{\bullet}} C_{H_2O_2T} \quad (29)$$

where,

$$r_{UV_H} = \phi_H I_0 \frac{\varepsilon_H C_H}{\sum_j \varepsilon_j C_j}\left[1 - exp\left(-2.303 L \sum \varepsilon_j C_j\right)\right] \quad (30)$$

with ϕ_H and ε_H as the quantum yield and absorptivity coefficient of hydrogen peroxide at 254 nm, respectively.

(c) For the hydroxyl radical:

$$r_{HO^{\bullet}} = \alpha\left(r_{UV_{ATL}} + r_{UV_{ACM}}\right) + 2r_{UV_H} - C_{HO^{\bullet}}\left(\sum k_{HO^{\bullet}i} C_i + k_H C_{H_2O_2T} + k_T + 2k_F C_{HO^{\bullet}}\right) \quad (31)$$

where,

$$k_H = \frac{k_{H1} + k_{H2} 10^{(pH - pK)}}{10^{(pH-pK)}} \quad (32)$$

(d) For the superoxide ion radical:

$$r_{O_2^{\bullet -}} = C_{HO} \sum k_{HOi} C_i + C_{1O_2} \sum k_{1O_{2i}} C_i - C_{O_2^{\bullet -}}\left(\sum k_{O_{2i}^{\bullet -}} C_i + k_{TT} + 2k_F C_{O_2^{\bullet -}}\right) \quad (33)$$

(e) For singlet oxygen:

$$r_{1O_{2i}} = \beta\left(r_{UV_{CLA}} + r_{UV_{ANT}}\right) - C_{1O_2} \sum k_{1O_{2i}} C_i \quad (34)$$

Secondary Effluent Water

When reactions occur in secondary effluent, the photolysis of natural organic matter (NOM) is incorporated into the mechanism:

$$NOM \xrightarrow{hv, \phi_{WWEf}} rP \quad (35)$$

A new balance equation corresponding to reaction (35) must then be considered. The absorbance of the water at 254 nm is used as a parameter to represent NOM so that its photolysis reaction rate is as follows:

$$r_i = \phi_{WWEf} e^a_{WWEf} = \phi_{WWEf}\ I_0 \frac{Abs_{WWEf}}{Abs_T}[-exp(-2.303\ L\ Abs_T)] \quad (36)$$

where

$$Abs_T = \sum \varepsilon_j C_j + Abs_{WWEf} \quad (37)$$

with Abs_{WWEf} representing the absorbance of the secondary effluent at 254 nm, excluding contributions of pharmaceuticals. Moreover, the NOM absorbance influences the photolysis reaction rates of pharmaceuticals (Equation (25)) and hydrogen peroxide (Equation (30)), which are now as follows:

$$r_{UVi} = \phi_i I_0 \frac{\varepsilon_i C_i}{Abs_T}[1 - exp(-2.303 L\ Abs_T)] \quad (38)$$

where i refers to any pharmaceutical or hydrogen peroxide.

Finally, the reaction rates of H_2O_2, $HO^\bullet$, $O_2^{\bullet-}$, and $^1O_2$ remain as defined in Equations (29), (31), (33), and (34), respectively.

The kinetic model comprising Equations (20)–(34) describes the photolysis of a mixture of the four pharmaceuticals in ultrapure buffered water (pH = 7) was first solved with the initial conditions:

$$t = 0 \quad C_M = C_{M0},\ C_H = 0,\ C_{HO^\bullet} = 0,\ C_{O_2^{\bullet-}} = 0 \quad (39)$$

For the kinetic model applied to the effluent wastewater, the following condition for water absorbance was added to Equation (39):

$$t = 0 \quad Abs_{WWEf} = Abs_{WWEf,0} = 0.16 \quad (40)$$

The kinetic models were solved using the Ode23s MATLAB 2024b code, incorporating the values of quantum yields and absorptivity coefficients given in Tables 2 and 3. For the rate constant of hydroxyl radical reactions, initial values were taken from the literature, as also referenced in the Section 1. For superoxide ion radical reaction rate constants a general value of $5 \times 10^6\ M^{-1}s^{-1}$ was initially taken for all pharmaceuticals while for the case of singlet oxygen reactions, values were experimentally determined and are shown in Table 4. Definitive rate constants for hydroxyl radical and superoxide ion radical reactions are provided in Table S2 along with others also used and present in the reaction rates of species. The used pharmaceutical-ROS reaction rate constants are of the expected order of magnitude for these kinds of processes [37,57].

Figure 4 presents the evaluation of the kinetic model applied to both ultrapure water (UP) and secondary effluent (WWEf).

As observed in Figure 4a,b, experimental and calculated results for ATL and ACM are in close agreement, while those for CLA and ANT (especially in UP water) present the greatest discrepancies, although their temporal evolution follows the same trend as the experimental results. These discrepancies are most likely attributable to the lack of data about the kinetics of intermediates formed (see Section 2.4). Moreover, the higher degradation rates of both compounds complicate the fitting of the model to the experimental results and the high number of parameters influencing the kinetic model (see Table S2). In any case, as observed in Figure 4, the fundamentals for establishing molar balances in the completely mixed batch photochemical reactor are valuable tools for predicting experimental results of processes such as photolysis. As also shown in Figure 4, the kinetic model provides a good fit for explaining differences due to the water content. As can be seen, the kinetic model predicts, even with higher precision, the concentrations in the effluent wastewater and the slower reaction rates observed in wastewater effluent compared to those in UP water. This is particularly clear in the case of the more refractory reacting compounds: ATL and ACM.

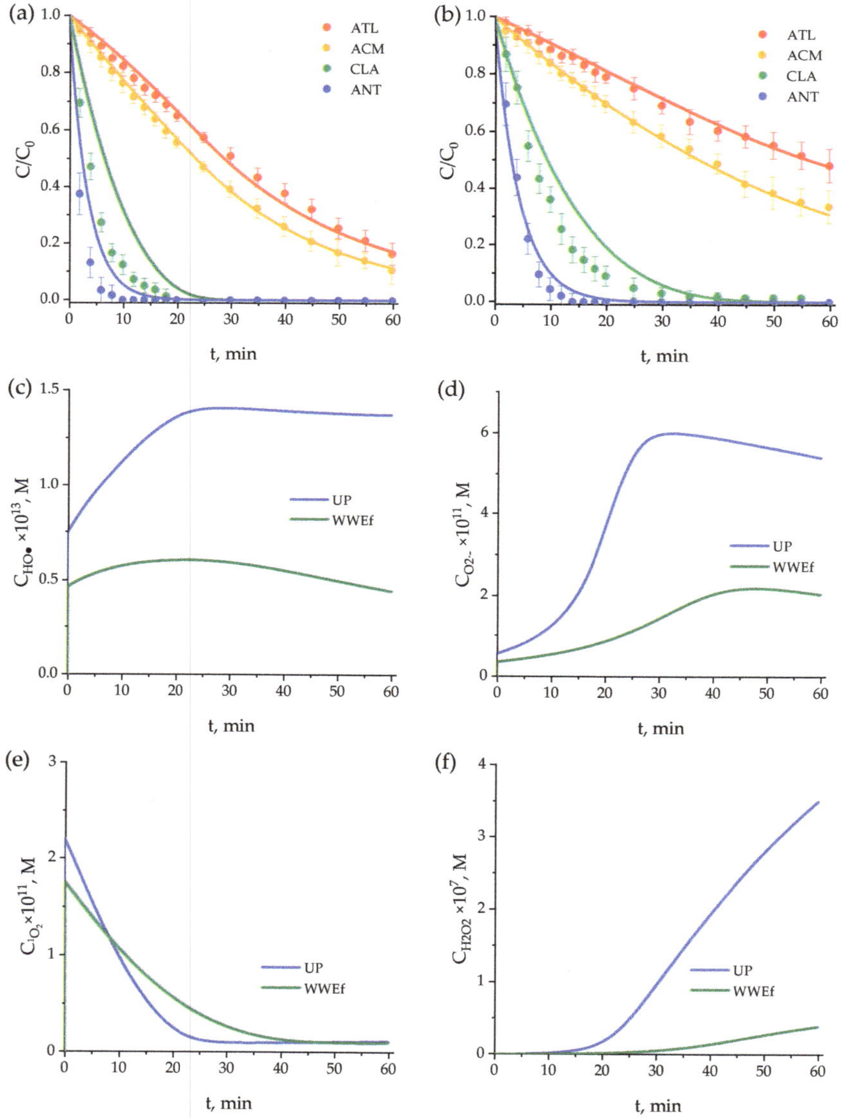

Figure 4. Evaluation of the kinetic model in (**a**) ultrapure water and (**b**) secondary effluent. Variation of normalized concentrations of pharmaceuticals over time: Symbols: experimental data; curves: calculated values. Evolution of the concentration of (**c**) hydroxyl radicals, (**d**) superoxide ion radical, (**e**) singlet oxygen, (**f**) hydrogen peroxide in ultrapure water (blue lines) and secondary effluent (green lines). Conditions: $C_{Pharm0} = 0.5$ mg·L^{-1}, $pH_0 = 7$. Other conditions, as in Figure 2.

On the other hand, Figure 4c,d depict the calculated concentrations of $HO^{\bullet}$ and $O_2^{\bullet-}$ over time, highlighting notable differences between the two water matrices. In ultrapure water, hydroxyl radicals increase steadily, indicating continuous generation with minimal interference, eventually reaching a near-plateau value. This behavior corresponds to the faster degradation of the compounds, as a higher concentration of reactive species is available. However, in secondary effluent, the concentration of hydroxyl radicals remains consistently lower and more stable. This can be explained by parallel reactions with NOM in the effluent, which limit the accumulation of radicals and reduce their overall effectiveness in degrading the compounds. During approximately the first 23 min, the formation trends of $HO^{\bullet}$ are similar in both ultrapure water and effluent wastewater. However, beyond this point, hydroxyl radical formation declines (more slowly in UP water), likely due to the generation of reaction intermediates that interact with these radicals [58] and with the NOM present in the water [59–61]. These interactions may inhibit further radical generation, leading to a slower degradation process in secondary effluent compared to ultrapure water [62].

Figure 4e shows the trend with time followed for the calculated concentration of singlet oxygen, which continuously decreases due to the consumption of CLA and ANT, whose photolysis generates this ROS. Singlet oxygen disappears after 20–25 min, which is approximately the time required for complete elimination of CLA and ANT.

Finally, Figure 4f shows the predicted hydrogen peroxide concentration over time. Experimental values could not be obtained due to the detection limits of the analytical method, which was unable to measure such low concentrations. Once again, concentrations in ultrapure water were higher than in wastewater, likely due to the greater presence of compounds (including NOM) that scavenge free radicals, thereby leading to lower concentrations of hydrogen peroxide.

2.3. UVC TiO_2 Photocatalysis

A series of TiO_2 photocatalytic reactions in secondary effluent wastewater was carried out using a mixture of the four pharmaceuticals, with the main objective of developing and validating a kinetic model. Figure 5 shows the results obtained at different catalyst concentrations. Adsorption experiments also indicated that none of the pharmaceuticals adhered to the catalyst surface, thereby reducing the complexity of the kinetic model, as surface-bound reactions can be disregarded.

Comparing Figures 4 and 5 reveals that the photocatalytic process significantly increases the degradation rates of pharmaceuticals, achieving complete removal in under 15 min while much longer times are needed in the absence of the catalyst, especially for ATL and ACM.

For different catalyst concentrations (see Figure 5a,b,d) beyond a certain threshold, increasing the TiO_2 concentration did not improve efficiency and, in some cases, even reduced it. This effect can be attributed to light shading and scattering [63–65], particle aggregation [66–68], electron-hole recombination [69], and limitations in the diffusion of reactants [70,71]. As in the case of photolysis, dissolved TOC was also monitored during the photocatalytic experiments. The results show that mineralization of the solution increases with higher catalyst concentrations (see Figure S5). Although complete removal of the parent compounds is achieved at early stages, the reduction in TOC does not follow the same trend. Specifically, TOC removal rates of 13%, 28%, and 51% were observed for catalyst concentrations of 0.05, 0.1, and 0.25 $g \cdot L^{-1}$, respectively, which is consistent with findings reported by other authors (see Table S1). These observations clearly highlight the importance of accounting for the formation of reaction intermediates, as well as the need to include an inhibition factor in the kinetic model (k_T in hydroxyl radical molar balance

Equation (31)) to reflect the presence of species that terminate the radical chains and hence slow down the degradation of the target pollutants.

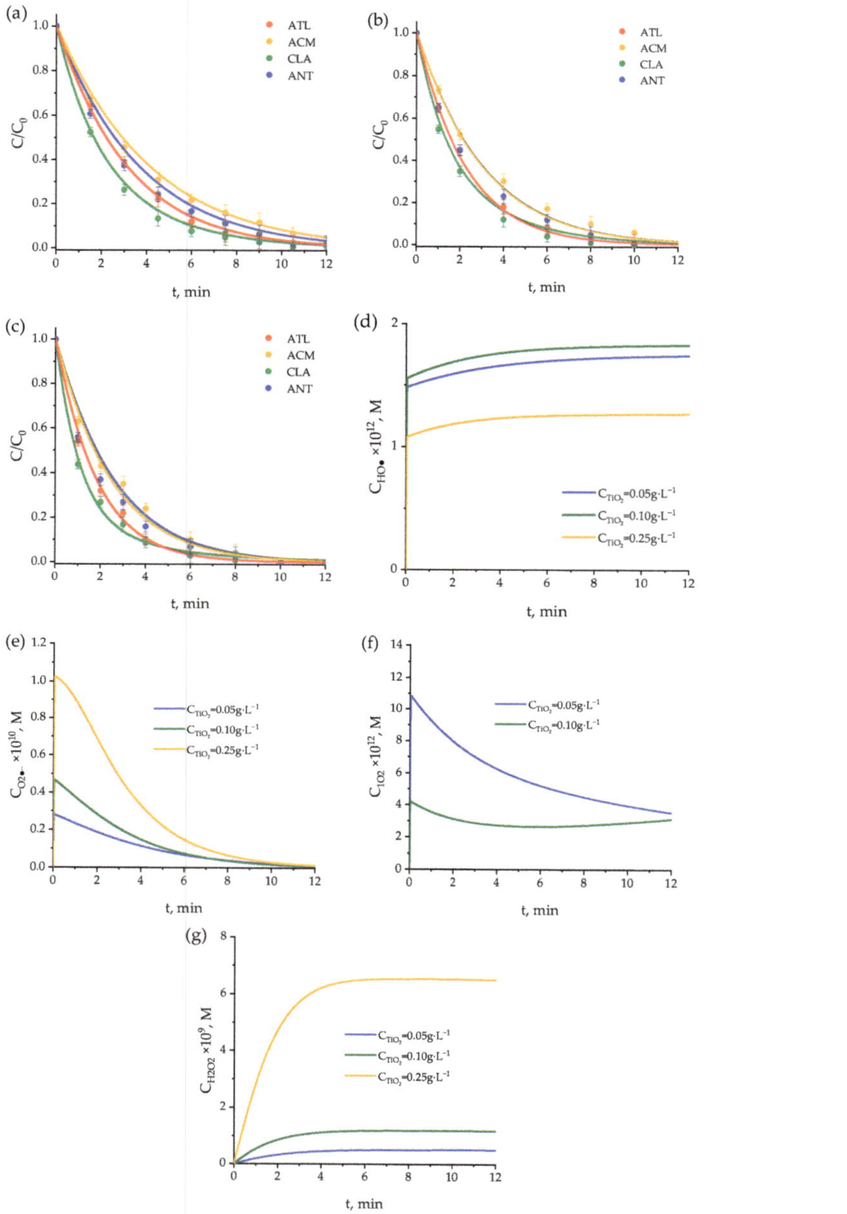

Figure 5. Evaluation of the kinetic model in TiO_2 photocatalytic experiments. Time evolution of the normalized concentration of pharmaceuticals for different TiO_2 concentrations: (**a**) 0.05 $g{\cdot}L^{-1}$, (**b**) 0.10 $g{\cdot}L^{-1}$, and (**c**) 0.25 $g{\cdot}L^{-1}$. Evolution of reactive species: (**d**) hydroxyl radicals ($HO^{\bullet}$), (**e**) superoxide radical ($O_2^{\bullet-}$), (**f**) singlet oxygen ($^1O_2$), and (**g**) hydrogen peroxide (H_2O_2). Conditions: $C_{Pharmac,0}$ = 0.5 $mg{\cdot}L^{-1}$, pH = 7.

2.3.1. Comparison with Photocatalytic Results from Literature

The photocatalytic degradation results obtained in this study are consistent with trends previously reported in the literature (see Table S1). For instance, the rapid and nearly complete removal of all compounds within 15 min under optimized TiO_2 conditions aligns with the high efficiencies reported for ATL and CLA using visible or UV-based photocatalysis. In particular, ATL removal values of 75–82% after 60 min have been observed with both UV (60 W) and visible light sources at higher catalyst doses (1.5–3.0 $g \cdot L^{-1}$) [72], while our system achieves full removal in a significantly shorter time with a lower dose. Similarly, although TOC reduction in our system remains moderate (13–51% depending on catalyst load), it matches or exceeds the values reported in pilot-scale studies such as that of TiO_2 P25 under solar CPCs (TOC removal of 40%) [73], or CLA degradation with Pd-doped TiO_2 under UV-A (TOC reductions of 32–65%) [74]. These comparisons confirm that the enhanced performance observed here may be attributed not only to the catalyst's activity, but also to optimized operational parameters such as pollutant concentration, mixing, and reactor design.

2.3.2. Mechanism and Kinetic Model

The photocatalytic kinetic model builds upon the photolysis reaction mechanism and incorporates additional steps arising from radiation absorption by the catalyst. First of the steps to add is the formation of charges (positive holes, h^+, and electrons, e^-) caused by excitation from the valence to the conduction band of TiO_2, due to the high energy associated to 254 nm wavelength (4.89 eV) that vastly exceeds the band gap of the catalyst (3.2 eV) [75]:

$$TiO_2 + hv \xrightarrow{r_0} h^+ + e^- \quad (41)$$

Oxidizing holes, h^+, subsequently react with water or hydroxyl ions adsorbed on the catalyst surface, producing adsorbed hydroxyl radicals that can subsequently react with pharmaceuticals [76]:

$$h^+ + H_2O - TiO_2 \xrightarrow{k_s} TiO_2 - HO^\bullet \quad (42)$$

and

$$TiO_2 - HO^\bullet + M \xrightarrow{k_{HO-M}} P \quad (43)$$

In turn, the electrons react with oxygen to generate adsorbed superoxide ion radicals [77]:

$$e^- + O_2 \xrightarrow{k_e} TiO_2 - O_2^{\bullet -} \quad (44)$$

Finally, recombination of charges regenerates the catalyst surface, releasing energy [75]:

$$h^+ + e^- \xrightarrow{k_{rec}} TiO_2 + energy \quad (45)$$

Also, hydrogen peroxide formed from free radical recombination, Equations (12) and (13), reacts with electrons to form more hydroxyl radicals [77]:

$$H_2O_2 + e^- \xrightarrow{k_{eH}} HO^- + OH \quad (46)$$

As described in the photolysis section, the photocatalysis kinetic model also requires solving a system of first-order ordinary differential equations of mass balances of species participating in the reacting system Equations (23) – (40). Now, to the reaction rate of the photolysis kinetic model, r_i, the following terms were added:

(a) For the balance of pharmaceuticals (Equation (24)):

$$r = k_{HOMs}C_{HOs}C_M \tag{47}$$

where k_{HOMs} and C_{HOs} are the rate constants of the reaction between any pharmaceutical M and adsorbed hydroxyl radicals and the concentration of the latter, respectively. However, adsorbed hydroxyl radicals are in equilibrium with hydroxyl radicals in solutions, that is, $C_{HOs} = K_{HO\cdot}\, C_{HO\cdot}$ [76], so that Equation (47) can be expressed as follows:

$$r = k_{HOs\text{-}M}C_{HO\bullet}C_M \tag{48}$$

with $k_{HOs\text{-}M} = k_{HO\text{-}M}\, K_{HO\cdot}$, the latter being the equilibrium constant [76].
Initial values of $k_{HOs\text{-}M}$ for the photocatalytic kinetic model at TiO_2 concentration of $0.1\ g{\cdot}L^{-1}$ were adopted from previous work [76] and initially considering the proportionality with TiO_2 concentration for other catalyst concentrations. Table S3 shows the final optimized values of $k_{HOs\text{-}M}$ applied in the kinetic model. A proportional relationship between $k_{HOs\text{-}M}$ and catalyst concentration was ultimately observed. For instance, with some exception, $k_{HOs\text{-}M}$ values for $0.1\ g{\cdot}L^{-1}$ TiO_2 concentration are twice the values for a $0.05\ g{\cdot}L^{-1}$ TiO_2 concentration, and the corresponding proportionality was found for $0.25\ g{\cdot}L^{-1}$ catalyst concentration. Something similar was applied in the case of reactions of pharmaceuticals with adsorbed superoxide ion radical. Nevertheless, the contribution of these reactions was found to be negligible, so that only the superoxide ion radical reactions in solution, as in the photolysis kinetic model, were considered.

(b) For the balance of hydroxyl radicals:

$$r = k_{1s}C_{h^+} + k_{eH}C_{H_2O_2}C_{e^-} - k_{HOs-M}C_{HO\bullet}C_M \tag{49}$$

where C_{h+} and C_{e-} are the concentrations of holes and electrons, which are considered equal [78] and expressed as follows:

$$C_{h^+} = C_{e^-} = \frac{-1 + (1 + 2\gamma r_0)^{0.5}}{k_{1s}\gamma} \tag{50}$$

where

$$\gamma = \frac{2k_{rec}}{k_{1s}^2} \tag{51}$$

with k_{1s} being a function of the rate constant of reaction (42), equilibrium constant between adsorbed and non-adsorbed hydroxyl radicals and catalyst concentration [76] and r_0 is the TiO_2 photon absorption rate calculated by solving the radiative transfer balance using the Six Flux Model (SFM) [79] (see Table S3 for r_0 values). It has to be noted that for $0.25\ g{\cdot}L^{-1}$ catalyst concentration, the r_0 value obtained from SFM was higher than the incident intensity of radiation, so this latter rate was taken as the photon absorption rate for this catalyst concentration. Therefore, at this catalyst concentration, the direct photolysis rate of pharmaceuticals, hydrogen peroxide, and WWEf was not included in the kinetic model. As a result, the participation of singlet oxygen was not considered either.

(c) For the balance of total hydrogen peroxide:

$$r = -k_{eH}C_{H_2O_2}C_{e^-} \tag{52}$$

(d) For the balance of superoxide ion radical [76]:

$$r = k_{HOs-M} C_{HO^\bullet} C_M - \omega \left((1 + 2\gamma r_0)^{0.5} - 1 \right) \tag{53}$$

where

$$\omega = \frac{k_{1s} k_e C^*_{O2}}{2k_{rec}} \tag{54}$$

with C^*_{O2} as oxygen solubility in water.

The kinetic model was then solved with initial conditions (39) and (40) with the Ode23s code of MATLAB 2024b and rate constants shown in Tables S2 and S3. Figure 5 also presents kinetic model calculated results of the photocatalytic degradation of the mixture of pharmaceuticals using different concentrations of TiO_2 (0.05, 0.10, and 0.25 $g \cdot L^{-1}$). Figure 5a–c show the temporal evolution of the dimensionless concentrations (C/C_0) of the pharmaceuticals, while Figure 5d–g display the simulated evolution of the main reactive species concentrations: $HO^\bullet$, $O_2^{\bullet-}$, $^1O_2$, and H_2O_2, respectively.

As shown in Figure 5a–c, increasing the TiO_2 concentration slightly enhances the degradation rate of all four pharmaceuticals. The fastest degradation occurs with 0.25 $g \cdot L^{-1}$ (Figure 5c), as indicated by the steeper decline in C/C_0 values relative to the other two concentrations. This behavior is attributed to the greater availability of active sites. However, the generation of hydroxyl radicals, which can be considered as the main oxidizing ROS, increases for 0.05 to 0.1 $g \cdot L^{-1}$ but it decreases at the highest photocatalyst concentration tested, 0.25 $g \cdot L^{-1}$, which can be due to light attenuation at high TiO_2 concentrations (Figure 5d). Also, likely for similar reason, the degradation curves tend to converge after 10 min, irrespective of the TiO_2 concentration applied.

The kinetic model accurately captures the degradation kinetics of the four pharmaceuticals, especially during the initial stages of the process, where differences among compounds are more evident. In later stages, slight deviations may be attributed to experimental limitations, the accumulation of intermediates, or partial deactivation of the photocatalyst. Nevertheless, overall agreement between experimental and simulated data confirms the ability of the model to accurately describe the system behavior under various operating conditions.

On the other hand, the evolution of $O_2^{\bullet-}$, $^1O_2$, and H_2O_2 concentration (Figure 5e–g) with time shows what seems to be a logical trend: the increase in catalyst concentration leads to an increase in these species concentrations. An opposite trend is observed in the case of hydroxyl radical concentration, which can explain the slight differences observed in pharmaceutical concentrations (both calculated and experimental) with varying TiO_2 concentrations.

In summary, the results confirm that increasing TiO_2 concentration improves photocatalytic performance up to a certain point. Beyond this threshold, excess TiO_2 may cause light scattering or shielding effects, which reduce process efficiency. The simulated ROS concentrations provide further insight into the degradation mechanisms, highlighting the predominant role of hydroxyl radicals and the complementary roles of other species, including the superoxide ion radical, singlet oxygen, and hydrogen peroxide.

2.4. Importance of End Radical Reactions

The comparison between experimental and simulated degradation curves in secondary effluent when intermediate reactions are not considered in the kinetic model (shown in Figure S6) highlights the importance of their inclusion in the model. Nonetheless, in UVC photolysis, from Figure S6a, it is seen that for the first 10–15 min, there is an acceptable agreement between experimental and calculated concentrations, even for the case of CLA and ANT. However, at higher reaction times, when these two compounds have already been completely removed, the deviation between experimental and calculated concentrations of ATL and ACM starts to be of increasing importance. There are two possible explanations for this behavior: (a) It is likely that due to the faster reaction rates of CLA and ANT, competition of intermediates could be negligible during the time these compounds need to be removed. The results obtained from the kinetic model when intermediate reactions (18) and (20) were not considered support this fact. (see Section 2.4 and Figure S6a,b. (b) The rate-terminating term (k_T for hydroxyl radicals and k_{TT} for superoxide ion radicals) should be a function of time, of practically negligible value during the initial minutes, as intermediates at such low concentrations do not yet compete for free radicals and radiation. The situation changes at higher reaction times, where competition becomes important as calculated concentrations of ACM and ATL start to be much lower than the experimental ones. However, during UVC photocatalysis, for k_T and $k_{TT} = 0$ (reactions (18) and (20)), removal of pharmaceuticals is almost instantaneous, highlighting the importance of free radical termination reactions. This indicates a higher formation of intermediate and end products in UVC photocatalytic oxidation.

3. Materials and Methods

3.1. Chemicals

The pharmaceuticals ATL (CAS 29122-68-7, 99%), CLA (CAS 882-09-7, >99.5%), ACM (CAS 103-90-2, 99%), and ANT (CAS 60-80-0, >99.5%) were purchased from Sigma-Aldrich (Burlington, MA, USA). Sodium hydroxide (CAS 1310-73-2, 99.5%) and hydrogen peroxide (CAS 7722-84-1, 30%) were purchased from PanReac AppliChem (Barcelona, Spain). Titanium (IV) oxysulfate (CAS 123334-00-9, >29%), sodium azide (CAS 26628-22-8, >99.5%), and tert-butanol (CAS 75-65-0, 99.5%), TIRON (4,5-Dihydroxy-1,3-benzenedisulfonic acid disodium salt monohydrate, CAS:270573-71-2, Sigma-Aldrich, 99%) were also obtained from Sigma-Aldrich, Furfural (Fischer Scientific, Waltham, MA, USA, CAS: 98-01-1), Rose Bengal (4,5,6,7-Tetrachloro-2′,4′,5′,7′-tetraiodofluorescein disodium salt, Sigma-Aldrich, CAS: 632-69-9). Ortho-phosphoric acid (PanReac AppliChem, CAS 7664-38-2) and acetonitrile (PanReac AppliChem, CAS 75-05-8). TiO_2 P25 Aeroxide® was supplied by Evonik Industries (Essen, Germany). This photocatalyst is widely used in photocatalytic studies due to its well-known physicochemical properties, including a high surface area (~50 m^2/g), average particle size of 30 nm, and a mixed-phase composition of approximately 70% anatase and 30% rutile [80]. Ultrapure water was sourced from a Milli-Q system with a resistivity of 18.2 MΩ·cm.

The wastewater effluent (WWEf) from a municipal wastewater treatment plant (WWTP) was used as the matrix to evaluate the process's performance. This effluent was pre-filtered using an S-PAC filter with a pore size of 0.45 μm (Merck, Darmstadt, Germany) to remove solid particles and stored in a refrigerator at 5 °C. The pre-treated effluent was used within five days of collection to ensure sample integrity. The physicochemical characterization of the WWEf is presented in Table S4 of the Supplementary section.

3.2. Photochemical Experiments

The photochemical experiments were performed in an 800 mL cylindrical glass reactor equipped with a 15 W low-pressure mercury vapor UV_C lamp (Heraeus, model TNN 15–32) emitting mainly at 254 nm. The lamp was placed in the middle of the reactor inside a quartz sleeve provided with a cooling jacket. The incident radiation (I_0) and the effective light path length of the reactor (L) were determined using hydrogen peroxide as an actinometer following standard procedures described elsewhere [81]. Values of 3.99×10^{-6} Einstein·L^{-1}·s^{-1} and 1.81 cm were found for I_0 and L, respectively. The apparent and direct quantum yields of the pharmaceuticals at 254 nm were determined by photolysis experiments under UVC radiation in the absence and presence of t-butanol (TBA) (0.01 M), TIRON, and sodium azide (NaN_3) (0.01 M) [40].

Simple photolysis experiments were conducted for each pharmaceutical individually and for a mixture of the four pharmaceuticals collectively, using both ultrapure water and wastewater buffered in advance with phosphate buffer at pH 7 (2 mM). The initial concentration of each pharmaceutical was set at 0.5 $mg \cdot L^{-1}$. Photocatalytic experiments were carried out following the same procedure, with TiO_2 P25 being added beforehand in varying concentrations ranging from 0.05 to 0.25 $g \cdot L^{-1}$. The aqueous mixture containing the pharmaceuticals and the powdered catalyst was kept under stirring in the dark for 30 min using a magnetic stirrer to allow the potential adsorption of the contaminants on the surface of the catalyst.

Before the experiment started, the UVC lamp was turned on to stabilize the emission and ensure consistent irradiance at the beginning of the test. After the stabilization period, the lamp was turned off, and the solution containing the pharmaceutical, with or without the catalyst, was introduced. The cooling system for the lamp was activated, and the lamp was switched on again, marking the start of the experiment. Samples were collected at intervals during the experiment and filtered (using 0.45 μm syringe filters, Millex-HA®, Millipore, Darmstadt, Germany) if TiO_2 P25 was present for subsequent analysis.

3.3. Analytical Methods

The absorption spectra of pharmaceuticals were recorded from 200 to 400 nm using a spectrophotometer (Thermo Scientific UV-Vis Evolution 201, Waltham, MA, USA) and a quartz cuvette (1 cm path length). The molar absorption coefficient was determined at 254 nm after applying the Beer–Lambert law to spectrometric measurements of stock solutions of the pharmaceuticals at different concentrations at pH 7. A TOC-VSCH analyzer (Shimadzu, Kyoto, Japan) was used to determine the dissolved organic carbon. The concentrations of pharmaceuticals in water were analyzed using an HPLC-DAD (Hitachi, Elite LaChrom, San Jose, CA, USA), using a C-18 Phenomenex column (3 × 150 mm, 5 μm) provided with a degassing unit, an automatic injector, an oven, and a diode array detector. The analysis was performed using a gradient method with a mixture of acidified ultrapure water (H_3PO_4 0.1% v/v) and acetonitrile at 0.6 $mL \cdot min^{-1}$. The method started with 5 min of initial stabilization with a water (A) and acetonitrile (B) ratio of 95:5. A linear gradient for 5 min to reach 80:20 A:B followed, and this ratio was maintained for 4 min. Subsequently, a second linear gradient for 8 min up to 5:95 A:B was applied; this ratio was maintained for 1 min. Finally, in order to return to the initial conditions, a 1 min linear gradient is applied to achieve a 95:5 initial ratio of A to B, which is then maintained for 6 min The pharmaceutical concentrations were measured at the following wavelengths and retention times: ATL (λ = 230 nm, t_r = 3.00 min), ACM (λ = 250 nm, t_r = 5.10 min), ANT (λ = 250 nm, t_r = 13.40 min), CLA (λ = 230 nm, t_r = 21.80 min). The concentration of hydrogen peroxide was tried to be spectrophotometrically evaluated following the formation of the yellowish pertitanic acid at 405 nm [82]. For the determination of the second-order rate constant of each pharmaceutical with singlet oxygen, the same HPLC analytical method was used, where F was detected at λ = 278 nm with a retention time (t_r) of 4.8 min.

4. Conclusions

This study proposes a comprehensive kinetic modeling approach for the UVC photolysis and UVC/TiO_2 photocatalysis of a mixture of four representative pharmaceuticals (ATL, ACM, CLA, and ANT) in aqueous media. Experimental results confirmed the generation of hydroxyl and, to a lesser extent, superoxide ion radicals and singlet oxygen during photolysis. Determination of quantum yields and scavenger experiments enables ROS identification of the reactive oxygen species (ROS) involved in each case.

The validated kinetic model, incorporating ROS generation and reaction pathways, demonstrated strong predictive capabilities, especially for a secondary effluent matrix. The presence of natural organic matter in the effluent reduced photodegradation efficiency by competing for radiation and reactive species. This matrix effect was successfully integrated into the model by accounting for effluent absorbance and ROS quenching reactions.

Under photocatalytic conditions, significant enhancement of degradation rates was observed, particularly for slower-reacting compounds (ATL and ACM). The model also captured the influence of catalyst concentration on ROS evolution and degradation kinetics, identifying an optimal TiO_2 loading beyond which performance declined due to scattering and recombination effects.

In summary, the proposed kinetic framework provides a detailed mechanistic understanding of pharmaceutical degradation in complex water matrices and offers a reliable tool for designing and optimizing advanced treatment processes under both photolytic and photocatalytic conditions.

Supplementary Materials: The following supporting information can be downloaded at: https://www.mdpi.com/article/10.3390/catal15050471/s1, Figure S1. Comparison of initial scavenger and pharmaceutical absorbance at 254 nm. Conditions: C_{Pharm} = 0.5 ppm, $C_{TBA,0}$ = 0.01 M, $C_{NAN3,0}$ = 0.01 M. Figure S2. Fit of absorbance at 254 nm of wastewater effluent with time using the analytical solution of Equation (5), employing the quantum yield value calculated from experiments with TBA and TIRON. Initial conditions: Abs_{2540} = 0.160 a.u., C_{Tiron0} = 0.001 M, $C_{TBA,0}$ = 0.01 M pH_0 = 7. Figure S3. Fitting Equation (3) to the experimental photolysis data: (a) ATL, (b) ACM, (c) CLA, (d) ANT. Conditions as in Figure 2. Green line: without scavengers; black line: with TBA and TIRON. Figure S4. Evaluation of Equation (4) based on the experiments with Rose Bengal. ATL, (b) ACM, (c) CLA, (d) ANT. Determination of the singlet oxygen-pharmaceutical reaction rate constant. Figure S5. Total organic carbon (TOC) removal during photolysis and photocatalysis of the mixture of pharmaceuticals in effluent wastewater: experimental results. Conditions: C_{Pharm0} = 0.5 $mg{\cdot}L^{-1}$, pH_0 = 7, Io = 3.99 × 10^{-6} Einstein$\cdot L^{-1}s^{-1}$, TOC_0 = 8.52 $mg{\cdot}L^{-1}$. Figure S6. Experimental degradation profiles (symbols) in secondary effluent under (a) photolytic (UVC) and (b) photocatalytic (UVC/TiO_2) conditions. Solid lines represent simulation results obtained without including termination and intermediate degradation reactions in the kinetic model. Table S1. Key results and experimental conditions from references related to the UVC photolysis and TiO_2-photocatalytic removal of the selected compounds. Table S2. Kinetic parameters and rate constants used in the photolysis model[a]. Table S3. Kinetic parameters and rate constants used in the photocatalysis model[a]. Table S4. Chemical characterization of secondary effluent. References [83–86] are cited in the Supplementary Materials.

Author Contributions: Conceptualization, M.A.J.-L. and F.J.B.; methodology, M.A.J.-L.; software, F.J.B.; validation, M.A.J.-L. and M.C.R.-G.; formal analysis, M.A.J.-L. and M.C.R.-G.; investigation, M.A.J.-L. and M.C.R.-G.; resources, F.J.B.; data curation, M.A.J.-L. and M.C.R.-G.; writing—original draft preparation, M.A.J.-L. and F.J.B.; writing—review and editing, M.A.J.-L. and F.J.B.; visualization, M.A.J.-L.; supervision, F.J.B.; project administration, F.J.B.; All authors have read and agreed to the published version of the manuscript.

Funding: This research was funded by the Spanish Ministerio de Ciencia e Innovación and Agencia Estatal de Investigación Grant PID2023-149919OB-I00 funded by MCIN/AEI/ 10.13039/501100011033 and by ESF+.

Data Availability Statement: The raw data supporting the conclusions of this article will be made available by the authors upon request.

Acknowledgments: The authors gratefully acknowledge financial support from Grant PID2023-149919OB-I00 funded by MCIN/AEI/ 10.13039/501100011033 and by ESF+. M.A. Jiménez-López thanks the Ministry of Science, Innovation, and Universities for granting the FPI PRE2020-091840 research training scholarship.

Conflicts of Interest: The funders had no role in the design of the study; in the collection, analyses, or interpretation of data; in the writing of the manuscript; or in the decision to publish the results.

References

1. Kanakaraju, D.; Glass, B.D.; Oelgemöller, M. Advanced Oxidation Process-Mediated Removal of Pharmaceuticals from Water: A Review. *J. Environ. Manag.* **2018**, *219*, 189–207. [CrossRef] [PubMed]
2. Kumar, R.; Sarmah, A.K.; Padhye, L.P. Fate of Pharmaceuticals and Personal Care Products in a Wastewater Treatment Plant with Parallel Secondary Wastewater Treatment Train. *J. Environ. Manag.* **2019**, *233*, 649–659. [CrossRef]
3. Peña-Guzmán, C.; Ulloa-Sánchez, S.; Mora, K.; Helena-Bustos, R.; Lopez-Barrera, E.; Alvarez, J.; Rodriguez-Pinzón, M. Emerging Pollutants in the Urban Water Cycle in Latin America: A Review of the Current Literature. *J. Environ. Manag.* **2019**, *237*, 408–423. [CrossRef]
4. Rivera-Utrilla, J.; Sánchez-Polo, M.; Ferro-García, M.Á.; Prados-Joya, G.; Ocampo-Pérez, R. Pharmaceuticals as Emerging Contaminants and Their Removal from Water. A Review. *Chemosphere* **2013**, *93*, 1268–1287. [CrossRef] [PubMed]
5. Dong, Z.; Senn, D.B.; Moran, R.E.; Shine, J.P. Prioritizing Environmental Risk of Prescription Pharmaceuticals. *Regul. Toxicol. Pharmacol.* **2013**, *65*, 60–67. [CrossRef] [PubMed]
6. Oates, R.P.; Longley, G.; Hamlett, P.; Klein, D. Pharmaceutical and Endocrine Disruptor Compounds in Surface and Wastewater in San Marcos, Texas. *Water Environ. Res.* **2017**, *89*, 2021–2030. [CrossRef]
7. Haro, N.K.; Del Vecchio, P.; Marcilio, N.R.; Féris, L.A. Removal of Atenolol by Adsorption—Study of Kinetics and Equilibrium. *J. Clean Prod.* **2017**, *154*, 214–219. [CrossRef]
8. Papageorgiou, M.; Kosma, C.; Lambropoulou, D. Seasonal Occurrence, Removal, Mass Loading and Environmental Risk Assessment of 55 Pharmaceuticals and Personal Care Products in a Municipal Wastewater Treatment Plant in Central Greece. *Sci. Total Environ.* **2016**, *543*, 547–569. [CrossRef]
9. Ulvi, A.; Aydın, S.; Aydın, M.E. Fate of Selected Pharmaceuticals in Hospital and Municipal Wastewater Effluent: Occurrence, Removal, and Environmental Risk Assessment. *Environ. Sci. Pollut. Res.* **2022**, *29*, 75609–75625. [CrossRef]
10. Folarin, O.S.; Otitoloju, A.A.; Amaeze, N.H.; Saliu, J.K. Occurrence of Acetaminophen, Amoxicillin, Diclofenac and Methylparaben in Lagos and Ologe Lagoons, Lagos, Nigeria. *J. Appl. Sci. Environ. Manag.* **2019**, *23*, 2143–2149. [CrossRef]
11. McLain, N.K.; Gomez, M.Y.; Gachomo, E.W. Acetaminophen Levels Found in Recycled Wastewater Alter Soil Microbial Community Structure and Functional Diversity. *Microb. Ecol.* **2023**, *85*, 1448–1462. [CrossRef]
12. Ayman, Z.; Işik, M. Pharmaceutically Active Compounds in Water, Aksaray, Turkey. *Clean* **2015**, *43*, 1381–1388. [CrossRef]
13. Pozdnyakov, I.P.; Zhang, X.; Maksimova, T.A.; Yanshole, V.V.; Wu, F.; Grivin, V.P.; Plyusnin, V.F. Wavelength-Dependent Photochemistry of Acetaminophen in Aqueous Solutions. *J. Photochem. Photobiol. A Chem.* **2014**, *274*, 117–123. [CrossRef]
14. Roveri, V.; Guimarães, L.L.; Toma, W.; Correia, A.T. Occurrence, Ecological Risk Assessment and Prioritization of Pharmaceuticals and Abuse Drugs in Estuarine Waters along the São Paulo Coast, Brazil. *Environ. Sci. Pollut. Res.* **2022**, *29*, 89712–89726. [CrossRef] [PubMed]
15. Cai, M.; Zhang, L.; Qi, F.; Feng, L. Influencing Factors and Degradation Products of Antipyrine Chlorination in Water with Free Chlorine. *J. Environ. Sci.* **2013**, *25*, 77–84. [CrossRef]
16. Gomes, I.B.; Simões, L.C.; Simões, M. The Effects of Emerging Environmental Contaminants on Stenotrophomonas Maltophilia Isolated from Drinking Water in Planktonic and Sessile States. *Sci. Total Environ.* **2018**, *643*, 1348–1356. [CrossRef]
17. Benotti, M.J.; Trenholm, R.A.; Vanderford, B.J.; Holady, J.C.; Stanford, B.D.; Snyder, S.A. Pharmaceuticals and Endocrine Disrupting Compounds in U.S. Drinking Water. *Environ. Sci. Technol.* **2009**, *43*, 597–603. [CrossRef]
18. Lu, J.; Huang, Q.; Mao, L. Removal of Acetaminophen Using Enzyme-Mediated Oxidative Coupling Processes: I. Reaction Rates and Pathways. *Environ. Sci. Technol.* **2009**, *43*, 7062–7067. [CrossRef]
19. Hwang, J.I.; Li, Z.; Andreacchio, N.; Ordonez Hinz, F.; Wilson, P.C. Potential Use of Floating Treatment Wetlands Established with Canna Flaccida for Removing Organic Contaminants from Surface Water. *Int. J. Phytoremediat.* **2020**, *22*, 1304–1312. [CrossRef]
20. Qutob, M.; Hussein, M.A.; Alamry, K.A.; Rafatullah, M. A Review on the Degradation of Acetaminophen by Advanced Oxidation Process: Pathway, by-Products, Biotoxicity, and Density Functional Theory Calculation. *RSC Adv.* **2022**, *12*, 18373–18396. [CrossRef]
21. Quesada, H.B.; Cusioli, L.F.; de O Bezerra, C.; Baptista, A.T.A.; Nishi, L.; Gomes, R.G.; Bergamasco, R. Acetaminophen Adsorption Using a Low-Cost Adsorbent Prepared from Modified Residues of Moringa Oleifera Lam. Seed Husks. *J. Chem. Technol. Biotechnol.* **2019**, *94*, 3147–3157. [CrossRef]

22. Feng, S.; Zhang, X.; Liu, Y. New Insights into the Primary Phototransformation of Acetaminophen by UV/H_2O_2: Photo-Fries Rearrangement versus Hydroxyl Radical Induced Hydroxylation. *Water Res.* **2015**, *86*, 35–45. [CrossRef] [PubMed]
23. Rivas, J.; Gimeno, O.; Borralho, T.; Sagasti, J. UV-C and UV-C/Peroxide Elimination of Selected Pharmaceuticals in Secondary Effluents. *Desalination* **2011**, *279*, 115–120. [CrossRef]
24. Ioannou, L.A.; Hapeshi, E.; Vasquez, M.I.; Mantzavinos, D.; Fatta-Kassinos, D. Solar/TiO_2 Photocatalytic Decomposition of β-Blockers Atenolol and Propranolol in Water and Wastewater. *Sol. Energy* **2011**, *85*, 1915–1926. [CrossRef]
25. Salgado, R.; Pereira, V.J.; Carvalho, G.; Soeiro, R.; Gaffney, V.; Almeida, C.; Cardoso, V.V.; Ferreira, E.; Benoliel, M.J.; Ternes, T.A.; et al. Photodegradation Kinetics and Transformation Products of Ketoprofen, Diclofenac and Atenolol in Pure Water and Treated Wastewater. *J. Hazard. Mater.* **2013**, *244–245*, 516–527. [CrossRef] [PubMed]
26. Su, C.C.; Bellotindos, L.M.; Chang, A.T.; Lu, M.C. Degradation of Acetaminophen in an Aerated Fenton Reactor. *J. Taiwan Inst. Chem. Eng.* **2013**, *44*, 310–316. [CrossRef]
27. Dewil, R.; Mantzavinos, D.; Poulios, I.; Rodrigo, M.A. New Perspectives for Advanced Oxidation Processes. *J. Environ. Manag.* **2017**, *195*, 93–99. [CrossRef]
28. Parsons, S. Advanced Oxidation Processes for Water and Wastewater Treatment. *Water Intell. Online* **2005**, *4*, 9781780403076. [CrossRef]
29. Xiao, R.; Wei, Z.; Chen, D.; Weavers, L.K. Kinetics and Mechanism of Sonochemical Degradation of Pharmaceuticals in Municipal Wastewater. *Environ. Sci. Technol.* **2014**, *48*, 9675–9683. [CrossRef]
30. Tang, Y.; Li, X.M.; Xu, Z.C.; Guo, Q.W.; Hong, C.Y.; Bing, Y.X. Removal of Naproxen and Bezafibrate by Activated Sludge under Aerobic Conditions: Kinetics and Effect of Substrates. *Biotechnol. Appl. Biochem.* **2014**, *61*, 333–341. [CrossRef]
31. Moreno, A.d.H.; Salgado, H.R.N. Stability Study and Degradation Kinetics of Ceftazidime in Pharmaceutical Preparations. *Adv. Anal. Chem.* **2012**, *2*, 1–5. [CrossRef]
32. Tazesh, S.; Tamizi, E.; Siahi Shadbad, M.; Mostaghimi, N.; Monajjemzadeh, F. Comparative Stability of Two Anti-Hyperpigmentation Agents: Kojic Acid as a Natural Metabolite and Its Di-Palmitate Ester, Under Oxidative Stress; Application to Pharmaceutical Formulation Design. *Adv. Pharm. Bull.* **2022**, *12*, 329–335. [CrossRef] [PubMed]
33. Bhangare, D.; Rajput, N.; Jadav, T.; Sahu, A.K.; Tekade, R.K.; Sengupta, P. Systematic Strategies for Degradation Kinetic Study of Pharmaceuticals: An Issue of Utmost Importance Concerning Current Stability Analysis Practices. *J. Anal. Sci. Technol.* **2022**, *13*, 1–13. [CrossRef]
34. Wang, C.; Xiong, M.C.; Zhao, X.; Liu, K.H. Kinetics Study on Reaction of Atenolol with Singlet Oxygen by Directly Monitoring the $1O_2$ Phosphorescence. *Chin. J. Chem. Phys.* **2021**, *34*, 406–412. [CrossRef]
35. Chen, P.; Wang, F.; Zhang, Q.; Su, Y.; Shen, L.; Yao, K.; Chen, Z.F.; Liu, Y.; Cai, Z.; Lv, W.; et al. Photocatalytic Degradation of Clofibric Acid by G-C3N4/P25 Composites under Simulated Sunlight Irradiation: The Significant Effects of Reactive Species. *Chemosphere* **2017**, *172*, 193–200. [CrossRef]
36. Zhang, X.; Liu, Z.; Kong, Q.; Liu, G.; Lv, W.; Li, F.; Lin, X. Aquatic Photodegradation of Clofibric Acid under Simulated Sunlight Irradiation: Kinetics and Mechanism Analysis. *RSC Adv.* **2018**, *8*, 27796–27804. [CrossRef] [PubMed]
37. Bielski, B.H.J.; Cabelli, D.E.; Arudi, R.L.; Ross, A.B. Reactivity of $HO_2/O^-{}_2$ Radicals in Aqueous Solution. *J. Phys. Chem. Ref. Data* **1985**, *14*, 1041–1100. [CrossRef]
38. Wang, J.; Zhang, X.; Fan, L.; Su, L.; Zhao, Y. Photolysis Mechanism of Eleven Insecticides under Simulated Sunlight Irradiation: Kinetics, Pathway and QSAR. *Chemosphere* **2023**, *334*, 138968. [CrossRef]
39. Rodríguez, E.M.; Márquez, G.; Tena, M.; Álvarez, P.M.; Beltrán, F.J. Determination of Main Species Involved in the First Steps of TiO2 Photocatalytic Degradation of Organics with the Use of Scavengers: The Case of Ofloxacin. *Appl. Catal. B* **2015**, *178*, 44–53. [CrossRef]
40. Beltrán, F.J.; Chávez, A.M.; Jiménez-López, M.A.; Álvarez, P.M. Kinetic Modelling of UVC and UVC/H_2O_2 Oxidation of an Aqueous Mixture of Antibiotics in a Completely Mixed Batch Photoreactor. *Environ. Sci. Pollut. Res.* **2024**, *31*, 55222–55238. [CrossRef]
41. Wang, L.; Xiao, K.; Zhao, H. The Debatable Role of Singlet Oxygen in Persulfate-Based Advanced Oxidation Processes. *Water Res.* **2023**, *235*, 119925. [CrossRef]
42. Nidheesh, P.V.; Boczkaj, G.; Ganiyu, S.O.; Oladipo, A.A.; Fedorov, K.; Xiao, R.; Dionysiou, D.D. Generation, Properties, and Applications of Singlet Oxygen for Wastewater Treatment: A Review. *Environ. Chem. Lett.* **2025**, *23*, 195–240. [CrossRef]
43. Katsumata, C.P.; Parizi, M.P.S.; Lastre-Acosta, A.M.; Teixeira, A.C.S.C. Low Pressure UV Photolysis of the Pharmaceutical Compounds Acetaminophen, Atenolol, Bezafibrate, Diclofenac and Ibuprofen. *Water* **2022**, *14*, 3165. [CrossRef]
44. Li, W.; Lu, S.; Qiu, Z.; Lin, K. Clofibric Acid Degradation in UV254/H_2O_2 Process: Effect of Temperature. *J. Hazard. Mater.* **2010**, *176*, 1051–1057. [CrossRef]
45. Tan, C.; Gao, N.; Deng, Y.; Zhang, Y.; Sui, M.; Deng, J.; Zhou, S. Degradation of Antipyrine by UV, UV/H_2O_2 and UV/PS. *J. Hazard Mater.* **2013**, *260*, 1008–1016. [CrossRef] [PubMed]

46. Gao, L.; Guo, Y.; Zhan, J.; Yu, G.; Wang, Y. Assessment of the Validity of the Quenching Method for Evaluating the Role of Reactive Species in Pollutant Abatement during the Persulfate-Based Process. *Water Res.* **2022**, *221*, 118730. [CrossRef]
47. Du, C.; Cheng, D.; Dai, J.; Du, Q.; Shao, A. Differences in the Ability of Dissolved Organic Matter from Different Sources to Produce Reactive Oxygen Species under Light Irradiation. *Huanjing Kexue Xuebao/Acta Sci. Circumstantiae* **2019**, *39*, 2279–2287. [CrossRef]
48. Fujii, M.; Otani, E. Photochemical Generation and Decay Kinetics of Superoxide and Hydrogen Peroxide in the Presence of Standard Humic and Fulvic Acids. *Water Res.* **2017**, *123*, 642–654. [CrossRef]
49. Vione, D.; Scozzaro, A. Photochemistry of Surface Fresh Waters in the Framework of Climate Change. *Environ. Sci. Technol.* **2019**, *53*, 7945–7963. [CrossRef]
50. Westerhoff, P.; Song, R.; Amy, G.; Minear, R. Applications of Ozone Decomposition Models. *Ozone Sci. Eng.* **1997**, *19*, 55–73. [CrossRef]
51. Rabani, J.; Matheson, M.S. The Pulse Radiolysis of Aqueous Solutions of Potassium Ferrocyanide. *J. Phys. Chem.* **1966**, *70*, 761–769. [CrossRef]
52. Baxendale, J.H.; Wilson, J.A. The Photolysis of Hydrogen Peroxide at High Light Intensities. *Trans. Faraday Soc.* **1957**, *53*, 344–356. [CrossRef]
53. Christensen, H.; Sehested, K.; Corfitzen, H. Reactions of Hydroxyl Radicals with Hydrogen Peroxide at Ambient and Elevated Temperatures. *J. Phys. Chem.* **1982**, *86*, 1588–1590. [CrossRef]
54. Staehelln, J.; Hoigne, J. Decomposition of Ozone in Water in the Presence of Organic Solutes Acting as Promoters and Inhibitors of Radical Chain Reactions. *Environ. Sci. Technol.* **1985**, *19*, 1206–1213. [CrossRef]
55. Fogler, H.S. *Elements of Chemical Reaction Engineering*; Prentice Hall Inc.: Upper Saddle River, NJ, USA, 1999; ISBN 0135317088.
56. Leifer, A. The Kinetics of Environmental Aquatic Photochemistry American Chemical Society. In *ACS Professional and Reference Book*; Oxford University Press: New York, NY, USA, 1988.
57. Buxton, G.V.; Greenstock, C.L.; Helman, W.P.; Ross, A.B. Critical Review of Rate Constants for Reactions of Hydrated Electrons, Hydrogen Atoms and Hydroxyl Radicals ($\cdot OH/\cdot O^-$ in Aqueous Solution. *J. Phys. Chem. Ref. Data* **1988**, *17*, 513–886. [CrossRef]
58. Brame, J.; Long, M.; Li, Q.; Alvarez, P. Trading Oxidation Power for Efficiency: Differential Inhibition of Photo-Generated Hydroxyl Radicals versus Singlet Oxygen. *Water Res.* **2014**, *60*, 259–266. [CrossRef]
59. Ahn, Y.; Lee, D.; Kwon, M.; Choi, I.h.; Nam, S.N.; Kang, J.W. Characteristics and Fate of Natural Organic Matter during UV Oxidation Processes. *Chemosphere* **2017**, *184*, 960–968. [CrossRef]
60. Song, C.; Zhang, K.X.; Wang, X.J.; Zhao, S.; Wang, S.G. Effects of Natural Organic Matter on the Photolysis of Tetracycline in Aquatic Environment: Kinetics and Mechanism. *Chemosphere* **2021**, *263*, 128338. [CrossRef]
61. Fu, H.; Zheng, W.; Duan, W.; Fang, G.; Duan, X.; Wang, S.; Feng, C.; Zhu, S. Overlooked Roles and Transformation of Carbon-Centered Radicals Produced from Natural Organic Matter in a Thermally Activated Persulfate System. *Environ. Sci. Technol.* **2024**, *58*, 14949–14960. [CrossRef]
62. Qin, W.; Wang, Y.; Fang, G.; Wu, T.; Liu, C.; Zhou, D. Evidence for the Generation of Reactive Oxygen Species from Hydroquinone and Benzoquinone: Roles in Arsenite Oxidation. *Chemosphere* **2016**, *150*, 71–78. [CrossRef]
63. Calza, P.; Minella, M.; Demarchis, L.; Sordello, F.; Minero, C. Photocatalytic Rate Dependence on Light Absorption Properties of Different TiO_2 Specimens. *Catal. Today* **2020**, *340*, 12–18. [CrossRef]
64. Yang, Q.; Ling Ang, P.; Ray, M.B.; Pehkonen, S.O. Light Distribution Field in Catalyst Suspensions within an Annular Photoreactor. *Chem. Eng. Sci.* **2005**, *60*, 5255–5268. [CrossRef]
65. Minero, C.; Vione, D. A Quantitative Evalution of the Photocatalytic Performance of TiO_2 Slurries. *Appl. Catal. B* **2006**, *67*, 257–269. [CrossRef]
66. Pellegrino, F.; Pellutiè, L.; Sordello, F.; Minero, C.; Ortel, E.; Hodoroaba, V.D.; Maurino, V. Influence of Agglomeration and Aggregation on the Photocatalytic Activity of TiO_2 Nanoparticles. *Appl. Catal. B* **2017**, *216*, 80–87. [CrossRef]
67. Luo, B.; Song, R.; Jing, D. Particle Aggregation Behavior during Photocatalytic Ethanol Reforming Reaction and Its Correlation with the Activity of H_2 Production. *Colloids Surf A Physicochem. Eng. Asp.* **2017**, *535*, 114–120. [CrossRef]
68. Ren, Y.; Jing, D. Study on Particle and Photonic Flux Distributions in a Magnetically Stirred Photocatalytic Reactor. *J. Photonics Energy* **2015**, *5*, 052097. [CrossRef]
69. Wang, X.; Shui, M.; Li, R.; Song, Y. Correlation of Lattice Distortion with Photocatalytic Activity of Titanium Dioxide. *Mater. Res. Bull.* **2008**, *43*, 2476–2484. [CrossRef]
70. Roushenas, P.; Ong, Z.C.; Ismail, Z.; Majidnia, Z.; Ang, B.C.; Asadsangabifard, M.; Onn, C.C.; Tam, J.H. Operational Parameters Effects on Photocatalytic Reactors of Wastewater Pollutant: A Review. *Desalin. Water Treat.* **2018**, *120*, 109–118. [CrossRef]
71. Li, X.; Chen, Y.; Tao, Y.; Shen, L.; Xu, Z.; Bian, Z.; Li, H. Challenges of Photocatalysis and Their Coping Strategies. *Chem. Catal.* **2022**, *2*, 1315–1345. [CrossRef]

72. Mehrabadi, Z.; Faghihian, H. Comparative Photocatalytic Performance of TiO_2 Supported on Clinoptilolite and TiO_2/Salicylaldehyde-NH2-MIL-101(Cr) for Degradation of Pharmaceutical Pollutant Atenolol under UV and Visible Irradiations. *J. Photochem. Photobiol. A Chem.* **2018**, *356*, 102–111. [CrossRef]
73. Radjenović, J.; Sirtori, C.; Petrović, M.; Barceló, D.; Malato, S. Solar Photocatalytic Degradation of Persistent Pharmaceuticals at Pilot-Scale: Kinetics and Characterization of Major Intermediate Products. *Appl Catal. B* **2009**, *89*, 255–264. [CrossRef]
74. Harja, M.; Sescu, A.M.; Favier, L.; Lutic, D. Doping Titanium Dioxide with Palladiun for Enhancing the Photocatalytic Decontamination and Mineralization of a Refractory Water Pollutant. *Rev. Chim.* **2020**, *71*, 145–152. [CrossRef]
75. Turchi, C.S.; Ollis, D.F. Photocatalytic Degradation of Organic Water Contaminants: Mechanisms Involving Hydroxyl Radical Attack. *J. Catal.* **1990**, *122*, 178–192. [CrossRef]
76. Beltrán, F.J.; Jiménez-López, M.A.; Álvarez, P.M.; Rivas, F.J. Kinetic Modeling of Photocatalytic Oxidation-Ozonation of Metronidazole Removal from Water. *J. Environ. Chem. Eng.* **2025**, *13*, 115869. [CrossRef]
77. Milosavijević, B.H.; Mićić, O.I. Solvated Electron Reactions in Water—Alcohol Solutions. *J. Phys. Chem.* **1978**, *82*, 1359–1362. [CrossRef]
78. Tolosana-Moranchel, A.; Casas, J.A.; Carbajo, J.; Faraldos, M.; Bahamonde, A. Influence of TiO_2 Optical Parameters in a Slurry Photocatalytic Reactor: Kinetic Modelling. *Appl. Catal. B* **2017**, *200*, 164–173. [CrossRef]
79. Beltrán, F.J.; Rivas, F.J.; García-araya, J.F. Six Flux Model for the Central Lamp Reactor Applied to an External Four-lamp Reactor. *Catalysts* **2021**, *11*, 1190. [CrossRef]
80. Wong, C.C.; Chu, W. The Hydrogen Peroxide-Assisted Photocatalytic Degradation of Alachlor in TiO_2 Suspensions. *Environ. Sci. Technol.* **2003**, *37*, 2310–2316. [CrossRef]
81. Rabani, J.; Mamane, H.; Pousty, D.; Bolton, J.R. Practical Chemical Actinometry—A Review. *Photochem. Photobiol.* **2021**, *97*, 873–902. [CrossRef]
82. O'Sullivan, D.W.; Tyree, M. The Kinetics of Complex Formation between Ti(IV) and Hydrogen Peroxide. *Int. J. Chem. Kinet.* **2007**, *39*, 457–461. [CrossRef]
83. Dalida, M.L.P.; Amer, K.M.S.; Su, C.C.; Lu, M.C. Photocatalytic Degradation of Acetaminophen in Modified TiO_2 under Visible Irradiation. *Environ. Sci. Pollut. Res.* **2014**, *21*, 1208–1216. [CrossRef] [PubMed]
84. Shahabi, H.; Allahrasani, A.; Naghizadeh, A. Photocatalytic Degradation of Acetaminophen in Aqueous Solution in the Presence of Montmorillonite Nanosheets Modified with Titanium Dioxide. *Desalin. Water Treat.* **2019**, *149*, 164–170. [CrossRef]
85. Silva, C.G.; Faria, J.L. Anatase vs. Rutile Efficiency on the Photocatalytic Degradation of Clofibric Acid under near UV to Visible Irradiation. *Photochem. Photobiol. Sci.* **2009**, *8*, 705–711. [CrossRef]
86. Peñas-Garzón, M.; Gómez-Avilés, A.; Álvarez-Conde, J.; Bedia, J.; García-Frutos, E.M.; Belver, C. Azaindole Grafted Titanium Dioxide for the Photodegradation of Pharmaceuticals under Solar Irradiation. *J. Colloid Interface Sci.* **2023**, *629*, 593–603. [CrossRef] [PubMed]

catalysts MDPI

Article

Green and Eco-Friendly Egg White–TiO_2 Hydrogel with Enhanced Antimicrobial, Adsorptive, and Photocatalytic Properties

Mei Zhang and Xu Wang *

School of Materials Science and Engineering, Yancheng Institute of Technology, Yancheng 224051, China
* Correspondence: wangxu@ycit.cn

Abstract: The design of multi-purpose decontaminants with environmentally friendly characteristics, low cost, and high efficiency in removing pollutants from the environment is an effective and economic strategy for maintaining the long-term development of the ecosystem. Based on the strategy of killing two birds with one stone, an egg white (EW)/TiO_2 hydrogel with a porous structure is devised as a bio-adsorbent using waste eggs nearing their expiration date for simultaneously achieving the efficient removal of organic dyes and the inactivation of microorganisms from industrial wastewater. The characterizations of its morphology and composition using scanning electron microscopy (SEM), the Brunauer–Emmett–Teller (BET) theory, energy-dispersive spectrometry (EDS), Fourier transform infrared spectroscopy (FTIR), and a thermogravimetric analyzer (TGA) validate the successful synthesis of EW/TiO_2. The maximum adsorption capacity of EW/TiO_2 is 333.172 $mg \cdot mL^{-1}$ according to the Langmuir model. The photodegradation of a methyl blue (MB) solution under irradiation via a xenon lamp is used to assess the photocatalytic behavior of EW/TiO_2. Among the different samples, the 5 wt% TiO_2-doped EW/TiO_2 hydrogel shows an efficiency of 99% for 120 min of irradiation. Finally, the antibacterial properties of the EW/TiO_2 hydrogel are evaluated by calculating its bacterial survival rate against *Escherichia coli* (*E. coli*). The EW/TiO_2 photocatalyst exhibits a photocatalytic inactivation efficiency of 90.4%, indicating that the EW/TiO_2 hydrogel possesses positive antibacterial activity via effectively inhibiting the growth of the bacteria, which is suitable for industrial wastewater treatment over a long period of time.

Keywords: egg white; TiO_2; hydrogel; wastewater treatment

1. Introduction

The deterioration of the ecological environment induced by water contamination is a great challenge for human beings [1–3]. Contaminated water mainly contains heavy metal ion contaminants, organic contaminants, and microbial contaminants. Among these contaminations, organic dye and microbial contaminations have attracted extensive attention owing to their high toxicity, stability, and pathogenicity in the natural environment [3–8]. Massive classical technologies such as chemical oxidation [9], adsorption [10–12], filtration [13,14], and photocatalysis [15,16] have been employed to remove organic dyes and pathogenic microorganisms. In recent decades, photocatalysis has proven to be an effective technology for organic dye removal and disinfecting industrial wastewater.

The photocatalytic degradation of organic dye is considered as a green technology that completely decomposes contaminants without secondary pollution [17,18]. Various semiconductor materials, such as ZnS [19], ZnO [20], MoS_2 [21], WO_3 [22,23], C_3N_4 [24], and TiO_2 [25,26], have been utilized in photocatalysis to degrade organic dyes. Significantly, nano-scaled TiO_2 is considered as a representative photocatalyst due to its satisfactory photoactivity, nontoxicity, low cost, and chemical stability. Additionally, it has been reported that TiO_2 crystals possess excellent antibacterial effects. However, TiO_2 nanoparticles are prone to agglomeration upon environmental changes, which can severely affect their photocatalytic activity [27]. Furthermore, the low adsorption capacity of TiO_2 nanoparticles in photocatalytic processes may reduce the contact of active sites with contaminants, which lowers the photocatalytic degradation efficiency. To solve the above issues, some researchers propose to immobilize TiO_2 on carbon-based materials, including carbon nanotubes, graphene, and carbon quantum dots, to enhance the photocatalytic efficiency via synergies of adsorption and photocatalysis [28,29]. However, the preparation conditions of carbon-based materials are strict, and the cost of these materials is high. Therefore, it is imperative to design a novel TiO_2-based photocatalyst with effective antibacterial performance, high adsorption capacity, and photocatalytic activity. Recently, biomass-derived materials, such as chitosan [30], cellulose [31], alginate [32], and protein [33–36], have received much attention as they are low-cost, environmentally friendly, and biodegradable, which makes them a momentous option for practical application.

Among these biomass-derived materials, egg-based materials have recently become a new research field. Eggs are regarded as a high-quality dietary source of protein for human beings and have been widely applied in biomedical and food engineering [37,38]. However, recently, the production of eggs in China has exceeded egg consumption [39]. To our knowledge, the quality guarantee period of eggs is short. As a result, identifying how to make reasonable use of superfluous eggs nearing their expiration date is a serious challenge. Considering the superiority of eggs, such as gelling characteristics, polymerization, low cost, eco-friendliness, and multiple other functions, there is much potential for different applications of these versatile materials.

For this purpose, we introduced a simple method to synthesize an EW/TiO_2 hydrogel with a hierarchical structure by combining TiO_2 and egg white (EW) protein nearing its expiration date to achieve effective bactericidal capability via photocatalytic inactivation and organic dye removal performance by the coordination of adsorption and photodegradation. The successful immobilization of TiO_2 nanoparticles in a 3D network nanostructured EW was confirmed with scanning electron microscopy, energy-dispersive spectrometry, and Fourier transform infrared spectroscopy. Methylene blue (MB) and *Escherichia coli* (*E. coli*) were employed as an organic dye model and biological model due to their abundant presence in wastewater and their adverse effects on various organisms. More importantly, the EW/TiO_2 hydrogel can be simply taken out and reused again. In a word, the fabricated hydrogel in this work provides new insight into simultaneous organic dye removal and the inactivation of pathogenic microorganisms using waste eggs nearing their expiration date based on "one stone kills two birds" tactics, thereby exhibiting a promising future in wastewater treatment.

2. Results and Discussion

2.1. Characterization of the EW/TiO_2 Hydrogel

To investigate the performances of the EW/TiO_2 hydrogel, its morphology was first investigated. As shown in Figure 1a–e, the microcosmic structures of EW, EW/TiO_2(2.5%), EW/TiO_2(5%), EW/TiO_2(7.5%), and EW/TiO_2(10%) were characterized. The surfaces of all the above hydrogels exhibit several cross-linking points with a relatively uniform distribution. In Figure 1a, the surface of the pure EW hydrogel containing massive lamella exhibits smoother morphology. In Figure 1b–e, the synthesized hydrogel displays a distinct 3D porous structure with a pore diameter from 0.15 to 0.45 μm. The pore wall is about 50 nm in thickness. Further, it is also observed that some particles are randomly distributed on the surface of the hydrogel. The addition of TiO_2 into the EW hydrogel makes the surface uneven and rougher, providing more active sites that improve the adsorption capacity. However, excess TiO_2 causes agglomeration inside the hydrogel, which will reduce the active sites of the EW/TiO_2 hydrogel. Therefore, 5% TiO_2 was selected for the following experiments. To investigate the performance of the EWP/TiO_2 hydrogel, EDS was applied to characterize the elemental composition of the beads. As shown in Figure 1f, the hydrogel mainly includes C, O, N, Au, S, and Ti elements (pink curve). Among these elements, the C,

O, N, and S elements originate from EW (green curve). Element Ti comes from TiO_2. The above results imply the successful synthesis of the EW/TiO_2 hydrogel.

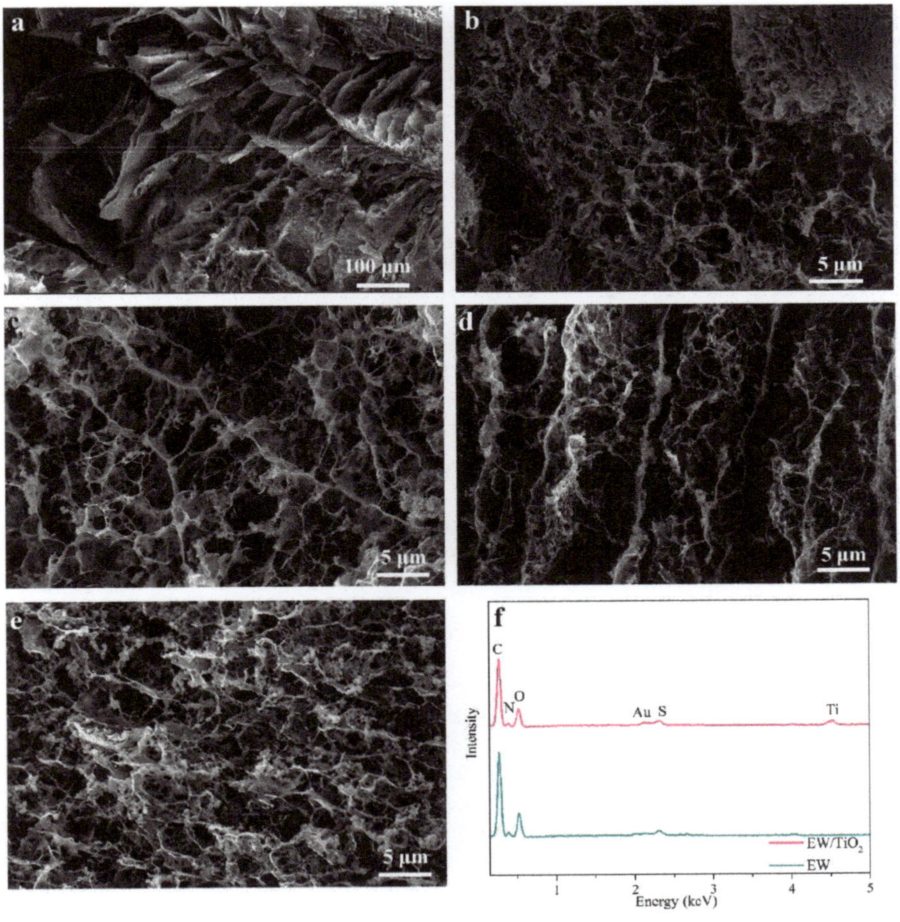

Figure 1. The morphology analysis and elemental characterization of the EW/TiO_2 hydrogel. SEM images of (**a**) EW hydrogel, (**b**) EW/TiO_2 (2.5%), (**c**) EW/TiO_2 (5%), (**d**) EW/TiO_2 (7.5%), and (**e**) EW/TiO_2 (10%). The addition of TiO_2 into the EW hydrogel makes the surface rougher, providing more active sites. (**f**) EDS of the EW and EW/TiO_2 hydrogels. Elemental characterizations verify the successful synthesis of the EW/TiO_2 hydrogel.

To illustrate the influence of the porous structure on the antibacterial, adsorption, and photocatalytic performance, we measured the specific surface areas of the as-synthesized EW/TiO_2 hydrogel via N_2 adsorption–desorption isotherms. As shown in Figure 2a, the EW/TiO_2 hydrogel exhibited a type III isotherm and visible H2 hysteresis loop in the isotherm, indicating irregular mesopores in the hydrogel. Additionally, the pore size distribution curve (Figure 2b) displays that mesopores were observed in the EW/TiO_2 hydrogel with a pore size from 20 to 50 nm. According to the obtained results, the specific surface area of the EW/TiO_2 hydrogel is 30.41 $m^2 \cdot g^{-1}$, attributed to the mesopores generated in the crosslinking of EW. The generated mesopores may provide more active sites in antibacterial, adsorption, and photocatalytic processes.

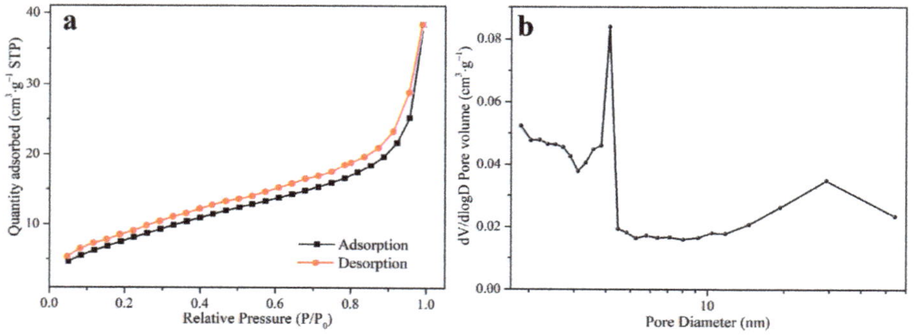

Figure 2. (**a**) N_2 adsorption–desorption isotherms of the EW/TiO_2 hydrogel. The EW/TiO_2 hydrogel with a specific surface area of 30.41 $m^2 \cdot g^{-1}$ exhibits a type III isotherm and visible H2 hysteresis loop in the isotherm. (**b**) The pore size distribution of the EW/TiO_2 hydrogel. The hydrogel with a pore size from 20 to 50 nm displays mesopores.

To further explore the performance of the EW/TiO_2 hydrogel, the functional groups of the hydrogel were characterized by employing FTIR (Figure 3a). A characteristic broadband at around 3347 cm^{-1} is attributed to the N-H stretching of EW [30]. The peaks at 1389 and 1068 cm^{-1} belong to the vibration of C-O and C-O-C in EW. The peaks at 1644, 1534, 1448, and 630 cm^{-1} belong to C=O stretching (amide I), N-H bending (amide II), C-N stretching (amide III), and Ti-O, certifying the successful synthesis of the EW/TiO_2 hydrogel [36].

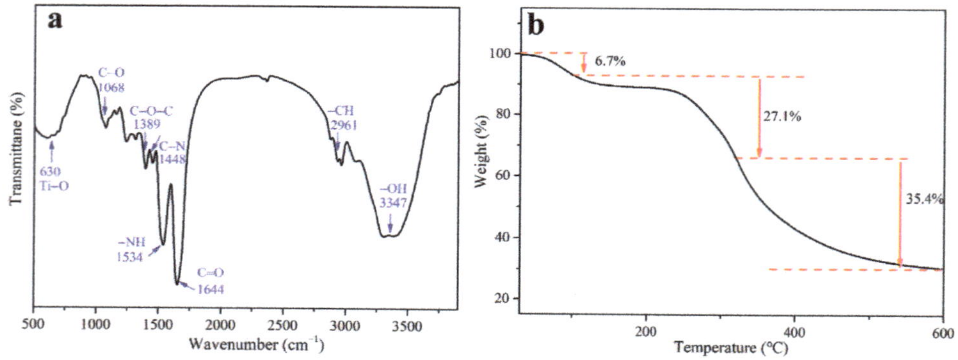

Figure 3. (**a**) FTIR spectra of the EW/TiO_2 hydrogel certify the successful synthesis of the EW/TiO_2 hydrogel. (**b**) TGA analysis confirms that the fabricated EW/TiO_2 possesses favorable thermal stability.

TGA was employed to check the thermal stability of the EW/TiO_2 hydrogel. As shown in Figure 3b, the TGA curves of the EW/TiO_2 hydrogel exhibit a weight loss of about 6.7% due to dehydration before 100 °C [36]. Between 100 and 320 °C, the EW/TiO_2 hydrogel displays a mass loss of 27.1% due to the dehydroxylation and decomposition of organic ingredients as well as the destruction of hydrogen bonds. From 320 to 600 °C, a conspicuous weight loss of 35.4% is observed due to carbonization and further decomposition of the organic components in the EW/TiO_2 hydrogel. According to the above results, we confirm that the fabricated EW/TiO_2 possesses favorable thermal stability.

2.2. *Adsorption Performance of the EW/TiO$_2$ Hydrogel*

2.2.1. Effects of TiO$_2$ on MB Adsorption Property

Figure 4 shows the MB removal rates by EW, EW/TiO$_2$(2.5%), EW/TiO$_2$(5%), EW/TiO$_2$ (7.5%), and EW/TiO$_2$(10%) for 60 min at 25 °C. As shown in Figure 4, EW exhibits a poor MB adsorption efficiency of ~40%. With the increase in the TiO$_2$ weight ratio from 2.5% to 5.0% (wt%), the MB removal rate rises from 52.1% to 70.0% as the addition of TiO$_2$ causes the surface to be rougher and provides more adsorption sites for MB. However, with the increase of TiO$_2$ in the weight ratio from 5% to 10%, the MB removal rate declines from 70.0% to 60.3%. The decline in the MB removal rate may be attributed to excess TiO$_2$, which gives rise to the agglomeration inside the hydrogel that has a negative impact on adsorption. The highest removal rate is achieved at 5% TiO$_2$. Therefore, 5% TiO$_2$ was selected for subsequent experiments.

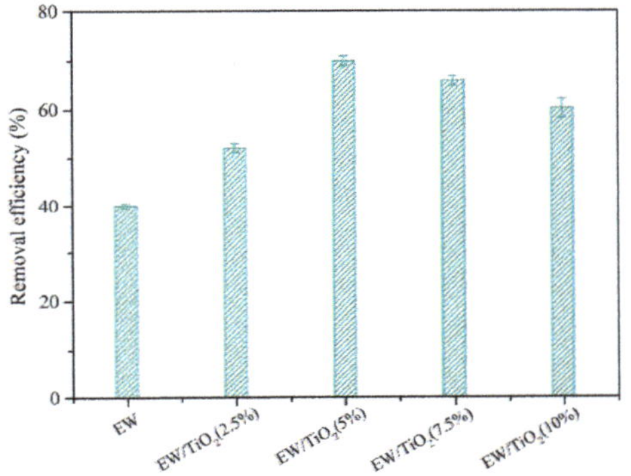

Figure 4. MB removal efficiency of different adsorbents for 60 min at 25 °C. The optimal dosage of TiO$_2$ in EW/TiO$_2$ is 5%. Error bars are based on the average of five measurements.

2.2.2. Effect of Adsorbent Dosage

To ensure cost-efficiency in wastewater treatment, the dosages of EW/TiO$_2$ were optimized to achieve high efficiency in MB adsorption. Figure 5 indicates an increase in the MB removal efficiency from 68 to 98.5% with the increased dosage of EW/TiO$_2$ from 5 to 25 mg. The maximum removal efficiency towards MB is 98.5%. However, excessive EW/TiO$_2$ leads to the overlapping of adsorption sites. Thus, the adsorbent dosage utilized in the following experiments was 10 mg.

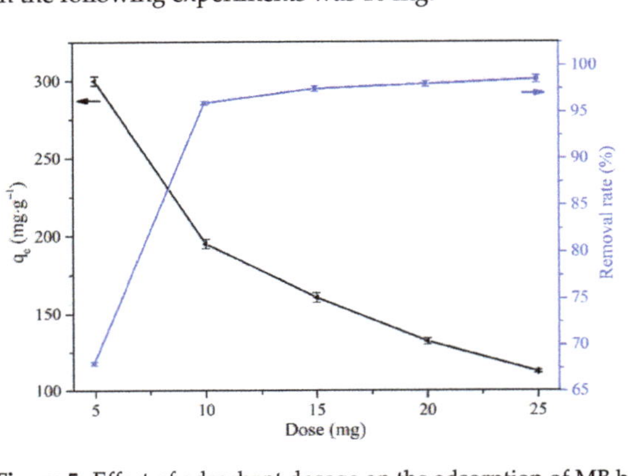

Figure 5. Effect of adsorbent dosage on the adsorption of MB by the EW/TiO$_2$ hydrogel at 25 °C. Error bars are based on the average of five measurements. The optimal EW/TiO$_2$ hydrogel dosage is selected as 10 mg.

To characterize the adsorption performance, adsorption kinetics was used to analyze the adsorption performance of the EW/TiO_2 hydrogel. The adsorption process was studied using the Elovich kinetic model, pseudo-first-order kinetic model, pseudo-second-order kinetic model, and intra-particle diffusion model. The pseudo-first-order model (Equation (1)), pseudo-second-order model (Equation (2)), Elovich kinetic model (Equation (3)), and intra-particle diffusion model (Equation (4)) are formulated as follows [40].

$$q_t = q_e\left(1 - e^{-k_1 t}\right) \tag{1}$$

$$q_t = \frac{q_e^2 k_2 t}{1 + k_2 q_e t} \tag{2}$$

$$q_t = \frac{\ln(\alpha\beta)}{\beta} + \frac{\ln(t)}{\beta} \tag{3}$$

$$q_t = k_i t^{1/2} + c_i \tag{4}$$

where q_t and q_e represent the adsorption capacity at time t and equilibrium adsorption capacity. k_1 (min^{-1}) and k_2 ($g{\cdot}mg^{-1}{\cdot}min^{-1}$) denote the rate constant of the pseudo-first-order model and the pseudo-second-order model. α ($mg{\cdot}g^{-1}{\cdot}min^{-1}$) refers to the initial adsorption rate, β ($g{\cdot}mg^{-1}$) stands for the parameters related to chemisorption activation energy and surface coverage extent. k_i ($mg{\cdot}g^{-1}{\cdot}min^{-1/2}$) represents the rate constant of the intra-particle diffusion model, and c_i ($mg\ mg^{-1}{\cdot}g$) refers a constant associated with the thickness of the boundary layer.

As seen in Figure 6a, the kinetic parameters and correlation coefficients are listed in Table 1. Compared with the pseudo-second-order model (R^2 = 0.991) and the Elovich kinetic model (R^2 = 0.964), the pseudo-first-order model (R^2 = 0.998) is more consistent with the kinetic adsorption results. To further discuss the interaction of the EW/TiO_2 hydrogel with MB, all results were fitted with the intra-particle diffusion model. As shown in Figure 6b, the fitting line is nonlinear and can be divided into two portions with different slopes. The slope of the first stage is 15.505, indicating a rapid adsorption process by transporting MB from the solution to the surface of the EW/TiO_2 hydrogel. The slope of second stage is 1.354, implying the adsorption process is slower than the first stage. In this stage, the adsorption process is dominated by the interaction between the EW/TiO_2 hydrogel and MB.

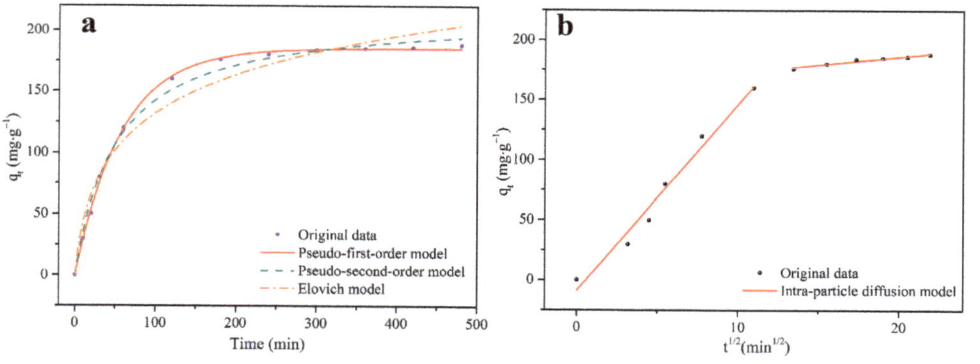

Figure 6. (**a**) Adsorption kinetics of the EW/TiO_2 hydrogel for MB. The pseudo-first-order model is more suitable for describing the adsorption kinetics of MB on EW/TiO_2. (**b**) The intra-particle diffusion model of the EW/TiO_2 hydrogel for MB. The adsorption process contains two steps: (I) MB in the solution diffuses to the surface of the EW/TiO_2 hydrogel; (II) the interaction between the EW/TiO_2 hydrogel and MB.

Table 1. Kinetics parameters of the EW/TiO_2 hydrogel for MB.

Kinetic Model	Parameter	Values
Pseudo-first-order kinetic model	k_1 (min^{-1})	0.0173
	q_e ($mg{\cdot}g^{-1}$)	185.220
	R^2	0.998
Pseudo-second-order kinetic model	$k_2 \times 10^5$ ($g{\cdot}mg^{-1}{\cdot}g^{-1}$)	9.335
	q_e ($mg{\cdot}g^{-1}$)	214.183
	R^2	0.991
Elovich kinetic model	α ($mg{\cdot}g^{-1}{\cdot}min^{-1}$)	7.264
	β ($g{\cdot}mg^{-1}$)	0.2116
	R^2	0.964
Intra-particle diffusion model	$k_{i,1}$ ($mg{\cdot}g^{-1}{\cdot}min^{-1/2}$)	15.505
	$c_{i,1}$ ($mg{\cdot}g^{-1}$)	−8.873
	R^2	0.973
	$k_{i,2}$ ($mg{\cdot}g^{-1}{\cdot}min^{-1/2}$)	1.354
	$c_{i,2}$ ($mg{\cdot}g^{-1}$)	158.876
	R^2	0.938

2.3. Adsorption Isotherm of the EW/TiO_2 Hydrogel for MB

To further explain the reaction between the solid EW/TiO_2 hydrogel and MB, the adsorption isotherm models (Langmuir (Equation (5)), Freundlich (Equation (6)), and Temkin (Equation (7)) were used to conduct nonlinear fitting to the experimental results and discuss the adsorption equilibrium during adsorption. The formulas of the adsorption isotherm models are listed below [40,41].

$$q_e = \frac{q_m K_L C_e}{1 + K_L c_e} \tag{5}$$

$$q_e = K_f C_e^{1/n} \tag{6}$$

$$q_e = B_T \ln(A_T C_e) \tag{7}$$

where K_L ($L{\cdot}g^{-1}$), K_f ($mg^{1-1/n}{\cdot}L^{1/n}{\cdot}g^{-1}$), and K_S [$(L{\cdot}mg^{-1})^{1/b}$] are the absorption constant of the Langmuir model, Freundlich model, and sips model, respectively. B_T ($J{\cdot}mol^{-1}$) represents the energy constant in the Temkin isotherm model. C_e refers to the concentration of MB at equilibrium. A_T is ($L{\cdot}mg^{-1}$) the equilibrium binding constant. n describes the degree of irregularity of the adsorbent surface.

The parameters of the adsorption isotherm are presented in Table 2. As shown in Figure 7, all the isotherm models have a good correlation with the experimental data. The correlation coefficient (R^2) of the isotherm models ranges from 0.918 to 0.975. Among these three isotherm models, the adsorption process expresses a better consistency with the Langmuir model (R^2 = 0.958) compared with the other two models. According to the above results, the adsorption of MB on the EW/TiO_2 hydrogel is speculated as a single-layer adsorption by the reaction of oxygen-containing functional groups in the hydrogel with MB molecules. The maximum adsorption capacity of the EW/TiO_2 hydrogel is 333.172 $mg{\cdot}g^{-1}$ using the Langmuir model.

Table 2. Isotherm parameters of the EW/TiO_2 hydrogel for MB.

Isotherm Models	Parameter	Values
Langmuir	q_m ($mg{\cdot}g^{-1}$)	333.172
	K_L ($L{\cdot}mg^{-1}$)	0.267
	R^2	0.958
Freundlich	K_f ($mg^{1-1/n}{\cdot}L^{1/n}{\cdot}g^{1}$)	129.88
	n	4.859
	R^2	0.918
Temkin	B_T ($J{\cdot}mol^{-1}$)	49.533
	A_T ($L{\cdot}mg^{-1}$)	8.333
	R^2	0.975

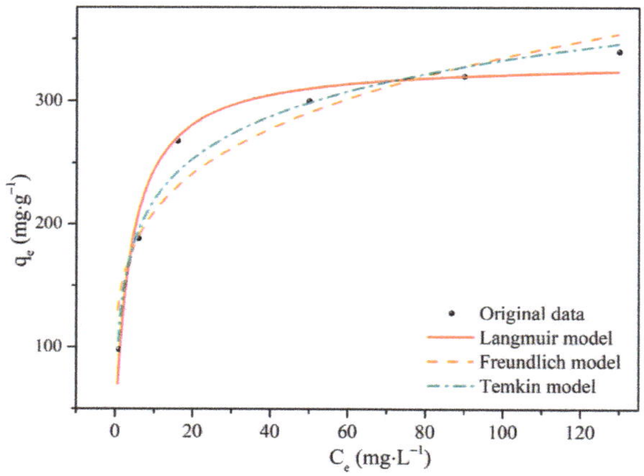

Figure 7. Adsorption isotherm of the EW/TiO_2 hydrogel for MB. The adsorption of MB on the EW/TiO_2 hydrogel is well-fitted with the Langmuir isotherm model.

2.4. Photocatalytic Performance of EW/TiO$_2$ Hydrogel for MB

The adsorption–photocatalytic degradation behavior was investigated by the degradation of the MB aqueous solution under visible light irradiation. As shown in Figure 8, MB in the control experiment decreased slightly (~9%) in the absence of the EW/TiO_2 hydrogel, indicating MB is stable and hardly degraded under an illumination of visible light via a 300 W xenon lamp. Under the same illumination of the xenon lamp, pure TiO_2 also has a low photocatalytic efficiency (47%) towards MB. As a comparison, MB decreased ~90% and 99% in the presence of the EW and EW/TiO_2 hydrogels after a visible light illumination of a xenon lamp for 120 min. These results demonstrate that the enhancement in degradation performance is mainly ascribed to the synergistic effect of a large surface area and excellent adsorption ability of the EW hydrogel towards MB and photocatalytic ability of TiO_2 towards MB via the generation of an electron–hole pair and radicals (·OH and ·O_2^-). A comparison of the performance of the EW/TiO_2 hydrogel with other photocatalysts is shown in Table 3.

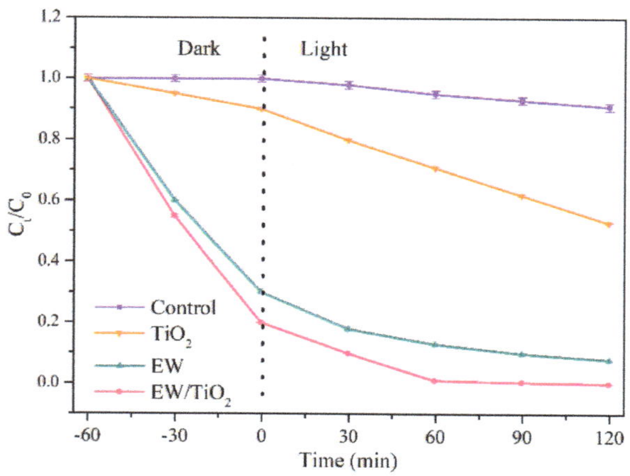

Figure 8. Photocatalytic behavior of MB over EW/TiO_2 under an irradiation of a 300 W xenon lamp. Error bars are based on the average of five measurements.

Table 3. Comparison of the performance of the EW/TiO_2 hydrogel with various photocatalysts.

Type of Photocatalyst	Photocatalytic Efficiency for MB	References
TiO_2/NC on cotton fibers	95.68%	[42]
Ag-TiO_2 nanofibers	94%	[43]
Xylan-based carbon dots@hydrogenated TiO_2	96%	[44]
Green production of TiO_2	96%	[45]
Nanocellulose–TiO_2/polyester fabric	90.02%	[46]
Plant-mediated Z-scheme ZnO/TiO_2-NCs	88.9%	[47]
EW/TiO_2 hydrogel	99%	This work

2.5. *Antibacterial Performance of the EW/TiO_2 Hydrogel*

The antimicrobial activity of the EW/TiO_2 hydrogel was studied by incubating *E. coil* with the hydrogel via the turbidity method under irradiation of a xenon lamp. The growth of bacteria was appraised by measuring the optical density of the suspension containing *E. coil*. As shown in Figure 9a, the bacteria display time-dependent growth in the control experiment due to the multiplication of *E. coil*. As a comparison, *E. coil* cultivated with the EW, TiO_2, and EW/TiO_2 hydrogels displays a distinct delayed growth. To further quantify the antibacterial activity of the EW/TiO_2 hydrogel, we estimated the survival rate of *E. coil* in the presence of the different samples in Figure 9b. The survival rate was calculated by comparing the value of OD_{600} for the different samples to that of the control. The survival rate of *E. coil* in the presence of the EW/TiO_2 hydrogel is 9.6%, which is lower than that incubated with pure TiO_2 nanoparticles (39%) and the EW hydrogel (64%), confirming that the EW/TiO_2 hydrogel has a significant antibacterial effect. The possible antibacterial mechanism of the EW/TiO_2 hydrogel is speculated as follows: the bacteria are easily contacted with the EW/TiO_2 hydrogel with a porous structure and then killed by cooperation between natural lysozymes in EW and the reactive oxygen species (ROS) released by TiO_2 via the degradation of the protein in the cell wall, inactivation of the enzyme, and destruction of DNA. As a result, the EW/TiO_2 hydrogel exhibits an obviously enhanced antibacterial activity compared with EW and pure TiO_2 nanoparticles.

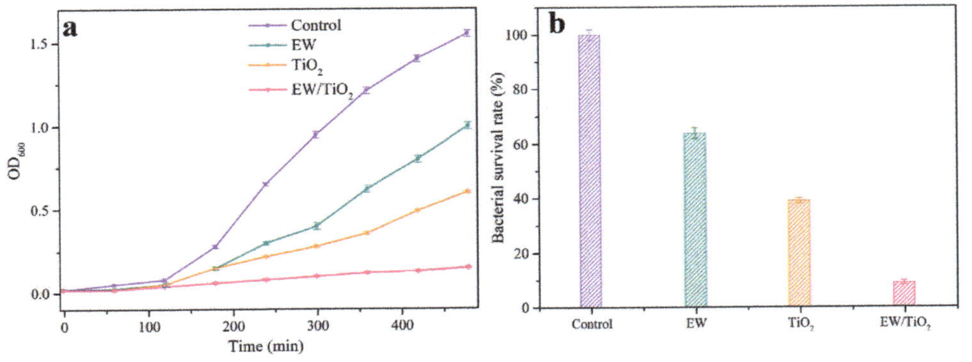

Figure 9. (**a**) OD_{600} of bacteria showing the distinctively different growth of bacteria in suspensions with different samples. (**b**) Comparison of the survival rate of bacteria in suspensions with different samples. Error bars are based on the average of five measurements.

2.6. *Cytotoxicity of the EW/TiO_2 Hydrogel*

To demonstrate the biosafety of the EW/TiO_2 hydrogel in practical applications, toxicity is a serious index that needs to be considered. Here, the cytotoxicity of the EW/TiO_2 hydrogel was appraised by choosing HEK 293 cells as a biological model. In Figure 10, the cell co-cultured with 5 mg EW/TiO_2 hydrogel sustains a survival rate of 90%. Even if the dose of EW/TiO_2 increases to 25 mg, the viability of HEK 293 is still above

70%. These results imply that EW/TiO_2 is a low-toxic hydrogel for mammalian cells in practical application.

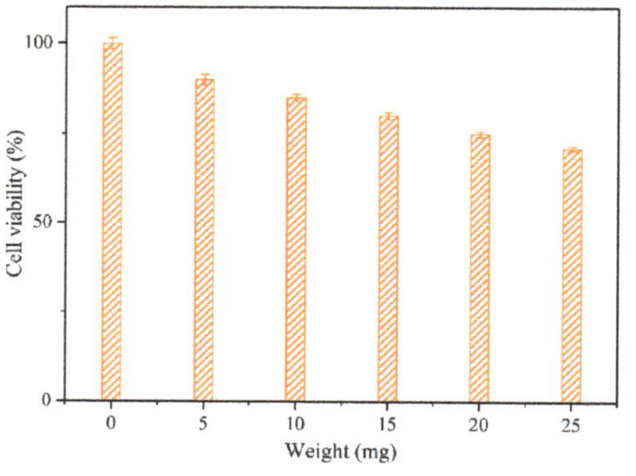

Figure 10. Viability of HEK 293 cells incubated for 24 h in the presence of the EW/TiO_2 hydrogel (5~25 mg) using the MTT assay. Error bars are based on the average of five measurements.

3. Materials and Methods

3.1. Materials

Eggs were obtained from a local market. TiO_2 (99.6%), dimethyl sulfoxide (DMSO, 99%), and MB (98%) were provided by Sinopharm Chemical Reagent Co., Ltd., Shanghai, China. Luria–Bertani (LB) broth, Miller was purchased from Becton, Dickinson and Company. Dulbecco's modified Eagle's medium (DMEM) and 3-(4,5-dimethylthiazol-2-yl) 2,5-diphenyl tetrazolium bromide (MTT, 98%) were provided by Thermo Fisher Scientific. All reagents in this experiment were analytical grade and were used directly without purification.

3.2. Preparation of the EW/TiO_2 Hydrogel

To prepare the EW/TiO_2 hydrogel, we firstly separated the EW from eggs [36,41]. Then, 0.25~1 g TiO_2 nanoparticles was dispersed into 10 g EW by simple physical commixing and stirring ultrasonication. Immediately after the mixture, the prepared EW/TiO_2 mixture was decanted into a prepared mold and heated to ~90 °C via a water bath. Finally, the EW/TiO_2 hydrogel was dehydrated using a lyophilizer.

3.3. Characterizations

The morphology and compositions of the as-fabricated EW/TiO_2 hydrogel were characterized using scanning electron microscopy (SEM)–energy-dispersive spectrometry (EDS). The functional groups were analyzed by Fourier transform infrared spectroscopy (FTIR). The specific surface area of the hydrogel was calculated using Brunauer–Emmett–Teller (BET) and Barrett–Joyner–Halenda (BJH) approaches. The concentrations of MB were determined using a UV-vis spectrometer (UV-vis). The thermal stability of the hydrogel was performed on a thermogravimetric analyzer (TGA) under a nitrogen atmosphere from 30 to 600 °C.

3.4. Adsorptive and Photocatalytic Experiments

3.4.1. Adsorptive Experiment

To investigate the adsorption performance of the EW/TiO_2 hydrogel, 5~25 mg hydrogel was added to 20 mL of MB (50 mg·L^{-1} to 300 mg·L^{-1}) at 200 rpm in a shaker. After reaching adsorption equilibrium, the adsorption performance of EW/TiO_2 was characterized by comparing the residual MB concentration with initial concentration using a UV-vis spectrophotometer. The adsorption capacity at time t (q_t, mg·g^{-1}, Equation (8)), equilibrium adsorption capacity (q_e, mg·g^{-1}, Equation (9)), and removal rate (%, Equation (10)) were calculated using the following formulas [40].

$$q_e = \frac{C_0 - C_e}{m} \times V \tag{8}$$

$$q_t = \frac{C_0 - C_t}{m} \times V \tag{9}$$

$$E = \frac{C_0 - C_t}{C_0} \times 100 \tag{10}$$

where C_0 (mg·L^{-1}) is the initial contaminant concentration, C_e (mg·L^{-1}) is MB concentration at equilibrium, C_t (mg·L^{-1}) is MB concentration at time t, V (L) is the volume of MB solution, and m is the weight of the EWP/TiO_2 hydrogel (g).

3.4.2. Photocatalytic Experiment

The photocatalytic activity of the EW/TiO_2 hydrogel was investigated by degrading MB under in a quartz vessel under visible light illumination using a 300 W xenon lamp. Typically, 10 mg different hydrogels were added into MB aqueous solutions and stirred for 60 min in the dark. In given time intervals of illumination, the residual concentration of MB was determined using a UV-vis spectrophotometer.

3.5. Antimicrobial Experiment

The antimicrobial performance of the EW/TiO_2 hydrogel was determined using the turbidity assay. The growth of bacteria was assessed by measuring the values of optical density values at 600 nm (OD_{600}). The hydrogel was put into a suspension containing *E. coli* with a concentration of ~10^6 CFU/mL with an irradiation of a 300 W xenon lamp. To estimate the bacteria viability co-cultured with the EW/TiO_2 hydrogel, we measured the OD_{600} of the *E. coli* suspension every hour for 8 h.

3.6. Cytotoxicity Characterization

The cytotoxicity of the EW/TiO_2 hydrogel was characterized through the MTT assay. To characterize the cytotoxicity of the EW/TiO_2 hydrogel, we first formed a monolayer of human embryonic kidney 293 (HEK 293) cells and exposed the cells to the prepared hydrogel for 24 h. Then, MTT was added to track the viable cells. After the addition of dimethyl sulfoxide (DMSO), the optical intensities were written down using a microplate reader. The cell viability was computed by comparing the measured optical intensity with that of the control sample, where the cells were incubated in fresh media.

4. Conclusions

In this study, a low-cost EW/TiO_2 hydrogel is developed using waste eggs nearing their expiration date and used as an antibacterial bio-adsorbent and degradant for the effective removal of organic dye and the inactivation of pathogenic bacteria in wastewater. The SEM, EDS, FTIR, BET, and TGA results confirm the successful immobilization of TiO_2 on the porous EW hydrogel. This prepared EW/TiO_2 hydrogel with a large specific surface area and abundant active sites provides a supportive structure and composition for antibacterial, adsorptive, and degradative applications. These results indicate the EW/TiO_2 hydrogel shows a favorable multifunctional ability in the water treatment field. Additionally, this study provides a new idea of cooperation of disinfection–adsorption–degradation and promotes the practical application of TiO_2-based hydrogels in the environmental field. To further reduce the consumption of energy, the utility of sunlight in the photocatalytic degradation of organic matter and the inactivation of pathogenic bacteria is an important and challenging direction in the waste treatment field.

Author Contributions: Investigation, X.W. and M.Z. methodology, validation, and writing. All authors have read and agreed to the published version of the manuscript.

Funding: This work was financially funded by the Natural Science Foundation of Jiangsu Province (grant number BK20201060) and the funding for school-level research projects of Yancheng Institute of Technology (grant number xjr2020003).

Data Availability Statement: The original contributions presented in this study are included in this article; further inquiries can be directed to the corresponding author.

Conflicts of Interest: The authors declare no conflicts of interest.

References

1. Wang, M.; Janssen, A.B.G.; Bazin, J.; Strokal, M.; Ma, L.; Kroeze, C. Accounting for interactions between Sustainable Development Goals is essential for water pollution control in China. *Nat. Commun.* **2022**, *13*, 730. [CrossRef] [PubMed]
2. Lin, L.; Yang, H.; Xu, X. Effects of water pollution on human health and disease heterogeneity: A review. *Front. Env. Sci-Switz.* **2022**, *10*, 880246. [CrossRef]
3. Rahmatpour, A.; Alijani, N.; Alizadeh, A.H. Preparation of chitosan-based ternary nanocomposite hydrogel film by loading graphene oxide nanosheets as adsorbent for enhanced methylene blue dye removal. *Int. J. Biol. Macromol.* **2023**, *253*, 126585. [CrossRef]
4. Georgouvelas, D.; Abdelhamid, H.N.; Li, J.; Edlund, U.; Mathew, A.P. All-cellulose functional membranes for water treatment: Adsorption of metal ions and catalytic decolorization of dyes. *Carbohyd. Polym.* **2021**, *264*, 118044. [CrossRef]
5. Xie, M.; Gao, M.; Yun, Y.; Malmsten, M.; Rotello, V.M.; Zboril, R.; Akhavan, O.; Kraskouski, A.; Amalraj, J.; Cai, X.; et al. Antibacterial nanomaterials: Mechanisms, impacts on antimicrobial resistance and design principles. *Angew. Chem. Int. Edit.* **2023**, *62*, 17.
6. Qi, X.; Xiang, Y.; Cai, E.; Ge, X.; Chen, X.; Zhang, W.; Li, Z.; Shen, J. Inorganic-organic hybrid nanomaterials for photothermal antibacterial therapy. *Coordin. Chem. Rev.* **2023**, *496*, 215426. [CrossRef]
7. Yao, H.; Wu, M.; Lin, L.; Wu, Z.; Bae, M.; Park, S.; Wang, S.; Zhang, W.; Gao, J.; Wang, D.; et al. Design strategies for adhesive hydrogels with natural antibacterial agents as wound dressings: Status and trends. *Mater. Today Bio.* **2022**, *16*, 100429. [CrossRef]
8. Wang, Z.; Liu, X.; Duan, Y.; Huang, Y. Infection microenvironment-related antibacterial nanotherapeutic strategies. *Biomaterials* **2022**, *280*, 121249. [CrossRef] [PubMed]
9. Chen, L.; Hu, X.; Cai, T.; Yang, Y.; Zhao, R.; Liu, C.; Li, A.; Jiang, C. Degradation of Triclosan in soils by thermally activated persulfate under conditions representative of in situ chemical oxidation (ISCO). *Chem. Eng. J.* **2019**, *369*, 344–352. [CrossRef]
10. Xue, H.; Wang, X.; Xu, Q.; Dhaouadi, F.; Sellaoui, L.; Seliem, M.K.; Ben Lamine, A.; Belmabrouk, H.; Bajahzar, A.; Bonilla-Petriciolet, A.; et al. Adsorption of methylene blue from aqueous solution on activated carbons and composite prepared from an agricultural waste biomass: A comparative study by experimental and advanced modeling analysis. *Chem. Eng. J.* **2021**, *430*, 132801. [CrossRef]
11. Zhao, R.; Shi, X.; Ma, T.; Rong, H.; Wang, Z.; Cui, F.; Zhu, G.; Wang, C. Constructing mesoporous adsorption channels and MOF-polymer interfaces in electrospun composite fibers for effective removal of emerging organic contaminants. *ACS Appl. Mater. Interfaces* **2021**, *13*, 755–764. [CrossRef] [PubMed]
12. Xiao, W.; Jiang, X.; Liu, X.; Zhou, W.; Garba, Z.N.; Lawan, I.; Wang, L.; Yuan, Z. Adsorption of organic dyes from wastewater by metal-doped porous carbon materials. *J. Clean. Prod.* **2021**, *284*, 124773. [CrossRef]
13. Atallah, C.; Mosadeghsedghi, S.; Kenari, S.L.D.; Hudder, M.; Morin, L.; Volchek, K.; Mortazavi, S.; Ben Salah, I. Removal of heavy metals from mine water using a hybrid electrocoagulation-ceramic membrane filtration process. *Desalin. Water Treat.* **2024**, *320*, 100730. [CrossRef]
14. Sefiddashti, F.T.; Homayoonfal, M. Nanostructure-manipulated filtration performance in nanocomposite membranes: A comprehensive investigation for water and wastewater treatment. *Heliyon* **2024**, *10*, e36874. [CrossRef]
15. Zeng, G.; He, Z.; Wan, T.; Wang, T.; Yang, Z.; Liu, Y.; Lin, Q.; Wang, Y.; Sengupta, A.; Pu, S. A self-cleaning photocatalytic composite membrane based on g-C_3N_4@MXene nanosheets for the removal of dyes and antibiotics from wastewater. *Sep. Purif. Technol.* **2022**, *292*, 121037. [CrossRef]
16. Wang, T.; Zheng, J.; Cai, J.; Liu, Q.; Zhang, X. Visible-light-driven photocatalytic degradation of dye and antibiotics by activated biochar composited with K^+ doped g-C_3N_4: Effects, mechanisms, actual wastewater treatment and disinfection. *Sci. Total Environ.* **2022**, *839*, 155955. [CrossRef] [PubMed]
17. Khan, M.K.; Farooq, U.; Naseem, K.; Khan, M.R.; Khan, M.E.; Ali, W.; Ali, S.K.; Alomar, M.S.; Arishi, M.; Bashiri, A.H.; et al. Solar-driven Pt free hydrogen production and successive degradation of sulfasalazine using CuO/ZnO binary nanocomposites: Reaction kinetics and determination of reaction parameters using response surface methodology. *J. Taiwan Inst. Chem. E.* **2024**, *165*, 105789. [CrossRef]
18. Raza, M.; Farooq, U.; Khan, S.A.; Ullah, Z.; Khan, M.E.; Ali, S.K.; Bakather, O.Y.; Alam, S.; Khan, M.Y.; Ali, W.; et al. Preparation and spectrochemical characterization of Ni-doped ZnS nanocomposite for effective removal of emerging contaminants and hydrogen production: Reaction kinetics, mechanistic insights. *Spectrochim. Acta A* **2024**, *318*, 124513. [CrossRef]

19. Guo, Y.; Chen, X.; Liu, Y.; Chen, Z.; Guo, P.; Luo, D.; Zhang, M.; Liu, X. Inorganic-organic dual-ligand-regulated photocatalysis of CdS@Zn_xCd_{1-x}S@ZnS quantum dots for lignin valorization. *ACS Appl. Mater. Interfaces* **2024**, *16*, 12624–12636. [CrossRef]
20. Doondani, P.; Panda, D.; Gomase, V.; Peta, K.R.; Jugade, R. Novel chitosan-ZnO nanocomposites derived from Nymphaeaceae fronds for highly efficient removal of Reactive Blue 19, Reactive Orange 16, and Congo Red dyes. *Environ. Res.* **2024**, *247*, 118228. [CrossRef]
21. Jiang, R.; Zhong, D.; Xu, Y.; Chang, H.; He, Y.; Zhang, J.; Liao, P. Chitosan derived N-doped carbon anchored Co_3O_4-doped MoS_2 nanosheets as an efficient peroxymonosulfate activator for degradation of dyes. *Int. J. Biol. Macromol.* **2024**, *265*, 130519. [CrossRef] [PubMed]
22. Subramani, T.; Nagarajan, S.K. Synergic effect of rare earth doped Sm doped WO_3 nanoparticles for enhanced MB dye photocatalytic activities of waste water treatment and antibacterial activities against Staphylococcus aureus. *Ceram. Int.* **2024**, *50*, 34226–34239. [CrossRef]
23. Feng, X.; Li, X.; Su, B.; Ma, J. Two-step construction of WO_3@TiO_2/CS-biochar S-scheme heterojunction and its synergic adsorption/photocatalytic removal performance for organic dye and antibiotic. *Diam. Relat. Mater.* **2023**, *131*, 109560. [CrossRef]
24. Alsulmi, A.; Mohammed, N.N.; Hassan, M.M.; Eltawil, M.A.; Amin, A.E.; Fahmy, M.; Sultan, A.; Ahmed, M.A. Rational engineering of S-scheme CeO_2/g-C_3N_4 heterojunctions for effective photocatalytic destruction of rhodamine B dye under natural solar radiations. *Colloid. Surface. A* **2024**, *689*, 133683. [CrossRef]
25. Li, M.; Zhang, R.; Zou, Z.; Zhang, L.; Ma, H. Optimizing physico-chemical properties of hierarchical ZnO/TiO_2 nano-film by the novel heating method for photocatalytic degradation of antibiotics and dye. *Chemosphere* **2024**, *346*, 140392. [CrossRef] [PubMed]
26. Habiba, U.; Islam, M.S.; Siddique, T.A.; Afifi, A.M.; Ang, B.C. Adsorption and photocatalytic degradation of anionic dyes on Chitosan/PVA/Na-Titanate/TiO_2 composites synthesized by solution casting method. *Carbohyd. Polym.* **2016**, *149*, 317–331. [CrossRef]
27. Zong, E.; Wang, C.; Yang, J.; Zhu, H.; Jiang, S.; Liu, X.; Song, P. Preparation of TiO_2/cellulose nanocomposites as antibacterial bio-adsorbents for effective phosphate removal from aqueous medium. *Int. J. Biol. Macromol.* **2021**, *182*, 434–444. [CrossRef] [PubMed]
28. Abdulhameed, A.S.; Jawad, A.H.; Ridwan, M.; Khadiran, T.; Wilson, L.D.; Yaseen, Z.M. Chitosan/Carbon-Doped TiO_2 Composite for Adsorption of two anionic dyes in solution and gaseous SO_2 capture: Experimental modeling and optimization. *J. Polym. Environ.* **2023**, *30*, 4619–4636. [CrossRef]
29. Alkorbi, A.S.; Javed, H.M.A.; Hussain, S.; Latif, S.; Mahr, M.S.; Mustafa, M.S.; Alsaiari, R.; Alhemiary, N.A. Solar light-driven photocatalytic degradation of methyl blue by carbon-doped TiO_2 nanoparticles. *Opt. Mater.* **2022**, *127*, 112259. [CrossRef]
30. Zhao, X.; Wang, X.; Lou, T. Preparation of fibrous chitosan/sodium alginate composite foams for the adsorption of cationic and anionic dyes. *J. Hazard. Mater.* **2021**, *403*, 124054. [CrossRef]
31. Dai, H.; Huang, Y.; Huang, H. Eco-friendly polyvinyl alcohol/carboxymethyl cellulose hydrogels reinforced with graphene oxide and bentonite for enhanced adsorption of methylene blue. *Carbohyd. Polym.* **2018**, *185*, 1–11. [CrossRef] [PubMed]
32. Bayat, M.; Javanbakht, V.; Esmaili, J. Synthesis of zeolite/nickel ferrite/sodium alginate bionanocomposite via a co-precipitation technique for efficient removal of water-soluble methylene blue dye. *Int. J. Biol. Macromol.* **2018**, *116*, 607–619. [CrossRef] [PubMed]
33. Zhang, T.; Chen, T.; Jiang, H.; Liu, J.; Wang, E.; Zhang, M.; Liu, X. pH-induced egg white protein foaming properties enhancement: Insight into protein structure and quantitative proteomic analysis at protein adsorption layer. *Food Hydrocolloid.* **2023**, *144*, 109060. [CrossRef]
34. Chen, Y.S.; Ooi, C.W.; Show, P.L.; Hoe, B.C.; Chai, W.S.; Chiu, C.Y.; Wang, S.S.S.; Chang, Y.K. Removal of ionic dyes by nanofiber membrane functionalized with chitosan and egg white proteins: Membrane Preparation and Adsorption Efficiency. *Membranes* **2022**, *12*, 63. [CrossRef]
35. Su, Y.; Wang, Y.; McClements, D.J.; Lu, C.; Chang, C.; Li, J.; Gu, L.; Yang, Y. Selective adsorption of egg white hydrolysates onto activated carbon: Establishment of physicochemical mechanisms for removing phenylalanine. *Food Chem.* **2021**, *364*, 130285. [CrossRef] [PubMed]
36. Jin, Y.; Du, Q.; Li, Y.; Zhang, Y.; Chen, B.; Wang, M.; Chen, K.; Sun, Y.; Zhao, S.; Jing, Z. Removal of methylene blue by crosslinked egg white protein/graphene oxide bionanocomposite aerogels. *Nanomaterials* **2022**, *12*, 2659. [CrossRef]
37. Farjami, T.; Babaei, J.; Nau, F.; Dupont, D.; Madadlou, A. Effects of thermal, non-thermal and emulsification processes on the gastrointestinal digestibility of egg white proteins. *Rends Food Sci. Tech.* **2021**, *107*, 45–56. [CrossRef]
38. Gharbi, N.; Labbafi, M. Effect of processing on aggregation mechanism of egg white proteins. *Food Chem.* **2018**, *252*, 126–133. [CrossRef]
39. Yang, Z.; Rose, S.P.; Yang, H.M.; Pirgozliev, V.; Wang, Z.Y. Egg production in China. World. *Poultry Sci. J.* **2018**, *74*, 417–426.
40. Sun, Y.; Li, Y.; Chen, B.; Wang, M.; Zhang, Y.; Chen, K.; Du, Q.; Wang, Y.; Pi, X. Methylene blue removed from aqueous solution by encapsulation of bentonite aerogel beads with cobalt alginate. *ACS Omega* **2022**, *7*, 41246–41255. [CrossRef]
41. Zhou, G.; Zhang, H.; Bai, Z.; Jiang, J.; Wang, Y.; Quan, F. Efficient and rapid adsorption of methylene blue dyes by novel metal-organic frameworks and organic/inorganic hybrid alginate-based dual network gel composites. *Int. J. Biol. Macromol.* **2023**, *253*, 127034. [CrossRef] [PubMed]

42. Chen, W.; Feng, X.; Zhang, D.; Lu, F.; Wang, H.; Tan, J.; Xu, Q.; Liu, Y.; Cao, Z.; Su, X. In situ synthesis of TiO_2/NC on cotton fibers with antibacterial properties and recyclable photocatalytic degradation of dyes. *RSC Adv.* **2022**, *12*, 19974–19980. [CrossRef] [PubMed]
43. Aravind, M.; Amalanathan, M.; Aslam, S.; Noor, A.E.; Jini, D.; Majeed, S.; Velusamy, P.; Alothman, A.A.; Alshgari, R.A.; Mushab, M.S.S.; et al. Hydrothermally synthesized Ag-TiO_2 nanofibers (NFs) for photocatalytic dye degradation and antibacterial activity. *Chemosphere* **2023**, *321*, 138077. [CrossRef] [PubMed]
44. Cong, S.; Cai, J.; Li, X.; You, J.; Wang, L.; Wang, X. Direct Z-Scheme Xylan-Based Carbon Dots@TiO_{2-x} Nanocomposites for Visible Light Driven Photocatalytic of Dye Degradation and Antibacterial. *Adv. Funct. Mater.* **2024**, *34*, 2401540. [CrossRef]
45. Shimi, A.K.; Ahmed, H.M.; Wahab, M.; Katheria, S.; Wabaidur, S.M.; Eldesoky, G.E.; Islam, M.A.; Rane, K.P. Synthesis and applications of green synthesized TiO_2 nanoparticles for photocatalytic dye degradation and antibacterial activity. *J. Nanomater.* **2022**, *2022*, 7060388. [CrossRef]
46. Tan, J.; Deng, H.; Lu, F.; Chen, W.; Su, X.; Wang, H. Antibacterial nanocellulose-TiO_2/polyester fabric for the recyclable photocatalytic degradation of dyes. *Polymers* **2023**, *15*, 4376. [CrossRef]
47. Negi, A.; Ringwal, S.; Pandey, M.; Yassin, M.T. Plant-mediated Z-scheme ZnO/TiO_2-NCs for antibacterial potential and dye degradation: Experimental and DFT study. *Sci. Rep. UK* **2024**, *14*, 7955. [CrossRef]

catalysts

MDPI

Article

Studies on Electronic Structure and Optical Properties of MoS_2/X ($X = WSe_2$, $MoSe_2$, AlN, and ZnO) Heterojunction by First Principles

Jibo Liu [1,2], Yuheng Jin [3], Bocheng Lei [1], Xucai Zhao [1], Yineng Huang [1], Lili Zhang [1,*] and Youliang Zhu [1,4,*]

1 Xinjiang Laboratory of Phase Transitions and Microstructures in Condensed Matters, College of Physical Science and Technology, Yili Normal University, Yining 835000, China; liujibo1207@163.com (J.L.); lbc0428@sina.com (B.L.); zxc85619876@sina.com (X.Z.); ynhuang@nju.edu.cn (Y.H.)
2 College of Information Science and Engineering, Xinjiang College of Science & Technology, Kuerle 841000, China
3 School of Physics, Sichuan University, Chengdu 610065, China; j1498922817@outlook.com
4 State Key Laboratory of Supramolecular Structure and Materials, College of Chemistry, Jilin University, Changchun 130023, China
* Correspondence: suyi2046@sina.com (L.Z.); youliangzhu@jlu.edu.cn (Y.Z.)

Abstract: The single-layer MoS_2 is a highly sought-after semiconductor material in the field of photoelectric performance due to its exceptional electron mobility and narrow bandgap. However, its photocatalytic efficiency is hindered by the rapid recombination rate of internal photogenerated electron–hole pairs. Currently, the construction of heterojunctions has been demonstrated to effectively mitigate the recombination rate of photogenerated electron–hole pairs. Therefore, this paper employs the first principles method to calculate and analyze the four heterojunctions formed by MoS_2/WSe_2, $MoS_2/MoSe_2$, MoS_2/AlN, and MoS_2/ZnO. The study demonstrates that the four heterojunctions exhibit structural stability. The construction of heterojunctions, as compared to a monolayer MoS_2, leads to a reduction in the band gap, thereby lowering the electron transition barrier and enhancing the light absorption capacity of the materials. The four systems exhibit II-type heterojunction. Therefore, the construction of heterojunctions can effectively enhance the optical properties of these systems. By forming heterojunctions MoS_2/WSe_2 and $MoS_2/MoSe_2$, the absorption coefficient in the visible light region is significantly increased, resulting in a greater ability to respond to light compared to that of MoS_2/ZnO and MoS_2/AlN. Consequently, MoS_2-based heterojunctions incorporating chalcogenide components WSe_2 and $MoSe_2$, respectively, exhibit superior catalytic activity compared to MoS_2 heterojunctions incorporating non-chalcogenide components ZnO and AlN, respectively. The absorption spectrum analysis reveals that $MoS_2/MoSe_2$ exhibits the highest light responsivity among all investigated systems, indicating its superior photoelectric performance.

Keywords: MoS_2; heterojunction; photocatalysis; first principles

1. Introduction

Due to their high electron mobility, surface activity, and narrow band gap, two-dimensional transition metal dichalcogenides (2D-TMDs) are extensively found in various applications such as photodetectors, optoelectronic devices, and photocatalysts [1,2]. The monolayer MoS_2 belongs to the family of 2D-TMDs and has a sandwich structure consisting of S-Mo-S [3–5]. The monolayer MoS_2 has significant potential in the fields of photocatalysts and optoelectronic devices, such as low power field effect transistors [6], valley electronics devices [7] and photodetectors [8], owing to its straightforward preparation process, favorable band gap structure, and exceptional magnetic and optical properties [9,10]. However, the limited application efficiency of the monolayer MoS_2 in the aforementioned optoelectronic devices stems from its low absorption of incident and high recombination rate of photogenerated electron–hole pairs [7,8]. Consequently, current research efforts are primarily directed towards identifying effective strategies to enhance the absorption of visible light by the monolayer MoS_2 and mitigate electron–hole pair recombination within the system.

Modulating the properties of two-dimensional materials through the construction of two-dimensional heterojunctions connected by van der Waals (vdW) force represents a commonly employed strategy. These heterojunctions give rise to electronic and optical properties that can surpass those of monolayer components, as evidenced by previous research [11–14]. In particular, the ones of the heterojunctions exhibit a type II band structure, which could induce an internal electric field that lowers the recombination probability of electron–hole pairs. As a result, constructing type II heterojunctions is a popular choice for photoelectric devices [15,16]. Deng et al. [17] used a first principles approach to investigate $CdTe/MoS_2$ heterojunctions. They found that constructing heterojunctions effectively promoted the separation of photogenerated carriers, which improved the photoelectron conversion efficiency and made the system suitable for high-efficiency photodetector material. The successful preparation of the MoS_2/CuO heterojunction [18], BP/MoS_2 heterojunction [19], and $MoS_2/MAPbBr_3$ [20] has demonstrated the effectiveness of type II heterojunctions in modulating the band gap and enhancing the system's response to light. Constructing a new type II heterojunction holds significant value. Despite previous experimental [13] and theoretical [21] studies on MoS_2-based heterojunctions, delving deeper into the microscopic mechanism of these heterojunctions remains a scientific pursuit of great importance.

Monolayer $MoSe_2$ and monolayer WSe_2 are semiconductors known for their high optical conversion efficiency, making them ideal materials for optoelectronic devices such as light-emitting diodes [22,23] and solder effect transistors [24,25]. Additionally, they share similar structures and properties with other broad-band semiconductors such as monolayers MoS_2, ZnO, and AlN, which are representative materials for constructing type II heterojunctions. Using MoS_2 as a base, this study will compare the difference in heterostructures built with other selenium-based compounds, such as $MoSe_2$ and WSe_2, to those compounds like ZnO and AlN [26,27]. In $MoS_2/MoSe_2$ and MoS_2/WSe_2, each monolayer is a sulfur-based compound, but in MoS_2/ZnO and MoS_2/AlN, ZnO and AlN are not sulfur-based compounds. To date, no theoretical investigation has been conducted on the differences in optical properties between these two types.

In this study, the heterojunction construction of WSe_2, $MoSe_2$, AlN, and ZnO based on the monolayer MoS_2 will offer additional options for the application of MoS_2-based heterojunctions and a theoretical reference for specific experimental preparation in materials for optoelectronic devices.

2. Computational Methods and Theoretical Model

The calculation program employed in this study is CASTEP (Cambridge Serial Total Energy Package) module in the Materials Studio 2017, which is based on density functional theory. The interaction between ions and electrons is described by the projected plane wave method using plane wave as the base group. The exchange correlation potential between electrons is described using the PBE (Perdew–Burke–Ernzerhof) exchange–correlation generalized gradient approximation [28,29]. The interlayer vans der Waals forces are corrected by the TS (Tkachenko–Scheffler) dispersion method in the calculations [30,31]. The electron configurations of atoms involved in this paper are Mo: $4d^5 5s^1$, S: $3s^2 3p^4$, Se: $4s^2 4p^4$, Al: $3s^2 3p^1$, N: $2s^2 2p^3$, Zn: $3d^{10} 4s^2$, and O: $2s^2 2p^4$. After conducting convergence tests, the cutoff energy of is set to 450 eV, and the Brillouin Zone-K grid point is chosen as $5 \times 5 \times 1$. To eliminate boundary effects and interface effects, periodic boundary conditions are adopted in the x-direction and y-direction and a fixed boundary is adopted in the z-direction. Additionally, to reduce the potential interlayer mutual coupling effects, a vacuum layer with a thickness of 20 angstroms is embedded along the Z-direction between the two layers of each heterojunction. The self-consistent convergence accuracy is set to 2.0×10^{-5} eV/atom, with a maximum stress limit of 0.1 GPa. Additionally, the maximum displacement is restricted to less than 2×10^{-3} Å, and the interatomic force field convergence accuracy is set to 0.05 eV/Å. All the calculations are based on the BFGS (Broyden–Fletcher–Goldfarb–Shanno) algorithm.

In the heterojunction structure designed in this paper, factors such as interface effects and boundary effects are considered. The two materials are designed in a vertical structure stacked up and down in the form of thin films. The relatively simple structure ensures the performance of the device. Based on the convergence test diagram (Figure 1), the supercell size of MoS_2 is set to $3 \times 3 \times 1$, which ensures calculation accuracy while conserving computing resources. To match with the substrate MoS_2, the supercell size of WSe_2, $MoSe_2$, AlN, and ZnO in the four heterojunctions is also set to $3 \times 3 \times 1$. The calculation of heterojunction energy for five stacking modes is as shown in the Supplementary Material (Figure S1), naming them from SP1 to SP5, respectively. The stacking mode SP1 exhibits the lowest ground state energy, as demonstrated by Figure S1. This finding indicates that mode SP1 possesses the most stable structure, thus justifying its selection in our paper. The constructed MoS_2/X (X = WSe_2, $MoSe_2$, AlN, and ZnO) four heterojunction models are shown in Figure 2.

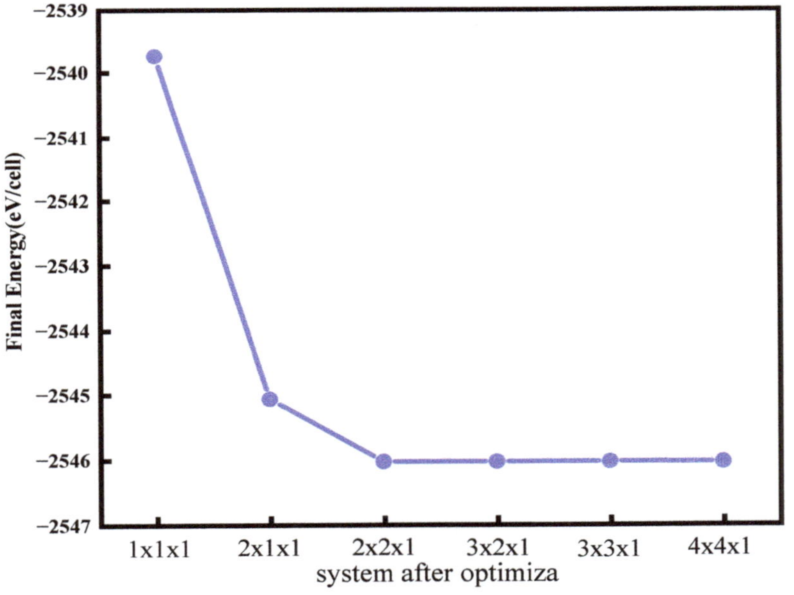

Figure 1. Energy of each monolayer system after optimization.

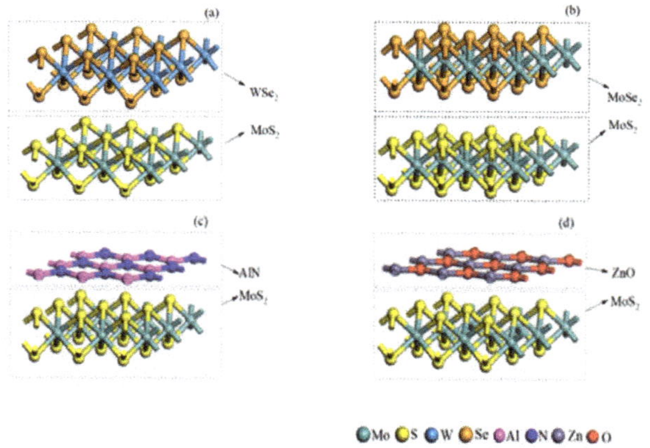

Figure 2. Heterojunction model diagrams: (**a**) MoS_2/WSe_2; (**b**) $MoS_2/MoSe_2$; (**c**) MoS_2/AlN; and (**d**) MoS_2/ZnO.

3. Geometric Parameters and System Stability

3.1. Geometric Parameters

Before constructing heterojunctions, we geometrically optimize the monolayers MoS_2, WSe_2, $MoSe_2$, AlN, and ZnO. During the optimization process, the c-axis was fixed, and the lattice constants and atomic positions in other directions were relaxed. The optimized lattice constant for each system is presented in Table 1. The results indicate that the optimized lattice constants agree well with theoretical values, confirming the reliability and validity of the selected calculation parameters.

Table 1. Lattice constants of each monolayer system after optimization.

	MoS_2	WSe_2	$MoSe_2$	AlN	ZnO
a = b/nm (this work)	0.317	0.331	0.330	0.312	0.329
a = b/nm (experimental)	0.316 [32]	0.327 [33]	0.329 [34]	0.311 [35]	0.325 [36]
Relative error	0.3%	1.2%	0.3%	0.3%	1.2%

3.2. Stability

To investigate the heterogeneous lattice matching, we assign the lattice constants of monolayers MoS_2 and X as a_1 and a_2, respectively. Next, we calculate the lattice mismatch rates in the four systems and represent them using Equation (1) [37].

$$\sigma = \frac{(a_1 - a_2)}{a_1} \tag{1}$$

The lattice mismatch rates of the four systems are calculated to be 4.02% for MoS_2/WSe_2, 3.83% for $MoS_2/MoSe_2$, 1.61% for MoS_2/AlN, and 2.43% for MoS_2/ZnO. These values satisfy the complete co-lattice condition ($\delta < 5\%$), indicating that all four heterojunctions are potential stable structures [38].

In order to quantitatively assess the lattice mismatch and structural stability of the four heterojunctions, we have conducted calculations for the lattice mismatch energy $\Delta(E_{mis})$, binding energy (E_b), and van der Waals energy (ΔE_{vdw}) using Equations (2)–(4) [39–41].

$$\Delta E_{mis} = \frac{1}{S}[E(MoS_2)_1 + E(X)_1 - E(MoS_2)_2 - E(X)_2] \tag{2}$$

$$E_b = E_{sum} - E(MoS_2)_2 - E(X)_2 \tag{3}$$

$E(MoS_2)_1$ and $E(X)_1$ are the total energies of the isolated monolayer MoS_2 and isolated monolayer X, respectively. $E(MoS_2)_2$ and $E(X)_2$ represent the total energy of the MoS_2 and X in a heterojunction, respectively. Additionally, S represents the area of a heterojunction layer. The stability of a heterojunction is indirectly assessed by both ΔE_{sum} and E_b. A smaller lattice mismatch energy indicates a more stable structure, while a smaller binding energy suggests tighter binding [38]. Table 2 displays the calculated results, which indicate that the lattice mismatch energy and binding energy of the four systems are negative. MoS_2/AlN exhibits the smallest lattice mismatch energy and binding energy, indicating that the heterojunction is the most stable system. It is found that the four systems, WSe_2, $MoSe_2$, AlN, and ZnO, can form a stable heterojunction with MoS_2.

Table 2. Lattice mismatch energy and binding energy of the four heterojunctions.

	MoS_2/WSe_2	$MoS_2/MoSe_2$	MoS_2/AlN	MoS_2/ZnO
$\Delta E_{mis}/meV{\cdot}Å^{-2}$	−0.332	−1.138	−8.675	−0.548
E_b/eV	−0.78	−0.79	−1.59	−1.05

In order to further verify the thermal stability of these, The DS - PAW [42] module in the Device Studio 2023 first principles plane wave calculation software was used to simulate the ab initio molecular dynamics (AIMDs) of the four systems at 300 K (equivalent to room temperature). The calculation is based on the canonical ensemble NVT (canonical ensemble with particles N, volume V, and temperature T) and is performed using the Nosé–Hoover method. Ref. [43], as shown in Figure 3, provides an example. It evident that the structures of the four structures remain stable even after 8000 steps of AIMDs with a step size of 1 fs, as no atomic bond is broken and the energy fluctuations are minimal. The results demonstrate that MoS_2/X exhibits exceptional high stability at ambient temperature (300 K).

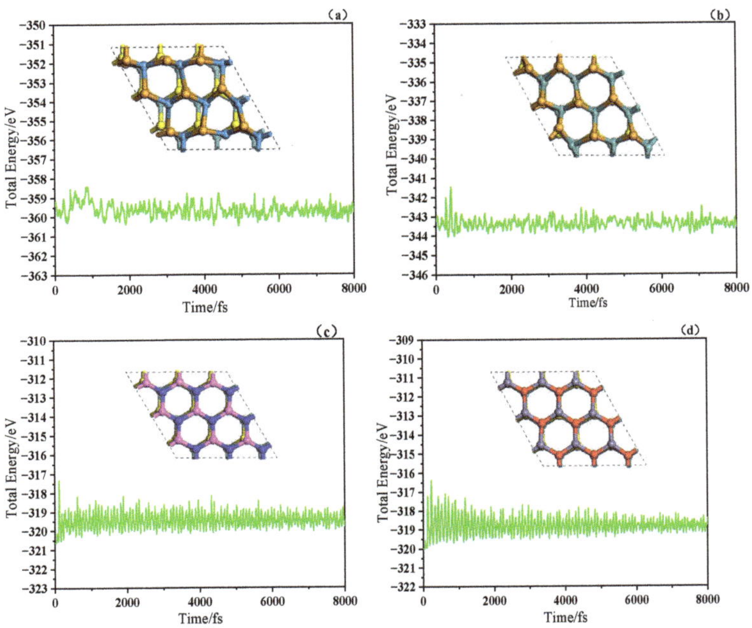

Figure 3. AIMDs (300 K) of heterojunctions: (**a**) MoS_2/WSe_2; (**b**) $MoS_2/MoSe_2$; (**c**) MoS_2/AlN; and (**d**) MoS_2/ZnO.

4. Electronic Structure

4.1. Energy Band Structure

To facilitate a comparative analysis of the electronic structures among various heterojunctions, this study computationally determines the energy band structure diagrams for the monolayer MoS_2 and four distinct heterojunctions, namely MoS_2/WSe_2, $MoS_2/MoSe_2$, MoS_2/AlN, and MoS_2/ZnO. The Fermi energy level (E_f) is set at 0 eV, as shown in Figure 4a–e. The diagram in Figure 4a illustrates that both the conduction band minimum (CBM) and the valence band maximum (VBM) correspond to the high symmetry point K, thereby demonstrating the direct band structure of single-layer MoS_2. The calculated band gap value of 1.75 eV exhibits a negligible deviation of only 0.5% from the experimental value of 1.76 eV [44], thereby affirming the reliability and reasonability of the parameters and methodologies employed in this study.

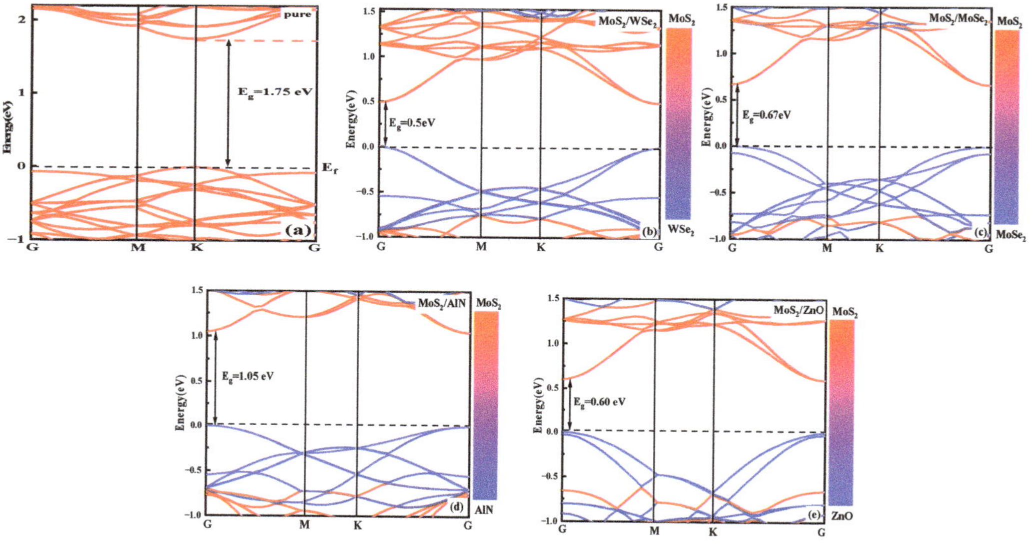

Figure 4. Energy band structure diagram: (**a**) monolayer MoS_2; (**b**) MoS_2/WSe_2 heterojunction; (**c**) $MoS_2/MoSe_2$ heterojunction; (**d**) MoS_2/AlN heterojunction; and (**e**) MoS_2/ZnO heterojunction.

The band gap of the four heterojunctions, namely MoS_2/WSe_2, $MoS_2/MoSe_2$, MoS_2/AlN, and MoS_2/ZnO, are 0.50 eV, 0.67 eV, 1.05 eV, and 0.60 eV, respectively (Figure 4b–e). Additionally, all heterojunctions still exhibit a direct band structure. The band gaps of all heterojunctions have been reduced compared to that of the monolayer MoS_2, facilitating the electron transitions and enhancing light absorption efficiency. Notably, the MoS_2/WSe_2 band gap experiences the most significant reduction.

The band structure in MoS_2/AlN and MoS_2/ZnO exhibits a relatively low density of states due to the significantly wider bandgaps of AlN and ZnO compared to that of single-layer MoS_2. Consequently, upon combination, there is a shift towards deeper energy levels in both conduction and valence bands. And it can be seen from the figure that the density on the conduction band of the two heterojunctions of MoS_2/WSe_2 and $MoS_2/MoSe_2$ is greater than that of MoS_2/AlN and MoS_2/ZnO, which may be because the components of both heterojunctions of MoS_2/WSe_2 and $MoS_2/MoSe_2$ are chalcogen compounds, and after combination, electrons are more likely to transfer from the valence band to the conduction band. The band diagrams reveal that the four heterojunctions demonstrate a type II band alignment. Further analysis on this aspect will be discussed subsequently.

4.2. Density of States

The density of state plots for the monolayer MoS_2 with four heterojunctions are presented in Figure 5. The conduction band of the monolayer MoS_2 is primarily constituted of Mo-4d, Mo-5s, and S-3p states, while the valence band predominantly consists of Mo-4d, S-3s, and S-3p states, as illustrated in Figure 5a. The Fermi energy level of the system intersects the valence band and is slightly situated below the VBM. The main reason for this phenomenon is attributed to the disruption of the periodic potential field in the system caused by cutting the bulk phase structure, leading to the formation of the monolayer MoS_2. This phenomenon leads to the generation of a continuous energy distribution at the uppermost region of the valence band, thereby inducing a band tail effect. The density of states of MoS_2/WSe_2 is presented in Figure 5b. The VBM of the heterojunction is found to arise from the spatial distribution of the Se-4p orbital within WSe_2, while CBM originates from the spatial distribution of the Mo-5s orbital within MoS_2. The density of states of $MoS_2/MoSe_2$ depicted in Figure 5c. The VBM of the heterojunction arises from the hybridization between Mo-5s orbitals and Se-4p orbitals in $MoSe_2$, while the CBM is primarily attributed to the Mo-5s orbitals of MoS_2. The density of states of the MoS_2/AlN is illustrated in Figure 5d. The VBM of the heterojunction is predominantly contributed to by the N-2p orbitals of AlN, while the CBM is mainly contributed to by the Mo-5s orbitals of MoS_2. The density of states of MoS_2/ZnO is depicted in Figure 5e. The O-2p orbital in ZnO predominantly contributes to the VBM of the heterojunction, whereas the Mo-5s orbital in the MoS_2 primarily governs the CBM.

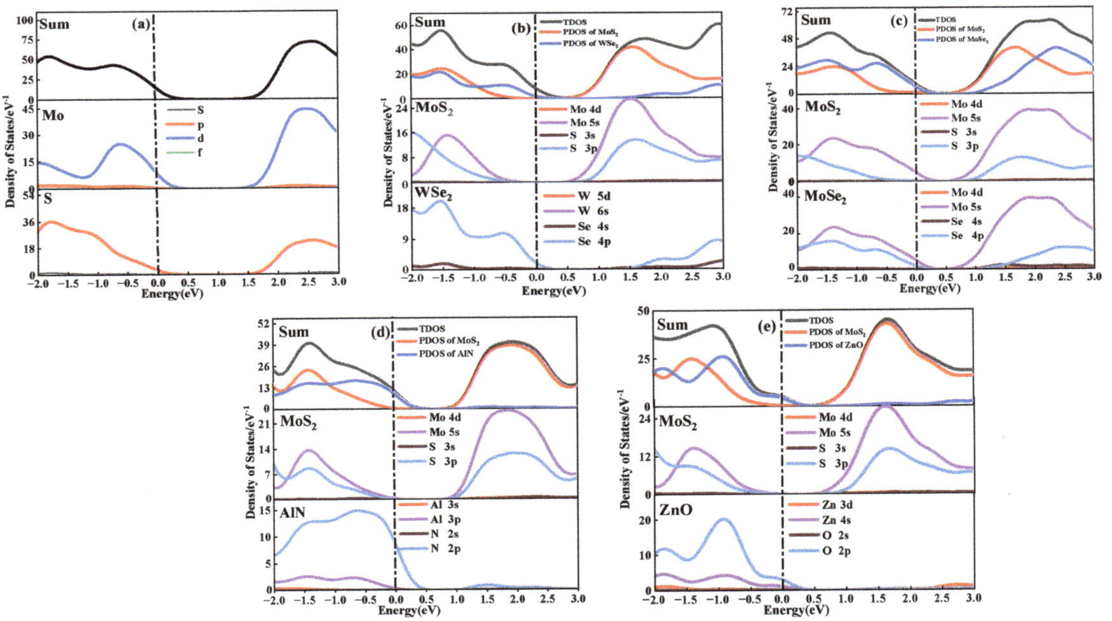

Figure 5. Density of state plots: (**a**) monolayer MoS_2; (**b**) MoS_2/WSe_2; (**c**) $MoS_2/MoSe_2$; (**d**) MoS_2/AlN; and (**e**) MoS_2/ZnO.

Combined with the energy band structure diagram, it can be observed that the VBMs of all four heterojunctions are predominantly contributed to by the X material, while the CBMs primarily arise from energy states associated with MoS_2. The VBM and CBM originate from distinct materials within the heterojunction, exhibiting characteristic features of typical type II heterojunctions [45,46]. It is evident that all four heterojunctions possess type II band alignments.

The conduction bands of all four heterojunctions exhibit a pronounced overlap in energy levels from the S-3p and Mo-5s orbitals of MoS_2. The strong orbital hybridization between S and Mo atoms is indicative of enhanced electron transfer from S-3p orbitals to Mo-5s orbitals, facilitating the movement of excited electrons. Consequently, the heterojunction facilitates a higher concentration of electrons at Mo atoms within the MoS_2 material. In heterojunctions, the degree of overlap between different orbitals in the density of state diagrams varies, resulting in a potential reduction in the band gap magnitude within the systems. This reduction in the band gap can enhance the light absorption efficiency of the system, as evidenced by the analysis of energy band diagrams.

4.3. Work Function

The work function(Φ), as defined by Equation (5), quantifies the energy required to overcome the surface potential barrier, enabling electrons to escape from the solid surface (E_F) and reach the vacuum energy level (E_{vac}). This function offers valuable insights into the photoelectric emission efficiency of the semiconductor, which is pivotal for comprehending its electronic characteristics and potential applications.

$$\Phi = E_{vac} - E_F \tag{4}$$

Table 3 demonstrates that the calculated work function is higher than the experimental value, which can be attributed to the tendency of GGA calculations to overestimate the work function. The calculated work function results presented in this study are generally consistent with recently published findings, thereby establishing the credibility of our calculations. Since the calculated results are solely utilized for comparative analysis of heterojunctions in this study, their relative trends can be considered reliable [47–56].

Table 3. Summary of the work functions of monolayers MoS_2, WSe_2, $MoSe_2$, AlN, and ZnO.

	MoS_2	WSe_2	$MoSe_2$	AlN	ZnO
Φ (this work)	5.78 eV	4.57 eV	4.64 eV	5.04 eV	5.09 eV
Φ (experimental)	4.80 eV [46]	3.70 eV [45]	4.40 eV [46]	4.45 eV [47]	4.85 eV [48]
Φ (calculation)	5.67 eV [53] GGA + PBE	4.36 eV [53] GGA + PBE	4.57 eV [54] PBE + SOC	4.92 eV [55] GGA + PBE	5.06 eV [56] GGA + PBE

The direction of electron transfer in the four heterojunctions can be tentatively determined as follows: electrons will migrate from the WSe_2 layer to the MoS_2 layer, from the AlN layer to the MoS_2 layer, and from the ZnO layer to the MoS_2 layer. The statement is derived from the analysis of work function images in various systems.

The simplified diagram illustrating the work function in Figure 6 reveals that the heterojunctions generate an internal electric field due to the disparity between each two constituents. The internal electric field serves to enhance the efficiency of separation of photogenerated electrons and holes, thereby augmenting the photoelectric functionality of the systems. This observation is consistent with the findings obtained from the electric structure analysis.

The analysis above indicates that electron transfer occurs at the interface of the heterojunction. To investigate the microscopic details of electron transfer, Figure 7 presents diagrams illustrating the lowest unoccupied molecular orbital (LUMO) and highest occupied molecular orbital (HOMO) of the four heterojunctions.

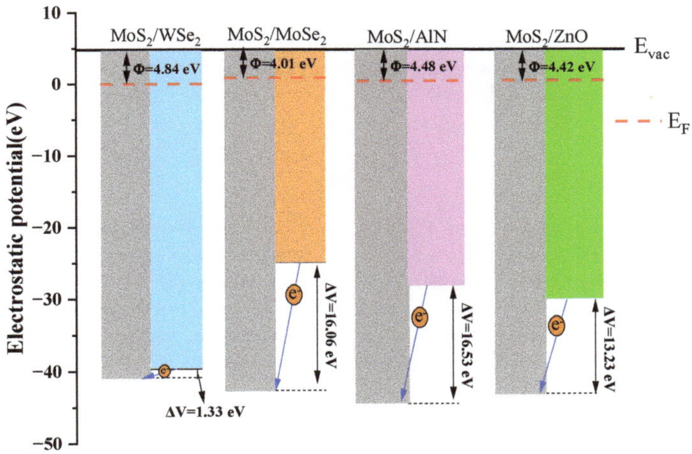

Figure 6. A simplified diagram illustrating the work function of four heterojunctions, namely MoS_2/WSe_2, $MoS_2/MoSe_2$, MoS_2/AlN, and MoS_2/ZnO, where each histogram represents the potential energy difference from the vacuum level to the minimum work function of each component in the respective heterojunction.

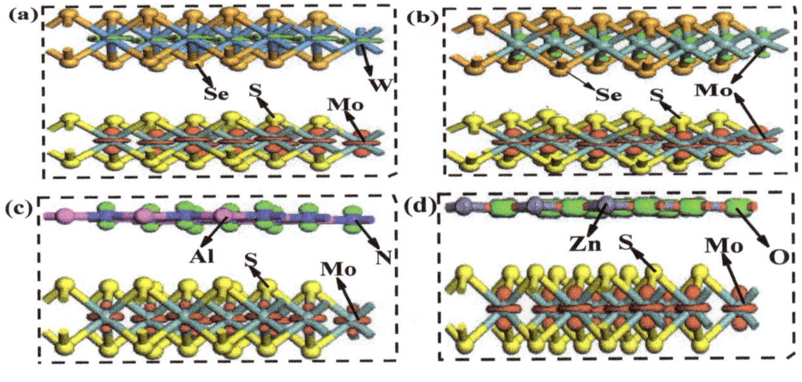

Figure 7. LUMO (red) and HOMO (green) diagrams: (**a**) MoS_2/WSe_2 heterojunction; (**b**) $MoS_2/MoSe_2$ heterojunction; (**c**) MoS_2/AlN heterojunction; and (**d**) MoS_2/ZnO heterojunction.

The electrons in the LOMO possess the minimum energy and exhibit enhanced accessibility to the external environment. Conversely, electrons residing in the HOMO demonstrate elevated energy levels but limited binding capacity, thereby facilitating facile electron loss. When subjected to solar irradiation, photoexcited electrons undergo a transition from the LUMO state to the HOMO state, thereby establishing the LUMO–HOMO transition as the primary pathway for the migration of photogenerated electrons [57–59]. Figure 7 indicates that the LUMO and HOMO of the four heterojunctions are located in distinct components. Therefore, the presence of an internal electric field within the heterojunction is observed to facilitate a reduction in the recombination rate of electron–hole pairs. This observation is also in line with the analysis of the internal electric field in type II heterojunctions.

Considering the results presented in Figure 7a–d, it is evident that electron transfer takes place from the W atoms in the WSe_2 layer to the Mo atoms in the MoS_2 layer within MoS_2/WSe_2. Similarly, in $MoS_2/MoSe_2$, the electrons transfer from Mo atoms in the $MoSe_2$ layer to Mo atoms in the MoS_2 layer. In MoS_2/AlN, the transfer flow is from N atoms in the AlN layer to Mo atoms in the MoS_2 layer. In MoS_2/ZnO, electrons transfer from O atoms in the ZnO layer to Mo atoms in the MoS_2 layer. Consequently, at the heterojunction interface, the transfer of electrons results in the formation of an internal electric field, and the field in the four heterojunctions are all directed towards MoS_2. As a result, the photoelectric properties of the four systems are enhanced, which is in line with the findings from the work function analysis.

5. Optical Property

5.1. Alignment with Edges

Typically, photocatalysts are assessed based on the semiconductor conduction band (CB) and valence band (VB) values relative to the standard hydrogen electrode (NHE). To determine the band edge potential on the NHE scale, Equations (5) and (6) [58] are commonly used. In semiconductor physics, X represents the absolute electronegativity of the material. The free electron energy at the NHE scale, which is approximately 4.5 eV, is denoted by E_{elce}. Meanwhile, E_g refers to the semiconductor band gap.

$$E_{CB} = X - E_{elec} - 0.5E_g \tag{5}$$

$$E_{VB} = \text{VBM} + E_g \tag{6}$$

The study calculated the E_{CB} and E_{VB} values for the monolayer MoS_2 and four heterojunctions, namely MoS_2/WSe_2, $MoS_2/MoSe_2$, MoS_2/AlN, and MoS_2/ZnO. The values obtained were (0.055 eV, 1.695 eV), (−0.75 eV, 0.25 eV), (0.005 eV, 0.675 eV), (−0.545 eV, 0.505 eV), and (−0.380 eV, 0.22 eV), respectively. The photocatalytic mechanism of the material involves the absorption of photon energy, which causes the transfer of electrons (e^-) in the VB to CB, leaving holes (h^+) in the VB. The resulting e^--h^+ pairs participate in the redox reaction to degrade pollutants. In this paper, it is observed that all the systems studied, namely the monolayer MoS_2, MoS_2/WSe_2, $MoS_2/MoSe_2$, MoS_2/AlN, and MoS_2/ZnO, have the capability to absorb photon energy and generate e^--h^+ for redox reactions. Additionally, the construction of heterojunctions has been found to alter the photocatalytic ability of the monolayer MoS_2. The monolayer MoS_2 has a higher E_{CB} than the $O^2/{\cdot}O^{2-}$ reduction potential of −0.33 eV [59]. Therefore, it cannot directly produce ${\cdot}O^{2-}$. However, through the construction of heterojunctions, the E_{CB} is effectively reduced. This reduction occurs in three heterojunctions, according to [60]; the monolayer MoS_2 has a higher E_{CB} than the $O^2/{\cdot}O^{2-}$ reduction potential of −0.33 eV, which means that it cannot directly produce O^{2-}. However, it is possible to reduce the E_{CB} effectively by constructing heterojunctions. This reduction can occur in three different heterojunctions.

$$E_{CB}^{vac} = -E_{CB} - 4.5 \tag{7}$$

$$E_{VB}^{vac} = -E_{VB} - 4.5 \tag{8}$$

$$E_{H^+/H_2}^{red} = -4.44\text{ eV} + \text{PH} \times 0.059\text{ eV} \tag{9}$$

$$E_{O_2/H_2O}^{oxd} = -5.67\text{ eV} + \text{PH} \times 0.059\text{ eV} \tag{10}$$

To ensure a photocatalyst's versatility, we evaluated the vacuum band edge potential of five systems against NHE and the redox potential of water at varying pH levels using Equations (7)–(10) [60]. The resulting data are presented in Figure 8. Research has indicated that the monolayer MoS_2 is not suitable for hydrogen precipitation in environments that are either acidic, neutral, or basic. In $MoS_2/MoSe_2$, the conduction band minimum (CBM) is similar to that of the monolayer MoS_2. However, the valence band (VB) is not lower than the potential of O_2/H_2O. This means that electrons in the CB can combine with O_2 to form ${\cdot}O_2{}^-$, which allows the heterojunction to retain some of the inherent photocatalytic properties of the monolayer MoS_2.

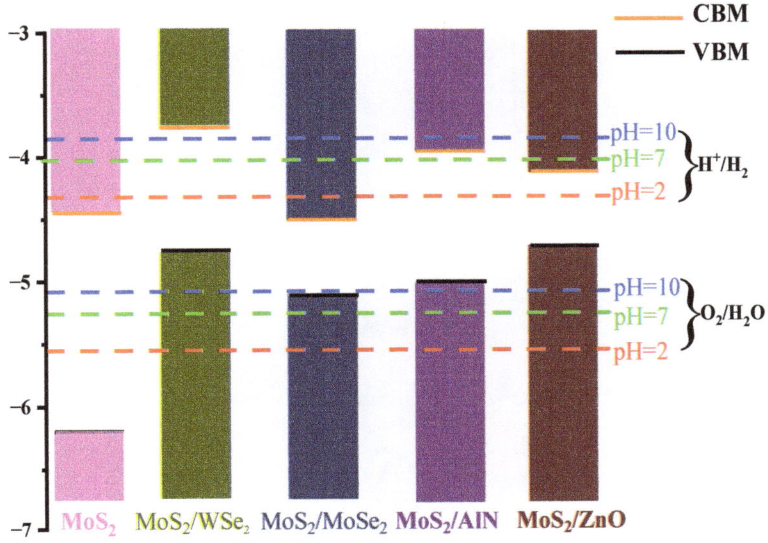

Figure 8. Trend in band edge potential of monolayer MoS_2 and MoS_2/WSe_2, $MoS_2/MoSe_2$, MoS_2/AlN, and MoS_2/ZnO under different pH conditions.

5.2. Complex Dielectric Function

The intricate dielectric functions for the monolayer MoS_2 and four heterojunctions are depicted in Figure 9. Specifically, Figure 9a illustrates the real component of the dielectric function, with the point of intersection on the vertical axis denoting the static dielectric constant of the five systems [61]. The values obtained for the systems are 5.80, 7.81, 9.30, 5.81, and 5.79, respectively. The static dielectric constants of the four heterojunctions are observed to surpass that of the monolayer MoS_2, thereby demonstrating enhanced polarization ability, intensified photogenerated electric field intensity, and accelerated carrier migration rates. These enhancements contribute to the efficient separation of photogenerated electron–hole pairs. $MoS_2/MoSe_2$ exhibits the highest static dielectric constant, indicating its superior polarization capability. In Figure 9b, the curves represent the imaginary component of the dielectric functions, and the first peaks of all the curves corresponds to electron transitions between the conduction band and valence band. The transverse coordinate value corresponding to the peak value is equivalent to the band gap width [62]. The primary peak of the monolayer MoS_2 is observed at 2.92 eV. Meanwhile, the primary peaks of the four heterojunctions, namely MoS_2/WSe_2, $MoS_2/MoSe_2$, MoS_2/AlN, and MoS_2/ZnO, are located at 2.80 eV, 2.77 eV, 2.83 eV, and 2.74 eV, respectively. In comparison to the monolayer MoS_2, the heterojunctions exhibit the shift of its main peaks towards the low-energy direction, indicating a smaller band gap value. Additionally, the figure illustrates that $MoS_2/MoSe_2$ exhibits the highest peak in the imaginary part of the dielectric functions, which can be attributed to a hybridization effect between Mo-5s and S-3p orbitals.

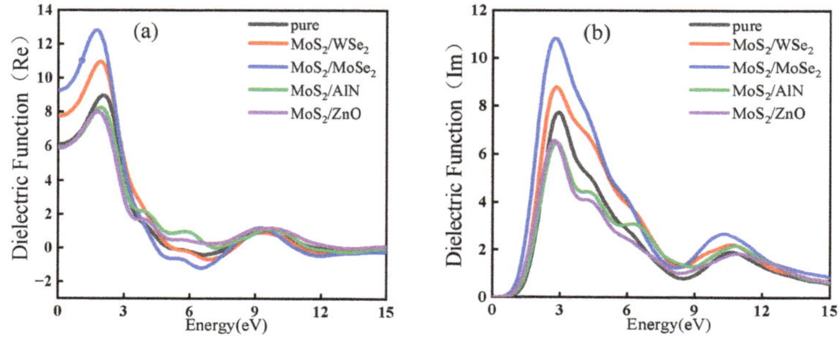

Figure 9. Complex dielectric function diagram: (**a**) real part and (**b**) imaginary part.

5.3. Absorption Spectrum

The semiconductor light absorption coefficient is denoted by α. This parameter quantifies the ability of the system to absorb light, with its magnitude reflecting the strength of this capability [61,62].

$$\alpha(\omega) = \frac{2\omega k(\omega)}{c} \tag{11}$$

In the equation, ω stands for angular frequency, $k(\omega)$ represents the complex refractive index, and c denotes the speed of light.

The absorption spectra of the monolayer MoS_2 and four heterojunctions are illustrated in Figure 10a,b, providing insights into the energy-dependent trend in the optical absorption coefficient within the photon energy range of 0–15 eV. The inset of Figure 10a reveals a red shift in the optical absorption band edges of the four heterojunctions compared to the monolayer MoS_2, indicating that the formation of heterojunctions can effectively broaden the system's spectral response range to light. The absorption spectra of all five systems show similar trends varying with photon energy, featuring more pronounced absorption peaks compared to the monolayer MoS_2 at intervals of 5–7 eV and 10–11 eV, while the $MoS_2/MoSe_2$ structure exhibited the highest peak.

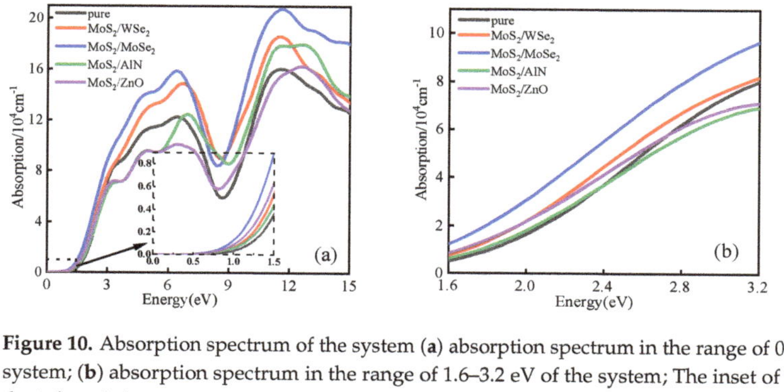

Figure 10. Absorption spectrum of the system (**a**) absorption spectrum in the range of 0–15 eV of the system; (**b**) absorption spectrum in the range of 1.6–3.2 eV of the system; The inset of Figure 10a is the infrared absorption spectrum.

To enhance the practicality of the data, we additionally generated absorption spectra for the five systems within the visible region (1.6–3.2 eV), as depicted in Figure 10b. The absorption coefficients of the four heterojunctions in the visible region have increased compared to that of the monolayer MoS_2. $MoS_2/MoSe_2$ demonstrates the highest absorption coefficient within the visible region, indicating its superior capacity for absorbing visible light.

6. Conclusions

The electronic structure and optical properties of four new heterostructures, MoS_2/WSe_2, $MoS_2/MoSe_2$, MoS_2/AlN, and MoS_2/ZnO, which are constructed with MoS_2 as the substrate, are calculated by the density functional theory. The constructed heterostructures have been observed to exhibit a stable existence, with a type II band structure and a band gap smaller than that of the monolayer. The generation of the MoS_2 type II band structure implies that the internal electric field generates at the interface of the four heterojunctions, enhancing the effective separation of photogenerated electron–hole pairs. The conclusion has been substantiated by means of an analysis conducted on the absorption spectrum and dielectric function. The heterostructures MoS_2/WSe_2, $MoS_2/MoSe_2$, and MoS_2/AlN exhibit a redshift absorption spectrum compared to the monolayer MoS_2, thereby expanding the systems' range of light response, and they have stronger static dielectric constants than that of the monolayer MoS_2. In addition, MoS_2/WSe_2 and $MoS_2/MoSe_2$ in the visible region have higher absorption coefficients and static dielectric constants than that of MoS_2/AlN and MoS_2/ZnO, indicating that the light absorption capacity of chalcogenide heterojunctions is superior. Chalcogenide heterojunctions MoS_2/WSe_2 and $MoS_2/MoSe_2$ have a stronger electronic transition ability and are more suitable for optical electronic devices.

Supplementary Materials: The following supporting information can be downloaded at: https://www.mdpi.com/article/10.3390/catal14100678/s1, Figure S1. Stacking pattern of heterojunctions. SP1 is the alignment of tungsten (W) and molybdenum (Mo); SP2 is the alignment of selenium (Se) and molybdenum (Mo); SP3, SP4, and SP5 are stacking patterns with different angles between selenium (Se) and molybdenum (Mo).

Author Contributions: Conceptualization, X.Z., L.Z. and Y.Z.; methodology, J.L., Y.H. and Y.Z.; validation, J.L. and Y.Z.; formal analysis, B.L.; investigation, J.L. and X.Z.; data curation, J.L., Y.J. and B.L.; writing—original draft preparation, J.L. and X.Z.; writing—review and editing, Y.H., L.Z. and Y.Z.; visualization, X.Z. and Y.J.; project administration, L.Z.; funding acquisition, L.Z. All authors have read and agreed to the published version of the manuscript.

Funding: This research was funded by the Open Project of Xinjiang Laboratory of Condensed Matter Phase Transition and Microstructure, grant number XJDX0912Y2409, the Science and Technology Plan Project of Yili Kazakh Autonomous Prefecture, grant number YZ2022B021, and the key project at the school level of Yili Normal University, grant number 2023YSZD003.

Data Availability Statement: The original contributions presented in the study are included in the article; further inquiries can be directed to the corresponding authors.

Acknowledgments: We gratefully acknowledge HZWTECH for providing computation facilities.

Conflicts of Interest: The authors declare no conflicts of interest.

References

1. Mak, K.F.; Shan, J. Photonics and optoelectronics of 2D semiconductor transition metal dichalcogenides. *Nat. Photonics* **2016**, *10*, 216–226. [CrossRef]
2. García-Dalí, S.; Paredes, J.I.; Villar-Rodil, S.; Martínez-Jódar, A.; Martínez-Alonso, A.; Tascón, J.M.D. Molecular Functionalization of 2H-Phase MoS_2 Nanosheets via an Electrolytic Route for Enhanced Catalytic Performance. *ACS Appl. Mater. Int.* **2021**, *13*, 33157–33171. [CrossRef] [PubMed]
3. Lu, S.; Li, C.; Zhao, Y.F. Tunable redox potential of nonmetal doped monolayer MoS_2: First principle calculations. *Appl. Surf. Sci.* **2016**, *384*, 360–367. [CrossRef]
4. Lin, S.Y.; Li, W.T.; Dong, Y.Y.; Wang, C.; Chen, Q.; Zhang, H. Two-dimensional metallic MoS_2: A DFT study. *Comput. Mater. Sci.* **2016**, *124*, 49–53. [CrossRef]
5. Shirazi, M.; Kessels, W.M.M.; Bol, A.A. Initial stage of atomic layer deposition of 2D-MoS_2 on a SiO_2 surface: A DFT study. *Phys. Chem. Chem. Phys.* **2018**, *20*, 16861. [CrossRef]
6. Yoon, Y.; Ganapathi, K.; Salahuddin, S. How Good Can Monolayer MoS_2 Transistors Be? *Nano Lett.* **2011**, *11*, 3768–3773. [CrossRef]
7. Mak, K.F.; He, K.; Shan, J.; Heinz, T.F. Control of valley polarization in monolayer MoS_2 by optical helicity. *Nat. Nanotechnol.* **2012**, *7*, 494–498. [CrossRef]

8. Lopez-Sanchez, O.; Lembke, D.; Kayci, M.; Radenovic, A.; Kis, A. Ultrasensitive photodetectors based on monolayer MoS_2. *Nat. Nanotechnol.* **2013**, *8*, 497–501. [CrossRef]
9. Zhuang, H.L.; Hennig, R.G. Computational Search for Single-Layer Transition-Metal Dichalcogenide Photocatalysts. *J. Phys. Chem. C* **2013**, *117*, 20440–20445. [CrossRef]
10. Schmidt, H.; Wang, S.; Chu, L. Transport Properties of Monolayer MoS_2 Grown by Chemical Vapor Deposition. *Nano Lett.* **2014**, *14*, 1909–1913. [CrossRef]
11. Li, S.S.; Sun, J.R.; Guan, J.Q. Strategies to improve electrocatalytic and photocatalytic performance of two-dimensional materials for hydrogen Strategies to improve electrocatalytic and photocatalytic performance of two-dimensional materials for hydrogen evolution reaction. *Chin. J. Catal.* **2021**, *42*, 511–556. [CrossRef]
12. Liu, Y.; Cao, Y.; Lv, H.; Li, S.; Zhang, H. Construction of a 2D nanoplate-like MoS_2/Bi_2WO_6 heterojunction with enhanced visible-light photocatalytic activities. *Mater. Lett.* **2017**, *188*, 99–102. [CrossRef]
13. Yu, Z.; Peng, J.; Lei, H.; Zhai, X.F.; Jiang, F.H.; Sun, C.J. Novel Z-scheme MoS_2/Bi_2WO_6 heterojunction with highly enhanced photocatalytic activity under visible light irradiation. *J. Alloys Compd.* **2021**, *8*, 157224.
14. Liao, J.; Zhong, Y.; He, Z.; Ding, H.; Chen, K.; Chen, D.M. One-step synthesis of MoS_2/Bi_2S_3 heterojunction with enhanced photocatalytic activity and high electrochemical performance. *J. Mater. Sci. Mater. Electron.* **2020**, *32*, 385–394. [CrossRef]
15. Zhang, X.; Xia, M.; Wang, F.; Lei, W. Cu_2O/MoS_2 composites: A novel photocatalyst for photocatalytic degradation of organic dyes under visible light. *Ionics* **2020**, *31*, 6359–6369. [CrossRef]
16. Yao, H.; Guo, X.; Bao, A.; Mao, H.Y.; Ma, Y.C.; Li, X.C. Graphene-based heterojunction for enhanced photodetectors. *Chin. Phys. B* **2022**, *31*, 38501. [CrossRef]
17. Deng, X.Z.; Zhang, J.R.; Zhao, Y.Q.; Yu, Z.L.; Yang, J.L.; Cai, M.Q. The energy band engineering for the high-performance infrared photodetectors constructed by $CdTe/MoS_2$ heterojunction. *J. Phys. Condens. Matter* **2020**, *32*, 65004. [CrossRef]
18. Um, D.; Lee, Y.; Lim, S.; Park, S.Y.; Lee, C.; Ko, H. High-Performance MoS_2/CuO Nanosheet-on-One-Dimensional Heterojunction Photodetectors. *ACS Appl. Mater. Interfaces* **2016**, *8*, 33955–33962. [CrossRef]
19. Abderrahmane, A.; Woo, C.; Ko, P.J. Black Phosphorus/Molybdenum Diselenide Heterojunction-Based Photodetector. *J. Electron. Mater.* **2021**, *50*, 5713–5720. [CrossRef]
20. Meng, F.X.; Xu, Z.Q.; Zeng, Y.H.; Wang, C.H.; Zou, Y.N.; Wageh, S.; Al-Ghamdi, A.; Guo, Z.N.; Zhang, H. Photodetectors Based on $MoS_2/MAPbBr_3$ van der Waals Heterojunction. *IEEE Electron Device Lett.* **2022**, *43*, 414–417. [CrossRef]
21. Xue, Z.; Zhang, X.; Qin, J.; Liu, R.P. Constructing $MoS_2/g\text{-}C_3N_4$ heterojunction with enhanced oxygen evolution reaction activity: A theoretical insight. *Appl. Surf. Sci.* **2019**, *5*, 254–268. [CrossRef]
22. Sundaram, R.S.; Engel, M.; Lombardo, A.; Krupke, R.; Ferrari, A.C.; Avouris, P.; Steiner, M. Electroluminescence in Single Layer MoS_2. *Nano Lett.* **2013**, *13*, 1416–1421. [CrossRef] [PubMed]
23. Ye, L.; Wang, P.; Luo, W.J.; Gong, F.; Liao, L.; Liu, T.D.; Tong, L.; Zang, J.F.; Xu, J.B.; Hu, W.D. Highly polarization sensitive infrared photodetector based on black phosphorus-on-WSe_2 photogate vertical heterostructure. *Nano Energy* **2017**, *37*, 53–60. [CrossRef]
24. Zhou, X.; Tian, Z.; Kim, H.J.; Wang, J.; Xu, B.R.; Pan, R.B.; Chang, Y.J.; Di, Z.F.; Zhou, P.; Mei, Y.F. Rolling up $MoSe_2$ Nanomembranes as a Sensitive Tubular Photodetector. *Small* **2019**, *15*, 1902528. [CrossRef] [PubMed]
25. Roy, T.; Tosun, M.; Hettick, M.; Ahn, G.H.; Hu, C.M.; Javey, A. 2D-2D tunneling field-effect transistors using $WSe_2/SnSe_2$ heterostructures. *Appl. Phys. Lett.* **2016**, *108*, 83111. [CrossRef]
26. Dybaa, F.; Polak, M.P.; Kopaczek, J.; Scharoch, P.; Wu, K.; Tongay, S.; Kudrawiec, R. crystals and semiconductor to metal transitions. *Sci. Rep.* **2016**, *6*, 26663.
27. Kopaczek, J.; Polak, M.P.; Scharoch, P.; Wu, K.; Chen, B.; Tongay, S.; Kudrawiec, R. Direct optical transitions at K- and H-point of Brillouin zone in bulk MoS_2, $MoSe_2$, WS_2, and WSe_2. *J. Appl. Phys.* **2016**, *119*, 1244. [CrossRef]
28. Long, L.; Guo, Y.P.; He, C.Z.; Tao, H.L.; Huang, J.T.; Yu, W.Y.; Chen, R.X.; Lou, M.S.; Yan, L.B. First-principles study of magnetism of 3d transition metals and nitrogen co-doped monolayer MoS_2. *Chin. Phys. B* **2020**, *29*, 517–524.
29. Weinhardt, L.; Blum, B.; Bär, M.; Heske, C.; Cole, B.; Marsen, B.; Miller, E.L. Electronic Surface Level Positions of WO_3 Thin Films for Photoelectrochemical Hydrogen Production. *J. Phys. Chem. C* **2008**, *112*, 3078–3082. [CrossRef]
30. Tkatchenko, A.; Scheffler, M. Accurate Molecular Van Der Waals Interactions from Ground-State Electron Density and Free-Atom Reference Data. *Phys. Rev. Lett.* **2009**, *102*, 69–72. [CrossRef]
31. Grimme, S. Semiempirical GGA-type density functional constructed with a long-range dispersion correction. *J. Comput. Chem.* **2010**, *27*, 1787–1799. [CrossRef] [PubMed]
32. Wang, Y.; Huang, S.; Zhao, H.; Liu, K.Y. First Principles Study on Properties of Monolayer MoS_2 Under Different Strains. *Braz. J. Phys.* **2021**, *51*, 1230–1236. [CrossRef]
33. Gustafsson, M.V.; Yankowitz, M.; Forsythe, C.; Rhodes, D.; Watanabe, K.; Taniguchi, T.; Hone, J.; Zhu, X.Y.; Dean, C.R. Ambipolar Landau levels and strong band-selective carrier interactions in monolayer WSe_2. *Nat. Mater.* **2018**, *17*, 4–11. [CrossRef] [PubMed]
34. Xiao, M.; Saki, N.; Lu, Z.; Goto, H.; Eguchi, R.; Ota, H.; Kambe, T.; Terashima, K.; Yokoya, K.; Huyen, T.L.; et al. Emergence of superconductivity in $(NH_3)_yM_xMoSe_2$ (M: Li, Na and K). *Sci. Rep.* **2016**, *6*, 29292.
35. Wu, H.; Qin, Z.; Yan, Z.; Sun, Z.H.; Li, B.K.; Zheng, R.S. First Principles Study on Li-doped and Li, O-codoped AlN. *Medžiagotyra* **2019**, *25*, 17–20.
36. Badreddine, K.; Kazah, I.; Rekaby, M.; Awad, R. Structural, Morphological, Optical, and Room Temperature Magnetic Characterization on Pure and Sm-Doped ZnO Nanoparticles. *J. Nanomater.* **2018**, 7096195. [CrossRef]

37. Guo, L.J.; Hu, J.S.; Ma, X.G.; Xiang, J. Interfacial interaction and Schottky contact of two-dimensional WS_2/graphene heterostructure. *Acta Phys. Sin.-Chin. Ed.* **2019**, *68*, 97101. [CrossRef]
38. Ma, H.H.; Zhang, X.B.; Wei, X.Y.; Cao, J.M. Theoretical study on Schottky regulation of WSe_2/graphene heterostructure doped with nonmetallic elements. *Acta Phys. Sin.-Chin. Ed.* **2020**, *69*, 117101. [CrossRef]
39. Hu, J.; Ji, G.; Ma, X.; He, H.; Huang, C.Y. Probing interfacial electronic properties of graphene/$CH_3NH_3PbI_3$ heterojunctions: A theoretical study. *Appl. Surf. Sci.* **2018**, *440*, 35–41. [CrossRef]
40. Zhang, J.R.; Yu, Q.; Chen, L.; Yin, S.F.; Cai, M.Q. Density functional theory calculation on facet-dependent photocatalytic activity of MoS_2/CdS heterostructures. *Appl. Surf. Sci.* **2019**, *469*, 27–33. [CrossRef]
41. Sun, M.; Nelson, A.E.; Adjaye, J. Ab initio DFT study of hydrogen dissociation on MoS_2, NiMoS, and CoMoS: Mechanism, kinetics, and vibrational frequencies. *J. Catal.* **2005**, *233*, 411–421. [CrossRef]
42. Blöchl, P.E. Projector augmented-wave method. *Phys. Rev. B* **1994**, *24*, 17953–17979. [CrossRef] [PubMed]
43. Bucher, D.; Pierce, L.C.T.; McCammon, J.A.; Markwick, P.R.L. On the Use of Accelerated Molecular Dynamics to Enhance Configurational Sampling in Ab Initio Simulations. *J. Chem. Theory Comput.* **2011**, *7*, 890–897. [CrossRef] [PubMed]
44. Kam, K.K.; Parkinson, B.A. Detailed photocurrent spectroscopy of the semiconducting group VIB transition metal dichalcogenides. *J. Phys. Chem.* **1982**, *86*, 463–467. [CrossRef]
45. Liu, J.; Hua, E. Photocatalytic properties and energy band offset of a tungsten disulfide/graphitic carbon nitride van der Waals heterojunction. *RSC Adv.* **2020**, *10*, 5260–5267. [CrossRef]
46. Chai, H.; Gao, L.; Wang, P.; Li, F.; Hu, G.W.; Jin, J. In_2S_3/F-Fe_2O_3 type-II heterojunction bonded by interfacial S-O for enhanced charge separation and transport in photoelectrochemical water oxidation. *Appl. Catal. B Environ.* **2022**, *305*, 121011. [CrossRef]
47. Zhou, F.; Ren, X.H.; Liu, J.; Liu, P. Development of Photocatalytic Degradation of Organic Pollutants in Water. *J. Mater. Eng.* **2018**, *46*, 9–19.
48. Liu, L.W.; Liu, C.S.; Huang, X.H.; Zeng, S.F.; Tang, Z.W.; Zhang, D.W.; Zhou, P. Tunable Current Regulative Diode Based on Van der Waals Stacked MoS_2/WSe_2 Heterojunction–Channel Field-Effect Transistor. *Adv. Electron. Mater.* **2022**, *8*, 2100869. [CrossRef]
49. Shimada, T.; Ohuchi, F.S.; Parkinson, B.A. Work function and photo threshold of layered metal dichalcogenides. *Jpn. J. Appl. Phys.* **1994**, *33*, 2696–2698. [CrossRef]
50. Sun, H.; Torres, C.G.; Liu, K.; Li, K.H.; Guo, W.Z.; Lin, R.H.; Liu, X.W.; Li, J.T.; Li, X.H. Valence and conduction band offsets of β-Ga_2O_3/AlN heterojunction. *Appl. Phys. Lett.* **2017**, *111*, 162105. [CrossRef]
51. Ozawa, K.; Edamoto, K. Photoelectron spectroscopy study of K adsorption on ZnO(1 0-1 0). *Surf. Sci.* **2003**, *524*, 78–88. [CrossRef]
52. Wei, D.; Li, Y.; Feng, Z.; Guo, G.F.; Ma, Y.Q.; Yu, H.; Luo, Q.Q.; Tang, Y.N.; Dai, X.Q. Electronic and optical properties of 3N-doped graphdiyne/MoS_2 heterostructures tuned by biaxial strain and external electric field. *Chin. Phys. B* **2021**, *30*, 438–445. [CrossRef]
53. Wang, Y.; Liu, T.; Tian, W.; Zhang, Y.; Shan, P.Y.; Chen, Y.J.; Wei, W.H.; Yuan, H.K.; Cui, H. Mechanism for hydrogen evolution from water splitting based on a MoS_2/WSe_2 heterojunction photocatalyst: A first-principle study. *RSC Adv.* **2020**, *10*, 41127–41136. [CrossRef] [PubMed]
54. Li, S.; Zang, W.; Liu, X.; Pennycook, S.J.; Kou, Z.K.; Yang, C.H.; Guan, G.; Wang, J. Heterojunction engineering of $MoSe_2$/MoS_2 with electronic modulation towards synergetic hydrogen evolution reaction and supercapacitance performance. *Chem. Eng. J.* **2019**, *359*, 1419–1426. [CrossRef]
55. Zou, H.; Peng, M.; Zhou, W.; Pan, J.L.; Ouyang, F.P. Type II GaS/AlN van der Waals heterostructure: Vertical strain, in-plane biaxial strain and electric field effect. *Phys. E Low-Dimens. Syst. Nanostruct.* **2021**, *126*, 114481. [CrossRef]
56. Chen, L.; Zhang, L.; Xiong, Z.; Cui, Y.Y.; Luo, H.J.; Gao, Y.F. Electronic and electron emission properties of X/ZnO monolayer (X = Li, Na, K, Rb, or Cs): A first-principles study. *J. Phys. Chem. Solids* **2020**, *147*, 109618. [CrossRef]
57. Li, B.; Nengzi, L.C.; Guo, R.N.; Cui, Y.Q.; Zhang, Y.X.; Cheng, X.W. Novel synthesis of Z-scheme α-Bi_2O_3/g-C_3N_4 composite photocatalyst and its enhanced visible light photocatalytic performance: Influence of calcination temperature. *Chin. Chem. Lett.* **2020**, *31*, 2705–2711. [CrossRef]
58. Tong, T.; He, B.; Zhu, B.H.; Cheng, B.; Zhang, L.Y. First-principle investigation on charge carrier transfer in transition-metal single atoms loaded g-C_3N_4. *Appl. Surf. Sci.* **2018**, *459*, 385–392. [CrossRef]
59. Wang, J.N.; Huang, Y.H.; Guo, J.X.; Zhang, J.M.; Wei, X.M.; Ma, F. Optoelectronic response and interfacial properties of BiOI/BiOX (X = F, Cl, Br) heterostructures based on DFT investigation. *Solid State Chem.* **2020**, *284*, 121181. [CrossRef]
60. Bartels, C.; Schultze, J.W.; Stimming, U.; Habib, M.A. The electrochemical behaviors of hafnium. *Electrochim. Acta* **1982**, *27*, 129–140. [CrossRef]
61. Zhang, L.L.; Xia, T.; Liu, G.A. Electronic and optical properties of n-pr co-doped anatase TiO_2 from first principles. *Acta Phys. Sin.* **2019**, *68*, 245–253.
62. Mahmood, Q.; Hassan, M.; Faridi, M.A. Study of magnetic and optical properties of Zn_(1−x)TM_xTe(TM = Mn,Fe,Co,Ni) diluted magnetic semiconductors: First principle approach. *Chin. Phys. B* **2017**, *26*, 456–464. [CrossRef]

catalysts

MDPI

Article

Modular Photocatalytic Reactor for the Removal of Estrogens from Aqueous Solutions

Liliana Bobirică [1], Cristina Orbeci [1], Cristian Pîrvu [2], Alexandra Constantinescu [2], Elena Iuliana Bîru [3], Giovanina Iuliana Ionică [1], Ecaterina Matei [4], Andrei Constantin Berbecaru [4] and Constantin Bobirică [1,*]

1 Department of Analytical Chemistry and Environmental Engineering, National University of Science and Technology POLITEHNICA Bucharest, 1–7 Polizu, 060042 Bucharest, Romania; liliana.bobirica@upb.ro (L.B.); cristina.orbeci@upb.ro (C.O.); giovanina.ionica@upb.ro (G.I.I.)
2 Department of General Chemistry, National University of Science and Technology POLITEHNICA Bucharest, 1–7 Polizu, 060042 Bucharest, Romania; cristian.pirvu@upb.ro (C.P.); alexandra.constantinescu.96@gmail.com (A.C.)
3 Department of Bioresources and Polymer Science, National University of Science and Technology POLITEHNICA Bucharest, 1–7 Polizu, 060042 Bucharest, Romania; iuliana.biru@upb.ro
4 Department of Metallic Materials Processing and Eco-Metallurgy, National University of Science and Technology POLITEHNICA Bucharest, 313 Spl. Independentei, 060042 Bucharest, Romania; ecaterina.matei@upb.ro (E.M.); andrei.berbecaru@upb.ro (A.C.B.)
* Correspondence: constantin.bobirica@upb.ro

Abstract: Estrogens, widely used for therapeutic or contraceptive purposes, act as endocrine disruptors in aquatic systems and have adverse effects on a wide range of living organisms. Wastewater insufficiently treated by conventional methods is the main way for estrogens to enter aquatic systems. Therefore, the purpose of this paper is to develop a novel photocatalytic system for the removal of the estrogenic mixture estradiol valerate/norgestrel from wastewater. The photocatalytic modules are operated in a plug flow reactor system under a UV-A radiation field, and the photocatalyst (TiO_2, ZnO or TiO_2/ZnO) is immobilized on an inert support of glass balls that are strung on stainless-steel wire and arranged in rows along the photocatalytic modules. The photocatalysts were synthesized by the sol–gel method and then deposited on the inert glass support by the hot method, after which it was calcined for two hours at a temperature of 500 °C. The experimental results showed that the efficiency of photocatalytic degradation largely depends on the dose of photocatalyst. The dose of photocatalyst can be adjusted by adding or removing photocatalytic modules, each of which have an approximately equal amount of photocatalyst. The best result was obtained for the TiO_2/ZnO photocatalyst, the organic substrate being practically mineralized in 120 min, for which only two photocatalytic modules are needed.

Keywords: estrogens; photocatalytic modules; immobilized photocatalyst; plug flow reactor

1. Introduction

Estrogens represent a group of hormones which can be natural or synthetic [1]. Natural estrogens, namely estrone (E1), estradiol (E2) and estriol (E3), are produced by the human body and help the female reproductive system to function properly. They are specifically involved in regulating the menstrual cycle of women as well as in the development of female secondary sexual characteristics [2]. Synthetic estrogens such as ethinylestradiol (EE2), estradiol valerate, diethylstilbestrol, hexetrol, dienestrol, etc., are obtained by chemical synthesis and are widely used for therapeutic or contraceptive purposes. Thus, in addition to the strict use for contraceptive purposes, they are often used to improve menopause symptoms, in the prevention of osteoporosis and hypoestrogenism, in the treatment of metastatic breast cancer, prostate cancer, etc. [3]. They are finally eliminated in partially metabolized or non-metabolized form through excretion in urine and feces, thus ending up in municipal wastewater or, accidentally, directly in natural surface waters [4].

Estrogens can also act as endocrine disruptors in aquatic systems and have adverse effects on a wide range of living organisms. Thus, sexual differentiation disorders have been observed in several aquatic species, antigonadotropic activity (inhibits the development of sexual glands), frequent intersexuality, vitellogenin production impairment, and other organ development effects in several fish species [5]. The Directive 2013/39/EU of the European Parliament established a maximum concentration allowed in surface waters for natural estrogens (E1, E2 and E3) of 0.4 ng/L, and for synthetic estrogen EE2 of 0.035 ng/L [6]. Subsequently, following numerous controversies regarding these values, Implementing Decision (EU) 2018/840 established a watch list of substances for monitoring across the Union under water policy, including estrogens E1, E2 and EE2 [7]. Moreover, Commission Implementing Decision (EU) 2022/679 established a watch list of substances and compounds of concern in water intended for human consumption, which includes estrogen E2 [8].

The main route through which estrogens reach natural surface waters is represented by the discharges of insufficiently treated wastewater in relation to them. Typical concentrations of natural estrogens in municipal wastewater are in the order of tens of ng/L for E1 and E2, and in the order of hundreds of ng/L for E3, while for the synthetic estrogen EE2, the typical concentration is around 1 ng/L [9]. The complete removal of these endocrine disruptors by classical methods for water treatment is practically impossible; thus, it is necessary to introduce additional stages into the technological treatment schemes that involve advanced procedures for removing them and other refractory organic compounds. Thus, at least in recent years, intense efforts have been made to identify, develop and test new methods for the advanced elimination of endocrine disruptors from wastewater such as physical methods (adsorption on different adsorbent materials, ultrafiltration, nanofiltration, degradation in the field of ultrasound, etc.), chemical methods (advanced oxidation methods such as Fenton oxidation, photo-Fenton, electro-Fenton, photocatalysis, etc.), and biological methods (microbiological degradation in classic or membrane nitrification/denitrification reactors) [10]. Of these, photocatalytic oxidation received special attention due to the multiple advantages that this method has, such as the complete mineralization of the organic substrate, a high degradation rate, the operation of photocatalytic reactors under normal conditions of temperature and pressure, low costs, no secondary waste generated, high availability, etc. [11].

There is currently a wide range of photocatalytic systems and photocatalysts that have been synthesized and tested under different working conditions for the degradation of several types of organic compounds, including compounds from the category of endocrine disruptors. However, serious efforts are still being made for the development of photocatalytic processes that can be transposed with good technical–economic efficiency from the laboratory scale to the industrial scale. This implies the possibility of the continuous operation of the photocatalytic reactor (reactors with photocatalysts deposited on the support), cheap photocatalysts, stability over time in terms of photocatalytic activity, etc. [12]. Currently, there is a wide range of photocatalytic reactors designed according to several criteria such as the configuration of the irradiation source, the mode of operation, the arrangement of the photocatalyst in the reactor, etc. [13]. For liquid–solid systems, the best-known configurations of photocatalytic reactors are slurry-bed (well-mixed) reactors, in which the photocatalyst is dispersed as a slurry, and fixed-bed (plug flow) reactors, in which the photocatalyst is immobilized over a support [14]. All configurations of photocatalytic reactors developed by plating from these common variants aim to increase the performances related to photocatalytic reactions, as well as to remove or mitigate the disadvantages related to their exploitation. However, the versatility of photocatalytic reactors represents one of the major problems to which it is desired to find a solution. The design of the photocatalytic reactors must allow for the simple adjustment of the operating parameters such as the photocatalyst dose according to the characteristics of the aqueous effluent undergoing processing.

Therefore, the objective of this work is to develop a novel modular photocatalytic system operated in a plug flow reactor system under a UV-A radiation field, equipped with photocatalysts based on TiO_2, ZnO or TiO_2-ZnO deposited on an inert support glass, for the removal of the estrogenic mixture estradiol valerate/norgestrel, found in pharmaceutical products widely used for contraceptive purposes, from aqueous solutions. This study highlights the innovative and very simple way to dose the photocatalysts in complex photocatalytic systems by adding or removing photocatalytic modules with a fixed amount of photocatalysts depending on the characteristics of the organic substrate to be degraded.

2. Results and Discussion

2.1. Characterization of the Synthesized Photocatalysts

2.1.1. X-ray Diffractometry

XRD data show the formation of anatase (JCPDF card no. 21-1272) and rutile (JCPDF card no. 21-1276) crystal structures in the TiO_2 sample (Figure 1a), zincite and zinc oxide (JCPDF card no. 36-1451) in the ZnO sample (Figure 1b) and all components in the TiO_2/ZnO sample (Figure 1c). XRD patterns show a very good crystalline nature of TiO_2, ZnO and TiO_2/ZnO photocatalysts. For particle size determination, Scherrer equation was used.

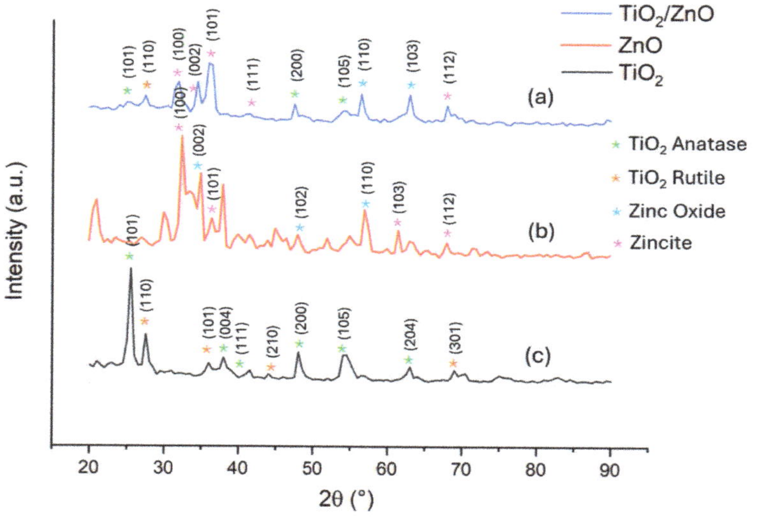

Figure 1. XRD patterns for the synthetized photocatalysts: (**a**) TiO_2, (**b**) ZnO and (**c**) TiO_2/ZnO.

For the TiO_2 powder, two predominant peaks are identified, anatase (101), which intersects the a and c axes, and also rutile (110), which intersects the a and b axes. They are found at $2\theta = 25.6°$ and 27.44°, respectively (Figure 1a). The average size of TiO_2 crystallites is 127 nm. The strong characteristic (100) and (002) peaks at $2\theta = 32.47°$ and 35.06° indicate the growth of ZnO nanostructures mainly along the a- (100) and c-axis (002) direction (Figure 1b). The average size of ZnO crystallites is 153 nm, slightly higher than that of the TiO_2 sample. The TiO_2/ZnO sample has the predominant peaks mentioned in Figure 1a,b for the TiO_2 and ZnO samples, with the exception of anatase (101). In addition, the most predominant peak in this sample is zincite (101), which is found at $2\theta = 36.30°$. This sample has the smallest crystallite average size (87 nm); thus, it has the best properties. TiO_2/ZnO is of high purity, as there are no unidentified peaks or peaks that correspond to other compounds.

2.1.2. X-ray Photoelectron Spectroscopy

The surface composition of the photocatalysts was assessed by carrying out X-ray photoelectron spectroscopy (XPS). XPS full spectra of the photocatalysts indicate the signals corresponding to Ti 2p, Zn 2p, O 1s and C 1s, with the last signal being associated with the preparation of the samples for analysis. The XPS surveys indicate pure synthetic photocatalysts, whereby the results obtained agree with those obtained by XRD analysis.

The valence states of the elements indicated by the XPS surveys are analyzed in the narrow spectra presented separately for each in Figures 2–4. Thus, the binding energy for Ti $2p_{3/2}$ and Ti $2p_{1/2}$ into the TiO_2 photocatalyst (Figure 2a) is 459.67 eV and 467.06 eV, respectively, which can most likely be attributed to Ti^{4+} ions in the oxide lattice of TiO_2 [15,16]. In comparison, the spectrum of the TiO_2/ZnO photocatalyst (Figure 2b) shows a shift towards lower binding energies for Ti $2p_{3/2}$, and towards higher binding energies for Ti $2p_{1/2}$, thus demonstrating a series of interactions between Ti and Zn in the oxide mixture [17]. The deconvolution of the peak corresponding to Ti $2p_{2/3}$ highlights two peaks, one at 458.54 eV and another at 456.68 eV, while the deconvolution of the peak corresponding to Ti $2p_{1/2}$ highlights two other peaks located at 466.11 eV and 469.23 eV. The peaks located at 458.54 eV and 469.23 eV were assigned to the $2p_{3/2}$ and $2p_{1/2}$ core levels of Ti^{4+}, while the peaks at 456.68 eV and 466.11 eV were assigned to the $2p_{3/2}$ and $2p_{1/2}$ core levels of Ti^{3+}. The peaks shifted (to lower binding energies of $2p_{3/2}$ core level and to higher binding energy of $2p_{1/2}$ core level) in the TiO_2/ZnO photocatalyst could be due to the existence of oxygen vacancies (O_V) around the Ti^{4+} ion, and therefore, the Ti^{3+} ion is most probably formed to compensate the charging requirement to establish the charge balance [18–20].

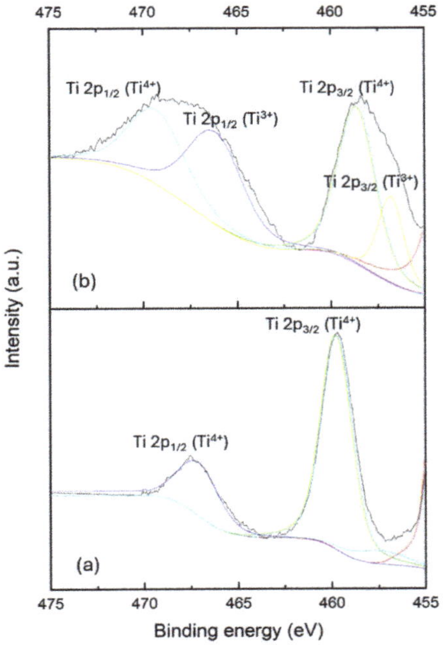

Figure 2. Deconvoluted XPS spectrum of Ti 2p: (**a**) Ti 2p into TiO_2; (**b**) Ti 2p into TiO_2/ZnO.

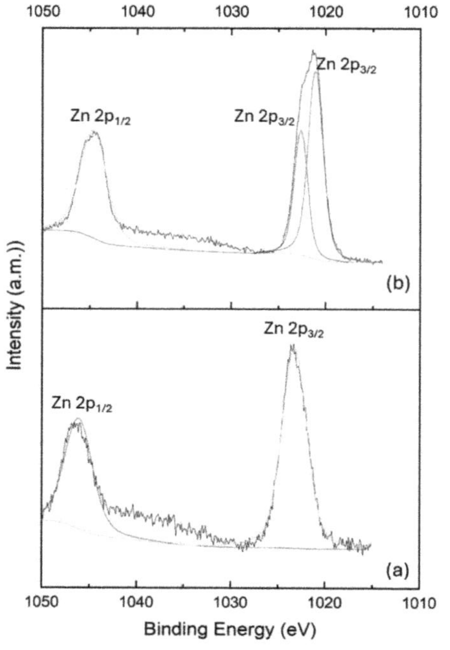

Figure 3. Deconvoluted XPS spectrum of Zn 2p: (**a**) Zn 2p into ZnO; (**b**) Zn 2p into TiO_2/ZnO.

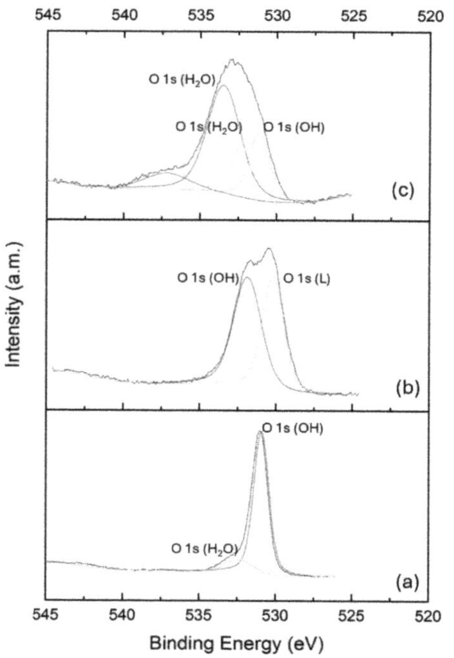

Figure 4. Deconvoluted XPS spectrum of O 1s: (**a**) O 1s into TiO_2; (**b**) O 1s into ZnO; (**c**) O 1s into TiO_2/ZnO.

Figure 3 shows the deconvoluted spectrum of Zn 2p into ZnO and TiO_2/ZnO photocatalysts. The peaks located at the binding energies of 1023 eV and 1046 eV (Figure 3a) are attributed to Zn $2p_{3/2}$ and Zn $2p_{1/2}$, respectively, which can be assigned to Zn^{2+} ions in oxide lattice of ZnO [21]. The peaks attributed to Zn $2p_{3/2}$ and Zn $2p_{1/2}$ in the TiO_2/ZnO photocatalyst are shifted to lower values, again indicating a series of interactions between the two elements, namely Ti and Zn, in their oxide mixture. Thus, the peaks located at the binding energies of 1021.1 eV and 1044.62 eV (Figure 3b) are attributed to Zn $2p_{3/2}$ and Zn $2p_{1/2}$, respectively, and can be assigned to Zn^{2+} ions in oxide lattice of ZnO. The shoulder appearing within the Zn $2p_{2/3}$ peak reveals a new peak located at the binding energy of 1023.1 eV after deconvolution, which could indicate the presence of interstitial zinc ions surrounded by oxygen atoms [22].

The deconvoluted spectrum of O 1s into TiO_2, ZnO and TiO_2/ZnO photocatalysts are shown in Figure 4. The main two peaks of O 1s in the TiO_2 photocatalyst (Figure 4a) are located at binding energies of 531.03 eV and 532.55 eV. They are attributed to oxygen in surface hydroxyl groups (Ti-OH) and to oxygen in adsorbed water molecules (Ti-H_2O), respectively. The deconvoluted spectrum of the ZnO photocatalyst (Figure 4b) highlights the peak attributed to lattice oxygen in ZnO (Zn-O) at 530.21 eV, and the peak attributed to week bonds of oxygen on the surface hydroxyl groups (Zn-OH) at 531.87 eV [23,24].

The deconvoluted XPS spectrum of O 1s into TiO_2/ZnO photocatalyst (Figure 4c) highlights three peaks located at the binding energies of 531.17 eV, 532.32 eV and 533.48 eV. The first peak is associated with the oxygen in surface hydroxyl groups (Ti-OH and Zn-OH), while the second and the third peaks are attributed to oxygen in adsorbed water molecules (Ti-H_2O and Zn-H_2O). Even if some of the peaks in the deconvoluted O 1s spectrum in the range of binding energies of 531–534 eV could be attributed to oxygen vacancies, which would justify the presence of Ti^{3+} in the deconvoluted spectrum of Ti 2p (the formation of surface Ti^{3+} is necessary for charge compensation in order to restore the charge balance of the lattice system), however, this attribution is uncertain because the XPS analysis was performed in ex-situ conditions, which expose the photocatalyst samples to ambient water vapor [25,26].

2.1.3. Scanning Electron Microscopy/Energy-Dispersive X-ray Spectroscopy

The morphology of the synthesized photocatalysts was examined by scanning electron microscopy (SEM); the results are shown in Figure 5. As can be seen, the TiO_2 particles (Figure 5a) seem to be a mixture of spherical and slight irregularly shaped particles with different granulometries. In contrast, the ZnO particles appear to be rod-shaped, arranging themselves in aggregated structures (Figure 5b). The morphology of the TiO_2/ZnO photocatalyst (Figure 5c) seems to be a combination of the two morphologies described above, with the mixture of spherical and slight irregularly shaped particles of TiO_2 being interspersed with the rod-shaped particles of ZnO. This arrangement of the particles leads to a more compact structure of the TiO_2/ZnO photocatalyst, in which the TiO_2 and ZnO particles seem to adhere to each other, a fact that could lead to the appearance of specific interactions between them.

The EDX analysis (Figure 6) indicates only the presence of the constituent elements of the three types of photocatalysts, indicating the purity of these materials, which is in close agreement with the results obtained from the XRD and XPS analyses.

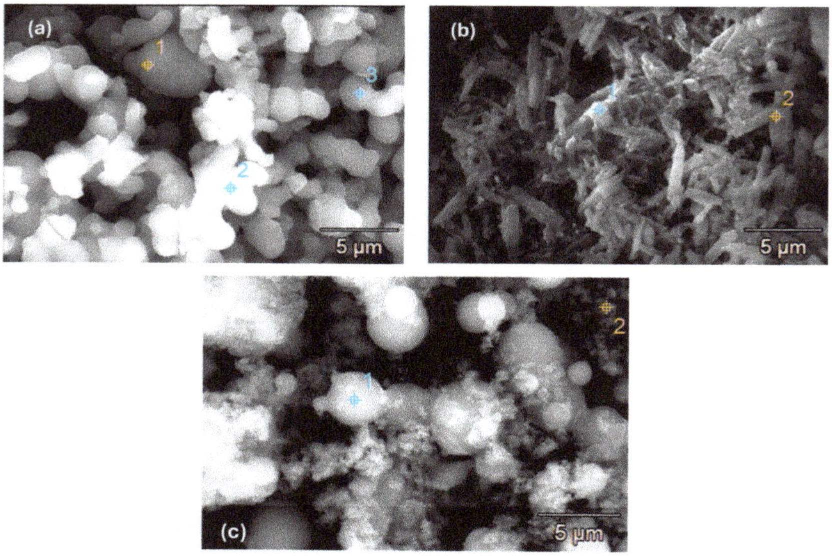

Figure 5. Scanning electron microscopy images (SEM) of: (**a**) TiO_2, (**b**) ZnO and (**c**) TiO_2/ZnO. The numbers and points of different colors in the SEM images correspond to the areas where the EDX analysis was performed.

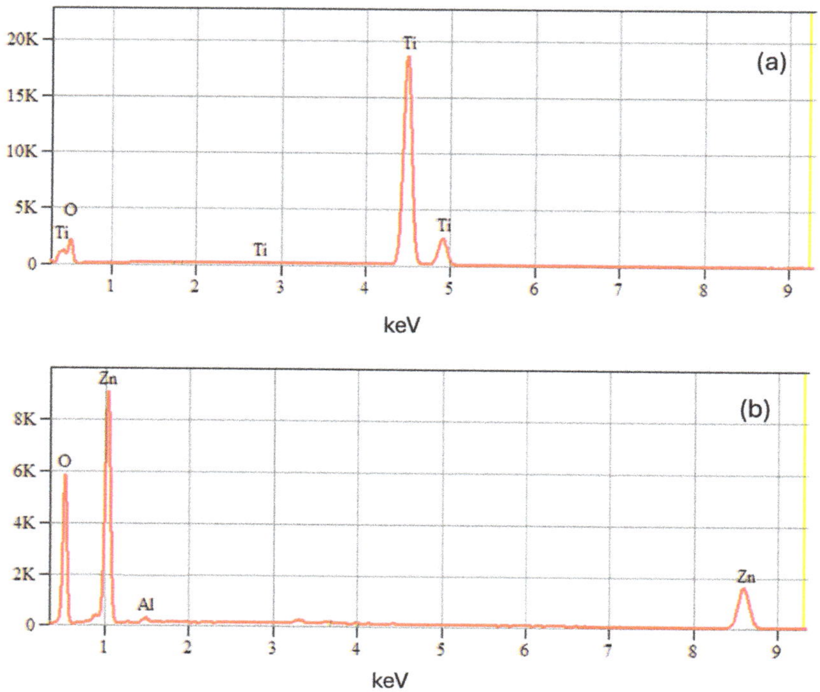

Figure 6. *Cont.*

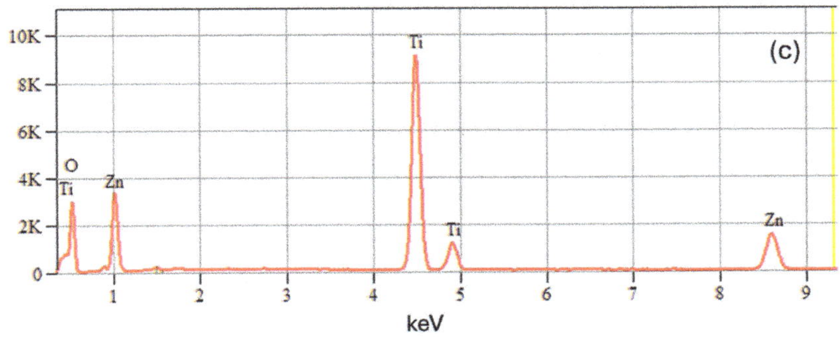

Figure 6. Elemental analysis (EDX) of: (**a**) TiO_2, (**b**) ZnO and (**c**) TiO_2/ZnO.

2.2. Photocatalytic Degradation of Estrogenic Mixture

The kinetics of the photocatalytic degradation of the estrogenic mixture estradiol valerate/norgestrel by using the three types of synthesized photocatalysts is shown in Figures 7–9. It should be specified that, hereinafter, the entire organic content made up of the initial organic mixture of estrogens and the organic degradation intermediates formed will be generically referred to as the organic substrate. Thus, Figure 7 shows the evolution of the photocatalytic degradation of the organic substrate during the two hours of irradiation, using TiO_2 as a photocatalyst. As can be seen, the overall degradation efficiency of the organic substrate increases with the increase in the number of photocatalytic modules (Figure 7a). It should be noted that the degradation of the organic substrate proceeds at a high speed during the first 10 to 30 min of the irradiation time (depending on the number of photocatalytic modules), after which, it decreases significantly and remains at low values throughout the time left of the irradiation time. These results indicate that the degradation kinetics of the organic substrate takes place in two stages. Most likely, in the first stage, the degradation of the organic substrate less resistant to the conditions in the reactor (strong oxidizing environment, acidic pH, and UV radiation) takes place, and later, the slow degradation of the organic substrate more resistant to these conditions occurs (probably, some organic intermediates formed). Thus, in the first 10 to 30 min of irradiation (the first stage of degradation), the degradation efficiency is approximately the same regardless of the number of photocatalytic modules, namely around 46% (46.0% for one photocatalytic module in 30 min of irradiation, 46.8% for two photocatalytic modules in 15 min of irradiation, and 46.4% for three photocatalytic modules in 10 min of irradiation), while, after the second stage of degradation (at the end of the irradiation time), the degradation efficiency is different depending on the number of photocatalytic reactors (54.0% for one photocatalytic module, 67.6% for two photocatalytic modules and 84.0% for three photocatalytic modules). The photocatalytic degradation of the organic substrate follows a pseudo-first-order kinetics (Equations (1) and (2)), which is valid for both degradation stages (Figure 7b) [27]:

$$C_t = C_0 e^{-kt} \quad (1)$$

$$\ln (C_0/C_t) = kt \quad (2)$$

where k is a pseudo-first-order rate constant (min^{-1}). The kinetic parameters of the photocatalytic degradation of the organic substrate by using TiO_2 photocatalyst are shown in Table 1. As can be seen, the rate constant corresponding to the first stage of degradation is clearly much higher than the second stage of degradation.

Table 1. Kinetic parameters of the photocatalytic degradation of the organic substrate.

Photocatalyst		One Module	Two Modules	Three Modules
		k, min^{-1}		
TiO_2	1st stage	0.0207	0.0418	0.0624
	2nd stage	0.0018	0.0045	0.0101
ZnO	1st stage	0.0324	0.0641	0.0978
	2nd stage	0.0027	0.0057	0.0124
TiO_2/ZnO	1st stage	0.0838	0.1317	-
	2nd stage	0.0060	0.0230	-

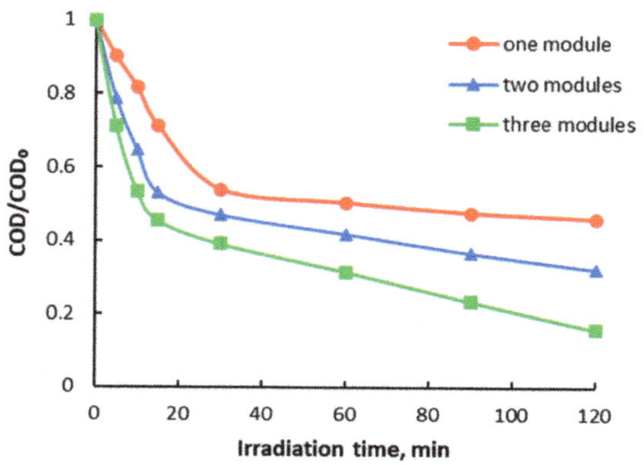

(**a**)

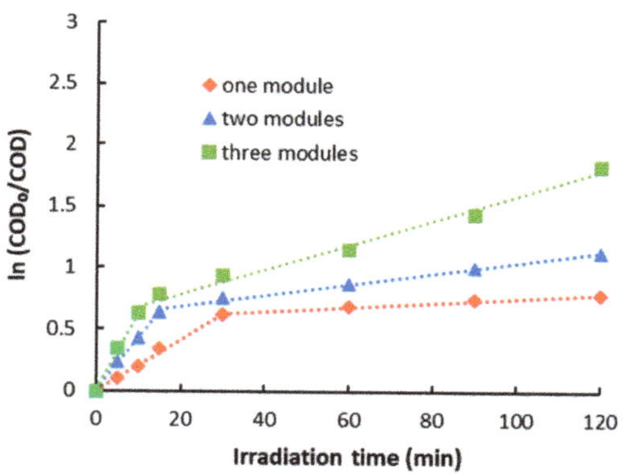

(**b**)

Figure 7. Kinetics of the photocatalytic degradation of the estrogen formulation by using TiO_2 photocatalyst: (**a**) degradation of the organic substrate during the irradiation time; (**b**) pseudo-first-order kinetic model. Experimental conditions: flow rate 20 L/h, pH 3, working solution volume 2 L, initial concentration of the organic substrate 250 mg O_2/L, estradiol valerate concentration 1 mg/L, norgestrel concentration 0.25 mg/L, photocatalyst concentration 1.5 g/L.

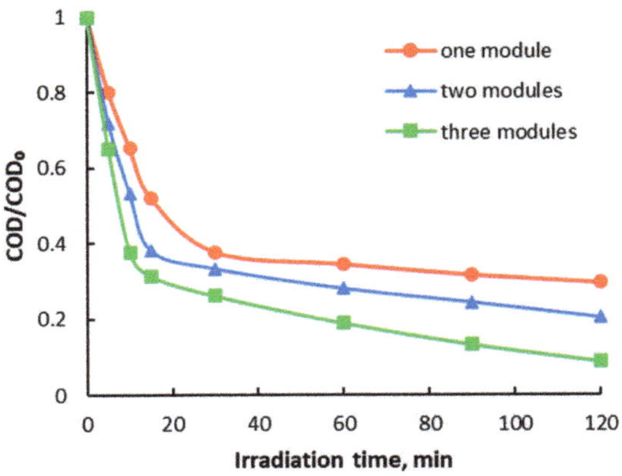

(**a**)

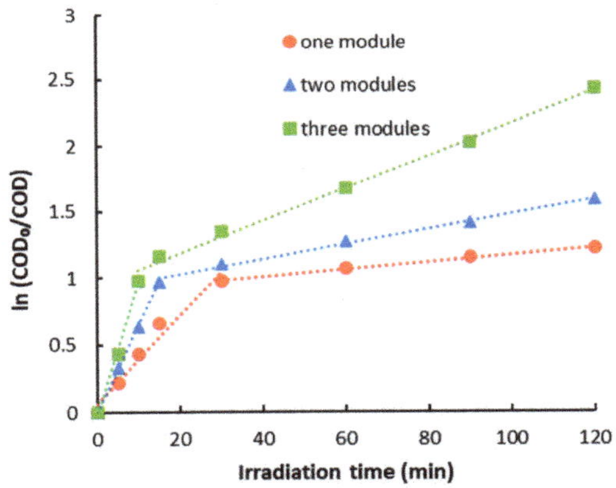

(**b**)

Figure 8. Kinetics of the photocatalytic degradation of the estrogen formulation by using ZnO photocatalyst: (**a**) degradation of the organic substrate during the irradiation time; (**b**) pseudo-first-order kinetic model. Experimental conditions: flow rate 20 L/h, pH 3, working solution volume 2 L, initial concentration of the organic substrate 250 mg O_2/L, estradiol valerate concentration 1 mg/L, norgestrel concentration 0.25 mg/L, and photocatalyst concentration 1.5 g/L.

Figure 8 shows the progress in the photocatalytic degradation of the organic substrate during the two hours of irradiation, using ZnO as a photocatalyst. The same trend is recorded in terms of the overall efficiency of photocatalytic degradation as in the case of the TiO_2 photocatalyst, which increases with the increase in the number of photocatalytic modules. In the first stage of degradation, the efficiency is around 62% (62.4% for a photocatalytic module in 30 min of irradiation, 62.0% for two photocatalytic modules in 15 min of irradiation and 62.4% for three photocatalytic modules in 10 min of irradiation), while, at the end of the irradiation period, the degradation efficiency is 70.4% for one photocatalytic module, 79.6% for two photocatalytic modules and 91.2% for three photocatalytic modules. The photocatalytic degradation of the organic substrate also follows a pseudo-first-order kinetics valid for both degradation stages. The kinetic parameters of the photocatalytic degradation of the organic substrate by using ZnO photocatalyst are shown in Table 1.

However, it should be noted that the degradation efficiency is higher than in the case of the TiO_2 photocatalyst for both degradation stages. Also, the kinetic constants have higher values, highlighting the higher degradation rate.

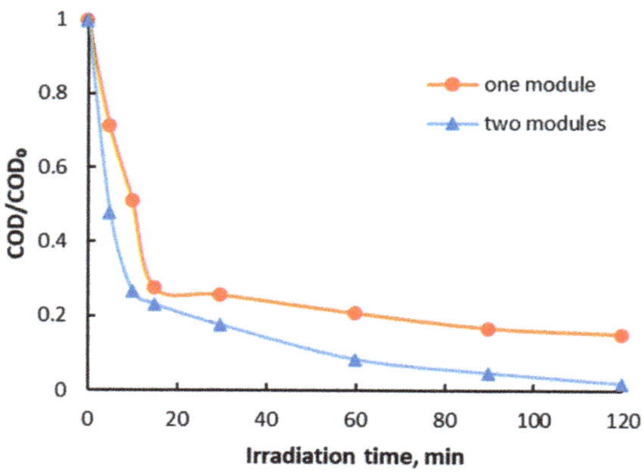

(a)

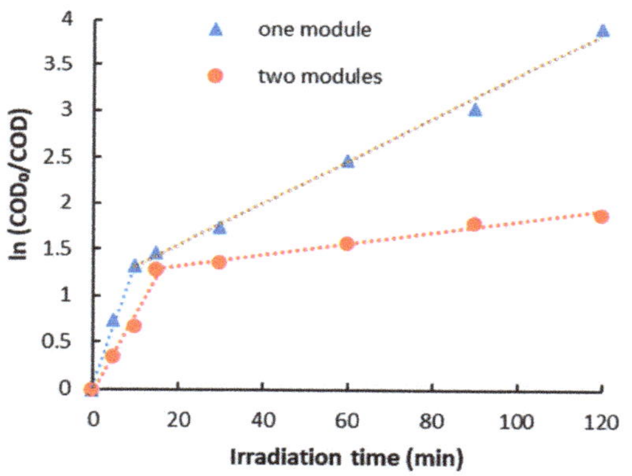

(b)

Figure 9. Kinetics of the photocatalytic degradation of the estrogen formulation by using TiO_2/ZnO photocatalyst: (**a**) degradation of the organic substrate during the irradiation time; (**b**) pseudo-first-order kinetic model. Experimental conditions: flow rate 20 L/h, pH 3, working solution volume 2 L, initial concentration of the organic substrate 250 mg O_2/L, estradiol valerate concentration 1 mg/L, norgestrel concentration 0.25 mg/L, and photocatalyst concentration 1 g/L.

In the case of the TiO_2/ZnO photocatalyst (Figure 9), although the same tendency is maintained in terms of efficiency and degradation speed, it should be noted that the organic substrate is almost mineralized during the two hours of irradiation by using only two photocatalytic modules. In the first stage of degradation, the efficiency is around 72.5% (72.4% for a photocatalytic module in 15 min of irradiation and 73.2% for two photocatalytic modules in 10 min of irradiation), and in the second stage of degradation (at the end of the irradiation period), the efficiency is 84.8% for one photocatalytic module and 98.0% for two photocatalytic modules. The photocatalytic degradation of the organic substrate follows a pseudo-first-order kinetics for both degradation stages. The kinetic parameters (Table 1) indicate a clearly higher degradation rate compared to the other two photocatalysts, at least at the level of two photocatalytic modules.

The obtained results highlight the fact that the photocatalytic degradation efficiency of the organic substrate is a function of the dose of photocatalysts used, regardless of the type of photocatalyst. Taking into account the fact that the dose of photocatalyst is kept approximately constant for each individual photocatalytic module, it is found that with each photocatalytic module added, the dose of photocatalyst in the photocatalytic system becomes twice as high. If after the first stage of photocatalytic degradation the efficiency is approximately the same for a certain type of photocatalyst, regardless of the number of photocatalytic modules, after the second stage of photocatalytic degradation (at the end of the 120 min of irradiation), efficiency increases constantly with the addition of photocatalytic modules. In this respect, for the TiO_2 photocatalyst, the efficiency is 1.14 higher after each added photocatalytic module; for the ZnO photocatalyst, it is 1.25 higher; and for the TiO_2/ZnO photocatalyst, it is 1.15 higher. Having this information available, the dose of photocatalyst can be determined in terms of the number of photocatalytic modules required for the complete mineralization of a certain type of organic substrate. Thus, the results clearly indicate that, for the initial organic substrate consisting of estrogenic mixture estradiol valerate/norgestre, the addition of a fourth photocatalytic module in the case of systems with TiO_2 or ZnO photocatalyst would result in complete mineralization of the organic substrate in the same irradiation time of 120 min, but using a higher dose of photocatalyst. Therefore, this result confirms the superior performance of the TiO_2/ZnO photocatalyst in terms of photocatalytic activity (complete mineralization with only two photocatalytic modules compared to the need for four photocatalytic modules for the other two photocatalytic systems).

Figure 10 shows the cycling test results for the photocatalyst for which the best degradation efficiency was obtained, namely TiO_2/ZnO. It should be mentioned that a fresh solution of estrogenic mixture was used in each run. As can be seen, the degradation efficiency remains approximately constant (98.0% for the first run, 97.2% for the second run, 97.6% for the third run and 98.4% for the fourth run), which highlights a high stability of the photocatalyst deposited on the glass spheres, as well as keeping its photocatalytic activity for a long period of irradiation.

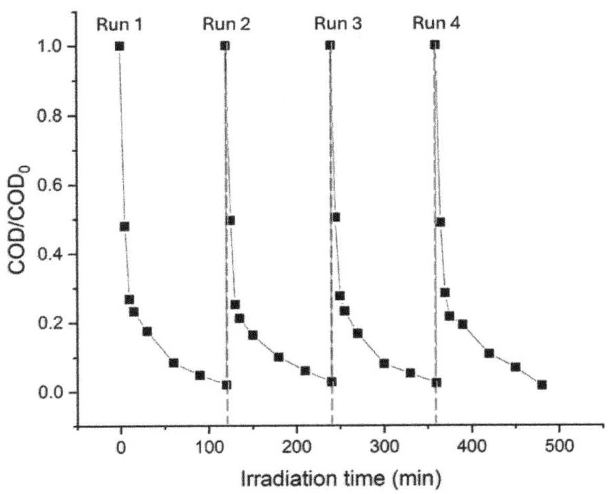

Figure 10. Cycling test of the photocatalytic degradation of the estrogen formulation by using TiO_2/ZnO photocatalyst. Experimental conditions: flow rate 20 L/h, pH 3, working solution volume 2 L, initial concentration of the organic substrate 250 mg O_2/L, estradiol valerate concentration 1 mg/L, norgestrel concentration 0.25 mg/L, and photocatalyst concentration 1 g/L.

3. Materials and Methods

3.1. Materials

The estrogenic mixture estradiol valerate/norgestrel derived from Cyclo-Progynova pharmaceutical formulation (pills containing 2 mg estradiol valerate and 0.5 mg norgestrel) from Bayer (Bayer AG, Leverkusen, Germany) was purchased from a local human pharmacy. Titanium(IV) isopropoxide (TTIP) 97%, zinc acetate dihydrate 98%, ethanol absolute 99.9%, and nitric acid 65%, all of analytical grade, were purchased from Sigma-Aldrich (Darmstadt, Germany), and used for photocatalyst synthesis. Hydrogen peroxide solution of 30% was purchased from Sigma-Aldrich and used in the photocatalysis tests.

The photocatalytic modules consist of quartz tubes with the dimensions 5 cm (d) × 10 cm (h), and the inert support on which the photocatalyst is deposited consists of perforated glass balls of 4 mm (d), so that they can be strung on stainless-steel wires. The photocatalytic modules are interconnected using the garden hose pipe fittings kit so they can be easily added or removed from the photocatalytic system. All these materials were purchased from various local companies (Bucharest, Romania).

3.2. Synthesis of Photocatalysts and Their Deposition on the Inert Support

3.2.1. TiO_2 Photocatalyst

TiO_2 photocatalyst was synthesized through the sol–gel method, which, in short, consists of the dropwise addition of an aqueous solution of nitric acid (1.3% by mass) over a solution of titanium(IV) isopropoxide (TTIP) in absolute ethanol (AE) at an AE/TTIP volume ratio of 3 (pre-mixed for one hour) under continuous stirring at a temperature of 60 °C (the temperature was chosen to be below the boiling point of the solvent) for two hours. It should be noted that the photocatalyst was deposited directly from the solution on the inert support, and therefore, the glass balls strung on rows of stainless-steel wires were immersed in the opaque and viscous mixture formed for one hour, after which they were kept in the oven at a temperature of 105 °C to remove the solvent for 24 h. Finally, the rows of glass balls were calcined at a temperature of 500 °C for four hours (the temperature was chosen to be below the limit of the glass transition temperature of the glass support).

3.2.2. ZnO Photocatalyst

ZnO photocatalyst was also synthesized by the sol–gel method, which, in short, consists of mixing 3 g of zinc acetate dihydrate with a solution consisting of absolute ethanol (EA) and nitric acid (NA) at an EA/NA volumetric ratio of 2, under continuous stirring for 10 min. Next, the glass balls strung on rows of stainless-steel wires were immersed in the obtained solution for one hour, kept in the oven at a temperature of 105 °C for 24 h, and calcined at a temperature of 500 °C for four hours.

3.2.3. TiO_2/ZnO Photocatalyst

For the synthesis of the TiO_2/ZnO photocatalyst, the opaque and viscous solution obtained in the synthesis of the TiO_2 photocatalyst is mixed with the solution obtained in the synthesis of the ZnO photocatalyst (both syntheses described above) under continuous stirring for two hours at a temperature of 60 °C. Next, the glass balls strung on rows of stainless-steel wires were immersed in the new mixture obtained for one hour, kept in the oven at a temperature of 105 °C for 24 h, and calcined at a temperature of 500 °C for four hours.

3.3. Photocatalytic Experiments

The photocatalytic experiments were carried out in modular photocatalytic systems designed so that the number of photocatalytic modules can be easily changed. The photocatalytic modules, which are made from quartz tubes with the dimensions 5 cm (d) × 10 cm (h), are operated in a plug flow reactor system under a UV-A radiation field, and the photocatalyst (TiO_2, ZnO or TiO_2/ZnO) is immobilized on an inert support of glass balls (4 mm in diameter) that are strung on stainless-steel wire and arranged in rows along the photocatalytic modules. It should be noted that the glass balls were initially cleaned by washing with an oxidizing mixture (potassium dichromate in concentrated sulfuric acid), rinsing with pure water several times, followed by washing with pure water in an ultrasonic field (37 kHz, Ultrasonic Bath Elmasonic Select 60 (Elma Schmidbauer GmbH, Singen, Germany). There are 65 glass balls on a string of stainless-steel wire, and a photocatalytic module contains 6 strings. The average amount of photocatalyst deposited on the inert support of a photocatalytic module is 1g (0.92 g–1.15 g). The photocatalytic modules are interconnected by using the garden hose pipe fittings kit and are continuously fed for two hours at room temperature through a feeding vessel using a peristaltic pump (LBX Pump-60J-001, Labbox Labware, S.L., Barcelona, Spain). The radiation source (Osram L Blue UV-A 18 W/78 G13 lamps, OSRAM GmbH, München, Germany) surrounds the modular photocatalytic system in such a way that each photocatalytic module receives an approximately equal amount of radiation (around 2400 μw/cm^2, measured with a Cole-Parmer UV-meter with mobile probe from General Tools & Instruments (General Tools & Instruments LLC, New York, NY, USA)). A schematic of the modular photocatalytic system is shown in Figure 11.

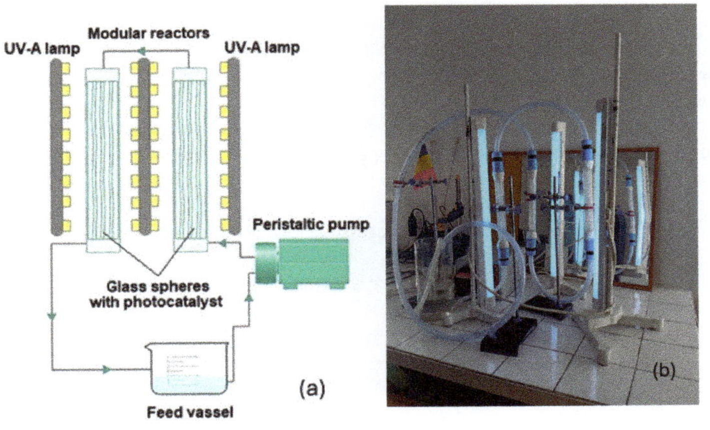

Figure 11. (**a**) Sketch of the modular photocatalytic reactor (two photocatalytic modules). (**b**) Photo of the real modular photocatalytic reactor (three photocatalytic modules).

The conditions in which the photocatalytic system was operated are the following: a flow rate of the working solution of 20 L/h, pH 3 of the working solution, the volume of the working solution was 2 L, the initial concentration of the organic substrate in the aqueous working solution was 250 mg O_2/L (Chemical Oxygen Demand—COD equivalent concentration), and a hydrogen peroxide/organic substrate mass ratio of 1. It should be noted that these working conditions were chosen after performing some preliminary tests. In order to monitor the degree of removal of the organic substrate during the two hours of irradiation, samples were taken from the feed vessel at predetermined times and subjected to COD analysis according to the APHA 5220 D standard method (closed reflux, colorimetric method) by using a Hach LT 200 thermostat and Hach DR 3800 spectrophotometer (Hach, Loveland, CO, USA).

3.4. Photocatalysts Characterization

The crystalline structure of the synthetized photocatalysts was investigated by XRD analysis using a Rigaku MiniFlex 600 X-ray diffractometer (Rigaku Holdings Corporation Tokyo, Japan), with a scan in the 20–90° 2θ range. Cu Kα monochromatic radiation was used as the X-ray source, with a characteristic wavelength of λ = 1.54 Å. Chemical compositions and valence-band spectra of the photocatalysis were investigated by XPS analysis using a Thermo Scientific K-Alpha instrument (Thermo Fisher Scientific Inc., Waltham, MA, USA) with a monochromate Al Kα source on a spot size of 400 μm. The binding energy scan range was 0–1361 eV in 1 eV steps, and the pass energy was 200 eV. Binding energies were calibrated by placing the C 1 s peak at 284.4 eV. The morphology and elemental analyses of the synthetized photocatalysts were performed by using a QUANTA 450 FEG electron microscope (FELMI-ZFE, Graz, Austria) coupled with an EDS detector operated at an accelerating voltage of 30 kV.

4. Conclusions

The experimental results of this work highlighted the following:

- The photocatalytic degradation efficiency of the organic substrate, initially formed from the estrogenic estradiol valerate/norgestrel mixture, largely depends on the dose of photocatalyst used regardless of its type. The dose can be easily modified by adding or removing photocatalytic modules that contain approximately the same amount of photocatalysts.
- The photocatalytic degradation of the organic substrate takes place in two stages with different speeds but following the same type of kinetics, namely pseudo-first-order kinetics. The number of photocatalytic modules required for the complete mineralization of a certain type of organic substrate can be determined based on the results obtained in the second stage of photocatalytic degradation in terms of degradation efficiency.
- The best results in terms of photocatalytic degradation efficiency were obtained for the photocatalytic system with a TiO_2/ZnO-type photocatalyst. The organic substrate was almost completely mineralized in 120 min of irradiation with only two photocatalytic modules.
- The morphological, mineralogical and spectroscopic analyses showed that the synthesized photocatalysts are of high purity, and that the high performance of the TiO_2/ZnO photocatalyst is due to its superior characteristics such as small crystallites, compact structure and highly reactive specific surface. The manifestation of specific interactions between the elements that form the oxide mixture leads to the appearance of surface defects beneficial to charge transfer, and thus, to photocatalytic activity.

Author Contributions: Conceptualization, C.B. and L.B.; methodology, L.B. and C.O.; formal analysis, G.I.I., E.I.B., A.C., E.M. and A.C.B.; investigation, C.O., C.P., C.B. and L.B.; resources, L.B. and C.O.; writing—original draft preparation, C.B., L.B., C.O. and C.P.; writing—review and editing, C.B., L.B. and C.O.; supervision, C.B. and C.O.; project administration, L.B.; funding acquisition, L.B. All authors have read and agreed to the published version of the manuscript.

Funding: This research was funded by the National Program for Research of the National Association of Technical Universities—GNAC ARUT 2023, grant no. 65/10.10.2023.

Data Availability Statement: The raw data supporting the conclusions of this article will be made available by the authors on request.

Conflicts of Interest: The authors declare no conflicts of interest. The funders had no role in the design of the study; in the collection, analyses, or interpretation of data; in the writing of the manuscript; or in the decision to publish the results.

References

1. Yang, X.; He, X.; Lin, H.; Lin, X.; Mo, J.; Chen, C.; Dai, X.; Liao, D.; Gao, C.; Li, Y. Occurrence and distribution of natural and synthetic progestins, androgens, and estrogens in soils from agricultural production areas in China. *Sci. Total Environ.* **2021**, *751*, 141766. [CrossRef]
2. Aubead, N.M. Role of Sex Hormones in Human Body. In *Reproductive Hormones*; Marsh, C., Ed.; IntechOpen: Kansas City, MO, USA, 2021. [CrossRef]
3. Amenyogbe, E.; Chen, G.; Wang, Z.; Lu, X.; Lin, M.; Lin, A.I. A Review on Sex Steroid Hormone Estrogen Receptors in Mammals and Fish. *Int. J. Endocrinol.* **2020**, *2020*, 5386193. [CrossRef] [PubMed]
4. Zhao, X.; Grimes, K.L.; Colosi, L.M.; Lung, W.-S. Attenuation, transport, and management of estrogens: A review. *Chemosphere* **2019**, *230*, 462–478. [CrossRef] [PubMed]

5. Li, N.; Li, J.; Zhang, Q.; Gao, S.; Quan, X.; Liu, P.; Xu, C. Effects of endocrine disrupting chemicals in host health: Three-way interactions between environmental exposure, host phenotypic responses, and gut microbiota. *Environ. Pollut.* **2021**, *271*, 116387. [CrossRef]
6. Directive 2013/39/EU of the European Parliament and of the Council of 12 August 2013 Amending Directives 2000/60/EC and 2008/105/EC as Regards Priority Substances in the Field of Water Policy Text with EEA Relevance. OJ L 226. 2013, pp. 1–17. Available online: http://data.europa.eu/eli/dir/2013/39/oj (accessed on 9 July 2024).
7. Commission Implementing Decision (EU) 2018/840 of 5 June 2018 Establishing a Watch List of Substances for Union-Wide Monitoring in the Field of Water Policy Pursuant to Directive 2008/105/EC of the European Parliament and of the Council and Repealing Commission Implementing Decision (EU) 2015/495 (Notified under Document C(2018) 3362). OJ L 141. 2018, pp. 9–12. Available online: http://data.europa.eu/eli/dec_impl/2018/840/oj (accessed on 9 July 2024).
8. Commission Implementing Decision (EU) 2022/679 of 19 January 2022 Establishing a Watch List of Substances and Compounds of Concern for Water Intended for Human Consumption as Provided for in Directive (EU) 2020/2184 of the European Parliament and of the Council (Notified under Document C(2022) 142). OJ L 124. 2022, pp. 41–43. Available online: http://data.europa.eu/eli/dec_impl/2022/679/oj (accessed on 9 July 2024).
9. Coello-Garcia, T.; Curtis, T.P.; Mrozik, W.; Davenpor, R.J. Enhanced estrogen removal in activated sludge processes through the optimization of the hydraulic flow pattern. *Water Res.* **2019**, *164*, 114905. [CrossRef] [PubMed]
10. Bilal, M.; Rizwan, K.; Adeel, M.; Barceló, D.; Awad, Y.A.; Iqbal, H.M.N. Robust strategies to eliminate endocrine disruptive estrogens inwater resources. *Environ. Pollut.* **2022**, *306*, 119373. [CrossRef]
11. Al-Nuaim, M.A.; Alwasiti, A.A.; Shnain, Z.Y. The photocatalytic process in the treatment of polluted water. *Chem. Pap.* **2023**, *77*, 677–701. [CrossRef]
12. Joseph, A.; Vijayanandan, A. Review on support materials used for immobilization of nano-photocatalysts for water treatment applications. *Inorganica Chim. Acta* **2023**, *545*, 121284. [CrossRef]
13. Huang, H.J.; Wu, J.C.-S.; Chiang, H.-P.; Chau, Y.-F.C.; Lin, Y.-S.; Wang, Y.H.; Chen, P.-J. Review of Experimental Setups for Plasmonic Photocatalytic Reactions. *Catalysts* **2020**, *10*, 46. [CrossRef]
14. Visan, A.; van Ommen, J.R.; Kreutzer, M.T.; Lammertink, R.G.H. Photocatalytic Reactor Design: Guidelines for Kinetic Investigation. *Ind. Eng. Chem. Res.* **2019**, *58*, 5349–5357. [CrossRef]
15. Erdem, B.; Hunsicker, R.A.; Simmons, G.W.; Sudol, E.D.; Dimonie, V.L.; El-Aasser, M.S. XPS and FTIR Surface Characterization of TiO_2 Particles Used in Polymer Encapsulation. *Langmuir* **2001**, *17*, 2664–2669. [CrossRef]
16. Kubiak, A.; Żółtowska, S.; Bartkowiak, A.; Gabała, E.; Sacharczuk, N.; Zalas, M.; Siwińska-Ciesielczyk, K.; Jesionowski, T. The TiO_2-ZnO Systems with Multifunctional Applications in Photoactive Processes—Efficient Photocatalyst under UV-LED Light and Electrode Materials in DSSCs. *Materials* **2021**, *14*, 6063. [CrossRef] [PubMed]
17. Saad, S.K.M.; Umar, A.A.; Nguyen, H.Q.; Dee, C.F.; Salleh, M.M.; Oyama, M. Porous (001)-faceted Zn-doped anatase TiO_2 nanowalls and their heterogeneous photocatalytic characterization. *RSC Adv.* **2014**, *4*, 57054. [CrossRef]
18. Ohno, T.; Tsubota, T.; Toyofukum, M.; Inaba, R. Preparation of S-doped TiO_2 Photocatalysts and their Photocatalytic Activities Under Visible Light. *Appl. Catal. A Gen.* **2004**, *265*, 255–258. [CrossRef]
19. Pandey, S.K.; Pandey, S.K.; Mukherjee, C.; Mishra, P.; Gupta, M.; Barman, S.R.; D'Souza, S.W.; Mukherjee, S. Effect of growth temperature on structural, electrical and optical properties of dual ion beam sputtered ZnO thin films. *J. Mater. Sci. Mater. Electron.* **2013**, *24*, 2541–2547. [CrossRef]
20. Fu, R.; Wang, Q.; Gao, S.; Wang, Z.; Huang, B.; Dai, Y.; Lu, J. Effect of different processes and Ti/Zn molar ratios on the structure, morphology, and enhanced photoelectrochemical and photocatalytic performance of Ti^{3+} self-doped titanium–zinc hybrid oxides. *J. Power Sources* **2015**, *285*, 449–459. [CrossRef]
21. Krylova, G.; Brioude, A.; Ababou-Girard, S.; Mrazeka, J.; Spanhel, L. Natural superhydrophilicity and photocatalytic properties of sol-gel derived $ZnTiO_3$-ilmenite/r-TiO_2 films. *Chem. Phys.* **2010**, *12*, 15101–15110. [CrossRef]
22. Kumar, A.; Ahmad, M.I. Role of defects and microstructure on the electrical properties of solution-processed Al-doped ZnO transparent conducting films. *Appl. Phys. A* **2020**, *126*, 598. [CrossRef]
23. Fan, C.; Chen, C.; Wang, J.; Fu, X.; Ren, Z.; Qian, G.; Wang, Z. Black Hydroxylated Titanium Dioxide Prepared Via Ultrasonication with Enhanced Photocatalytic Activity. *Sci. Rep.* **2015**, *5*, 11712. [CrossRef]
24. Claros, M.; Setka, M.; Jimenez, Y.; Vallejos, S. AACVD Synthesis and Characterization of Iron and Copper Oxides Modified ZnO Structured Films. *Nanomaterials* **2020**, *10*, 471. [CrossRef]
25. Idriss, H. On the wrong assignment of the XPS O1s signal at 531–532 eV attributed to oxygen vacancies in photo- and electro-catalysts for water splitting and other materials applications. *Surf. Sci.* **2021**, *712*, 121894. [CrossRef]
26. Frankcombe, T.J.; Liu, Y. Interpretation of Oxygen 1s X-ray Photoelectron Spectroscopy of ZnO. *Chem. Mater.* **2023**, *35*, 5468–5474. [CrossRef]
27. Li, B.; Yuan, D.; Gao, C.; Zhang, H.; Li, Z. Synthesis and characterization of TiO_2/ZnO heterostructural composite for ultraviolet photocatalytic degrading DOM in landfill leachate. *Environ. Sci. Pollut. Res.* **2022**, *29*, 85510–85524. [CrossRef] [PubMed]

catalysts

MDPI

Article

Fe_3O_4-CdO Nanocomposite for Organic Dye Photocatalytic Degradation: Synthesis and Characterization

Ahlam Albeladi [1], Zaheer Khan [1], Shaeel Ahmed Al-Thabaiti [1], Rajan Patel [2], Maqsood Ahmad Malik [3,*] and Shilpa Mehta [4]

1 Chemistry Department, Faculty of Science, King Abdulaziz University, P.O. Box 80203, Jeddah 21589, Saudi Arabia; ahlamalbeladi@gmail.com (A.A.); drkhanchem@yahoo.co.in (Z.K.); sthabiti@hotmail.com (S.A.A.-T.)
2 Biophysical Chemistry Laboratory, Centre for Interdisciplinary Research in Basic Sciences, Jamia Millia Islamia, New Delhi 110025, India; rpatel@jmi.ac.in
3 Department of Chemistry, Faculty of Sciences, Jamia Millia Islamia, New Delhi 110025, India
4 Department of Chemistry, Ramjas College, University of Delhi, Delhi 110007, India; shilpa.dr@gmail.com
* Correspondence: mamalik@jmi.ac.in

Abstract: In this study, pure CdO nanoparticles, magnetic Fe_3O_4 nanoparticles, and Fe_3O_4-CdO nanocomposites were prepared via a solution combustion method using cetyltrimethylammonium bromide (CTAB) as a template. These prepared nanomaterial samples were characterized by X-ray diffraction (XRD), ultraviolet-visible diffuse reflectance spectroscopy (UV-vis DRS), Fourier-transform infrared spectroscopy (FTIR), Raman spectroscopy, X-ray photoelectron microscopy (XPS), transmittance electron microscopy (TEM), and scanning electron microscopy (SEM) analysis. XRD patterns confirmed the purity and the crystalline nature of the prepared samples. FTIR and Raman spectra observed the metal-oxygen (M-O) bond formation. UV-vis DRS studies were performed to investigate the optical properties and the bandgap energy determination. The surface morphology and the size of the pure CdO nanoparticles, magnetic Fe_3O_4, and nanocomposites of Fe_3O_4-CdO were determined via TEM and SEM analysis. Under optimum experimental conditions, the Fe_3O_4-CdO nanocomposites were applied for photocatalytic activity against Methylene blue dye. Under visible light irradiation, Fe_3O_4-CdO nanostructures showed an efficient photocatalytic degradation of 92% against Methylene blue organic dye and showed excellent stability for multiple cycles of reuse.

Keywords: nanocomposite; magnetic nanoparticles; surface morphology; photodegradation

1. Introduction

In recent years, human poisoning and the contamination of aquatic life have occurred as significant consequences of the production of poisonous organic pigments and their wastewater by-products in several industries, including the textile, plastic, paper, pesticides, leather, and petrochemical sectors [1]. Because of their carcinogenic properties and limited biodegradability [2], these dyes pose a substantial risk to humans, livestock, and aquatic ecosystems. Due to their non-biodegradability and adverse effects on humans due to their high carcinogenic potential [3], organic dyes have received special attention as important environmental pollutants [4]. In addition, they have a degrading influence on the nature and quality of water and inhibit the penetration of sunlight, which is detrimental to photosynthetic aquatic plants. For example, trace amounts of organic dyes such as Methylene blue (MB) in water can cause gastrointestinal issues, irritations, anemia due to hemolysis, and other diseases [3,4]. MB is a positively charged dye that finds utility in diverse applications, including but not limited to the wood and textile sectors [5,6]. A major problem in the industry is the remediation of water from MB, as failure to do so may lead to health hazards for living organisms and the environment.

In recent years, various techniques have been developed for removing pollutants from aqueous media, including adsorption [7], biosorption mechanisms [6], membranes [8], Fenton [9], and photocatalytic processes [10]. As an alternative approach, photocatalysis has attracted much attention due to its ability to degrade the organic dye in wastewater into environmentally friendly compounds such as CO_2 and water [11]. The significance of photocatalytic degradation of hazardous compounds has grown in importance [12,13] due to its distinct benefits that do not generate secondary pollutants. During the degradation process, the photocatalyst facilitates the conversion of noxious compounds into innocuous constituents, namely carbon dioxide and water [12]. In recent times, numerous research investigations have been dedicated to advancing photocatalysts that are responsive to visible light or sunshine to utilize them for environmental cleanup purposes. Photocatalysts can operate well in ambient conditions and achieve complete mineralization of organic molecules. Another essential factor of a catalyst that is efficient is the separation and recycling of the catalyst to prevent catalyst loss. In this regard, fabricating a photocatalyst with a magnetic component would be of great interest. The unique properties of semiconductor photocatalysts may be achieved by combining them with different materials. A bi-functional material that exhibits magnetic and photocatalytic activity makes it valuable for developing environmentally benign catalytic processes [13].

Nanomaterials (NMs) are considered one of the special categories of materials that have received exceptional attention and interest due to their explicit use in various applications, such as sensors [14], photocatalysis [12], catalysis [15] and material chemistry, etc. [16]. NMs have several advantages over similar nanostructures, including the ability to regulate the thickness of their shell material, which allows morphology-dependent properties to be tuned to the desired level [16]; in addition, they possess dual properties in contrast to single nanoparticles (NPs). The functionality of NPs is contingent upon the bandgap and the relative positioning of the electronic energy levels of the specific semiconductor [17] As a result, it may aid in reducing the consumption of expensive materials by controlling the release of the core. Numerous differently shaped nanocomposites (NCs) exist, but metal oxide is one of the most effective materials due to its improved properties. It has been demonstrated that transition metal oxide or hydroxides NPs exhibit diverse properties and functionalities due to modifications in their surface properties at the nanoscale [18].

In recent years, there has been a significant expansion in the field of study about iron oxide NPs, namely magnetite NPs (Fe_3O_4 NPs), regarding their potential application in removing both organic and inorganic contaminants. Fe_3O_4 NPs have emerged as a promising option for various technological applications, particularly, environmental remediation [19]. This is because of their tiny size, biological compatibility, non-toxicity, and superior full-activity magnetism [13,19].

Cadmium oxide (CdO) is a significant semiconductor in the II–IV group, characterized by a straight bandgap of 2.2–2.5 eV. It has high optical transmittance in the visible range and low resistivity, making it an attractive material for optoelectronic applications; it is suitable for use as a photocatalyst.There are also numerous applications for CdO, such as gas sensors, solar cells, photodiodes, lithium batteries, etc. [20–22]. Various researchers have already discovered that when Fe and CdO are combined as photocatalytic hybrids, the performance can be enhanced for the removal of target pollutants. For instance, the Fe-CdO combination demonstrated high photodegradation effectiveness of 92.85% against the breakdown of MB dye when exposed to visible-light radiation [23]. In addition, CdO and Fe_3O_4-based NCs have shown promising and excellent degradation results against other organic dyes [23,24].

Heterojunction photocatalysts have gained significant interest in recent years as a potential environmentally acceptable solution for eliminating organic contaminants in wastewater. These photocatalysts utilize abundant sunlight to convert pollutants into innocuous chemicals [25–28]. The literature extensively discusses heterojunction photocatalysts for wastewater treatment due to their distinctive characteristics, including simple production, affordability, chemical durability, and lack of toxicity [29–32]. Heterojunction building is an approach that is widely used to improve the photocatalytic efficiency of a single semiconductor. It involves modifying the surface properties of the catalyst and optimizing the crystal structure. It can significantly enhance the surface catalytic reaction and efficiently modify the energy-band structure [33–35]. So, in this work, we highlighted the synthesis of Fe_3O_4, CdO NPs, and Fe_3O_4-CdO NPs by using a simple and fast method with CTAB as a template, and their photocatalytic activities were performed under visible light.

2. Results and Discussion

2.1. UV-Visible Spectroscopy Analysis

The research entailed using UV-visible diffuse reflectance spectroscopy (UV-Vis DRS) techniques to investigate the light-absorption characteristics of CdO, Fe_3O_4 NPs, and Fe_3O_4-CdO NC materials (Figure 1a). The UV-vis DRS spectra of CdO and Fe_3O_4 display a conspicuous and wide absorption band in the ultraviolet and visible area, which is potentially attributable to their narrow bandgap energy. This pertains to the direct bandgap phenomenon resulting from the electron transition occurring between the valence band (VB) and the conduction band (CB) [36]. The Fe_3O_4-CdO NC, formed by the heterojunction between CdO and Fe_3O_4, demonstrates an absorption band ranging from 400 to 550 nm. This leads to a notable enhancement in the absorption of visible light [37].

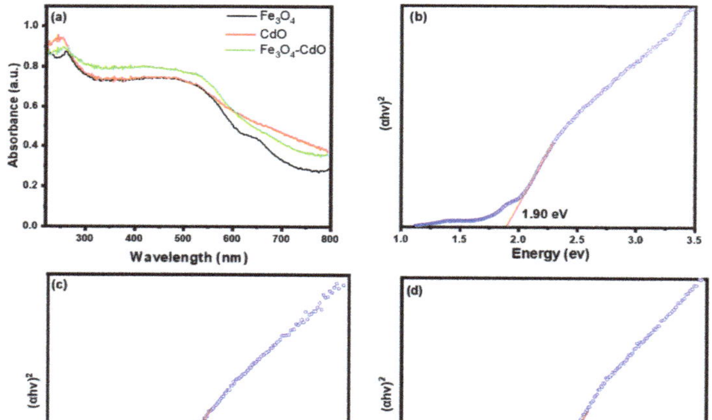

Figure 1. (**a**) UV-DR spectra and the Tauc plots for Fe_3O_4, CdO, Fe_3O_4-CdO, (**b–d**) bandgap of Fe_3O_4, CdO, Fe_3O_4-CdONPs, respectively.

Furthermore, the presence of CdO-NPs on the surface of Fe_3O_4-NPs mitigates the dissipation of light energy caused by the intense light scattering resulting from the interaction between the incident light and CdO-NPs [38,39]. The results obtained from the UV-DRS analysis suggest a minor redshift in the Fe_3O_4-CdO NPs, as compared to the pure CdO NPs. This redshift implies that the Fe_3O_4-CdO NPs have a higher energy absorption capability, which makes them more suitable for a wide range of applications in photocatalysis. The observed alteration in the bandgap energy within the Fe_3O_4-CdO NCs further validates the effective deposition of CdO-NPs onto the Fe_3O_4 NPs' surface, as reported in reference [40]. The observed phenomenon can be attributed to the significant electron–electron interaction occurring within the d-orbital of Cadmium (Cd), as well as its interaction with the s- and p-orbitals of the host material Iron (II,III) oxide (Fe_3O_4). The renormalization effect in pure Fe_3O_4 is caused by the exchange of sp-d band electrons between Fe_3O_4 and localized d-electrons of Cd^+ [41]. This observation provides more evidence supporting the enormous potential of Fe_3O_4-CdO NPs in the field of photocatalysis, since they demonstrate effective use of visible light. Figure 1b–d illustrates the respective bandgap values of 1.90 eV, 1.83 eV, and 1.80 eV for Fe_3O_4, CdO, and NPs, as well as the Fe_3O_4-CdO NC. The findings of the study indicate that the presence of CdO NPs on Fe_3O_4 NPs resulted in a shift in the bandgap of Fe_3O_4 from 1.90 eV to 1.80 eV. The observed transition can be ascribed to the existence of oxygen vacancies, which enhance the mobility of electrons from the valence band (VB) to the conduction band (CB) with increased efficiency [42].

2.2. XRD Analysis

The XRD analysis was ascribed to deduce the crystallinity, phase, crystallite size, and composition of Fe_3O_4, CdO, and Fe_3O_4-CdO NMs [43]. The XRD pattern of Fe_3O_4 reveals the formation of magnetite with well-defined crystallinity. The observed different XRD patterns in Fe_3O_4, CdO, and Fe_3O_4-CdO NMs are shown in Figure 2. In Fe_3O_4 NPs at an angle (2θ), the observed major diffraction peaks were 30.38°, 35.57°, 40.88°, 43.24°, 53.68°, 57.45°, 63.04°, 64.11°, and 74.58° with corresponding crystal planes at (200) (311), (220), (400), (422), (511), (440), (441), and (533), respectively, attributing to cubic structure in accordance with Fe_3O_4 (JCPDS card no. 89-0691) [44]. The additional minor diffraction peaks were observed at 2θ values of 24.17, 33.21, and 49.55, corresponding to Miller indices (012), (104), and (024), respectively, which resemble Fe_2O_3. However, the major and highly intense diffraction peaks observed confirm the formation of Fe_3O_4. The XRD pattern of pure CdO annealed at 450 °C in angular range 2θ from 10–80° is shown in Figure 2. The diffraction patterns of CdO NPs showed crystalline behavior with five distinguishable peaks centered at 2θ = 32.84°, 38.29°, 55.29°, 65.90° and 69.27° corresponding to reflections from (111), (200), (220), (311), and (222) crystal planes, respectively (JCPDS card no. 05-0640) [45]. The purity of the materials obtained was further confirmed by the absence of any further impurity peaks observed in the X-ray diffraction (XRD) spectra. The XRD patterns of Fe_3O_4-CdO NC shown in Figure 2 exhibit no extra major peaks except diffraction peaks of Fe_3O_4 and CdO NPs. In addition, there was no shifting of the diffraction peaks of as-prepared Fe_3O_4-CdO NC, thus determining the well-organized Fe_3O_4-CdO heterojunction without altering the structure of Fe_3O_4 and CdO NPs. The crystallite size of Fe_3O_4, CdO NP, and Fe_3O_4-CdO was calculated using the Scherrer equation ($D = 0.9\lambda/\beta\cos\theta$), where λ is the wavelength of the radiation (0.154056 nm), β is the full width of half maximum (FWHM), and θ is the diffraction angle. The average crystallite size for Fe_3O_4, CdO, and Fe_3O_4-CdO was 24.86 nm, 31.53 nm, and 28.30 nm, respectively.

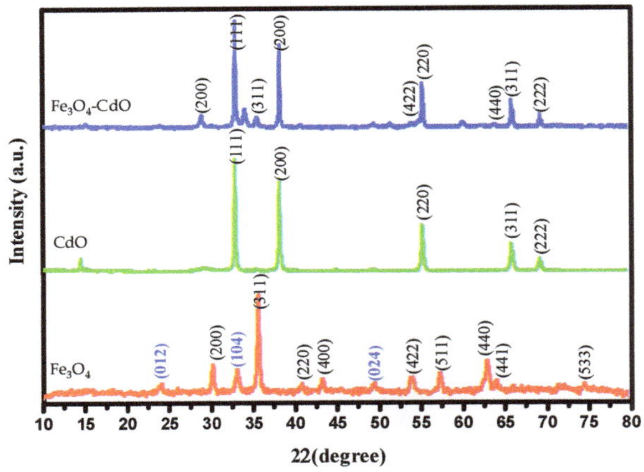

Figure 2. XRD patterns for Fe_3O_4, CdO, and Fe_3O_4-CdO NPs samples.

2.3. FTIR Spectroscopy Analysis

Fourier Transform Infrared (FTIR) spectroscopy is a highly dependable and sensitive spectroscopic method utilized to identify and detect prominent functional groups on the surface of synthetic materials. The involvement of corresponding functional groups is observed following the synthesis of CdO-NPs, forming a Fe_3O_4-supported CdO NC. The FTIR spectral analysis was conducted in accordance with the methodology illustrated in Figure 3. Figure 3 displays the primary peak intensities of Fe_3O_4, CdO, and Fe_3O_4-CdO NPs at specific wavenumbers: 3454.1 cm^{-1}, 2962.0 cm^{-1}, 1723.1 cm^{-1}, 1630.3 cm^{-1}, 1384.7 cm^{-1}, 1270.3 cm^{-1}, 1117.9 cm^{-1}, 731.5 cm^{-1}, 545.3 cm^{-1}, and 457.4 cm^{-1}. The peak intensity found at a wavenumber of 3454.1 cm^{-1} can be attributed to the presence of the hydroxyl (OH) functional group [46]. The broad spectral band corresponds to the stretching vibration of O-H bonds within H_2O molecules that are adsorbed onto the surface of NPs. The observed peak at a wavenumber of 1630.3 cm^{-1} can be related to the bending and stretching frequency of the hydroxyl (OH) group. Similarly, the peak observed at 1384.7 cm^{-1} can be attributed to the bending vibrations of carbon–carbon (C–C) bonds.

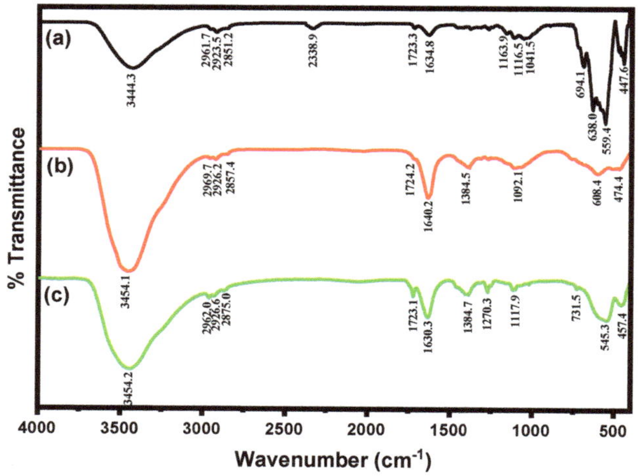

Figure 3. FT-IR spectra for (**a**) Fe_3O_4, (**b**) CdO, and (**c**) Fe_3O_4-CdO NPs samples.

Additionally, the shift in peak intensities seen in the Fe_3O_4 NPs and Fe_3O_4-CdO NC towards lower (blue) and higher (red) wavenumbers, respectively, provides further confirmation of the successful synthesis of both CdO NPs and the Fe_3O_4-CdO NC. The stretching modes associated with the metal–oxygen connection can also be seen in the spectral region of 400–600 cm^{-1}. These observations provide evidence for the presence of CdO NPs and the creation of a Fe_3O_4-CdO NC. The frequencies observed at values below 500 cm^{-1} in the case of CdO NPs have provided evidence of the presence of chemical bonding between the metal and oxygen atoms on the surfaces of the NPs [24]. Additionally, these findings provide insights into the oxidation processes occurring on the surface of the NPs.

2.4. Raman Spectra Analysis

Raman spectroscopy analysis reveals clear differences in peak patterns between the two materials by examining specific molecular vibrations and rotations [47]. In this study, the Raman spectra for the Fe_3O_4, CdO, and Fe_3O_4-CdO NCs are shown in Figure 4, which shows the characteristics Raman peaks of these as-prepared samples, respectively. Based on a comparative analysis of Raman spectra of the composite (Fe_3O_4-CdO) with virgin samples (Fe_3O_4 and CdO), the composite prepared in the present study can be confirmed as follows: Figure 4 shows the characteristic Raman bands at 672 and 346 cm^{-1} for Fe_3O_4, whereas the peak at 477 cm^{-1} suggested the Fe–O vibration mode in Fe_3O_4. The Raman band at 1381 cm^{-1} indicated that the D band of Fe_3O_4 may originate from defects on the surface of Fe_3O_4 [48]. The two peaks at 193 and 255 cm^{-1} may be due to the oxidation reaction during the Raman experiment [49]. As reported in earlier studies, there is Raman spectrum inactivity for CdO. Therefore, the Raman explanation for CdO is quite difficult. However, many studies have shown that the CdO also shows Raman spectra because of structural disorder and resonance caused in the Raman process [50]. For current CdO, the Raman spectra profile (Figure 4) shows three Raman bands at 260, 567, and 1094 cm^{-1}. These bands were attributed to second-order Raman scattering, as expected in CdO [51]. A broad peak at 260 cm^{-1} has been assigned as the 2TA(L) mode for CdO, a prominent feature in the Raman spectrum of CdO [52]. The Raman band at 567 cm^{-1} was assigned for E1 (LO). The peak around 1100 cm^{-1} was reported earlier with CdO for 2LO phonon [53]. The Raman spectrum for the Fe_3O_4-CdO NC is shown in Figure 4. In this spectrum, both CdO and Fe_3O_4 peaks are visible; these confirm the formation of Fe_3O_4-CdO NC by interacting with each other. Not much change was observed in the peaks of CdO as well as Fe_3O_4, which indicates that doping of CdO in Fe_3O_4 does not cause any change in their structure.

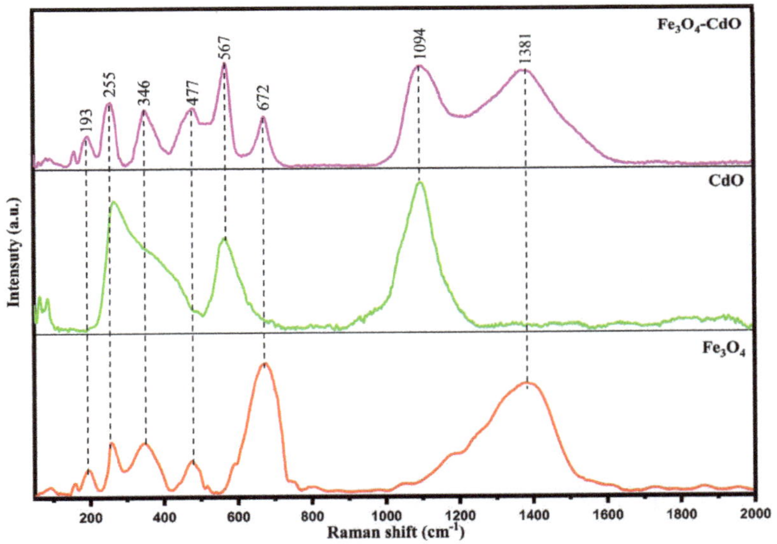

Figure 4. Raman spectra for Fe_3O_4, CdO, and Fe_3O_4-CdO NPs samples.

2.5. SEM-EDX and TEM Analysis

The surface, morphology, and elemental composition of the prepared materials were confirmed by SEM and EDX analyses and are depicted in Figure 5a–c, respectively. SEM imaging analysis is effectively used to characterize heterogeneous solid inorganic and organic materials [54]. For present study, the SEM images (Figure 5a,b) showed the presence of aggregated spherical structures resembling mushrooms in the Fe_3O_4 NPs. However, there was only a slight reduction in the accumulation of CdO on the surface of the Fe_3O_4 NPs.

The elemental composition of the pure Fe_3O_4 NPs indicates that they primarily consist of iron (61.40%) and oxygen (38.60%). This significant presence of iron and oxygen is evidence of the great purity of the synthesized Fe_3O_4 NPs [55]. According to the elemental composition analysis of pure CdO NPs, it was determined that the NPs consisted of 77.86% Cd and 15.40% O. The presence of a Cd peak in the Fe_3O_4-CdO NC indicates the production of a FeO-CdO NC, with elemental compositions of Fe (21.02%), O (35.63%),

Cd (31.34%), and C (10.19%) [24]. The close examination of Figure 5 reveals the presence of spherical clumps consisting of irregular-edged aggregated masses of Fe_3O_4 NPs. CdO NPs may be observed through their adsorption onto the surface of Fe_3O_4 NPs, forming a spherical mass. This adsorption occurs along an uneven edge in the CdO-supported Fe_3O_4 NCs. The ImageJ software was utilized to determine the average particle size of CdO, Fe_3O_4 NP, and Fe_3O_4-CdO, yielding measurements of around 32 nm, 53 nm, and 60 nm, respectively. The successful creation of CdO, Fe_3O_4 NPs, and Fe_3O_4-CdO NC is confirmed by the comprehensive results obtained by investigating elemental composition and surface morphology using EDX and SEM.

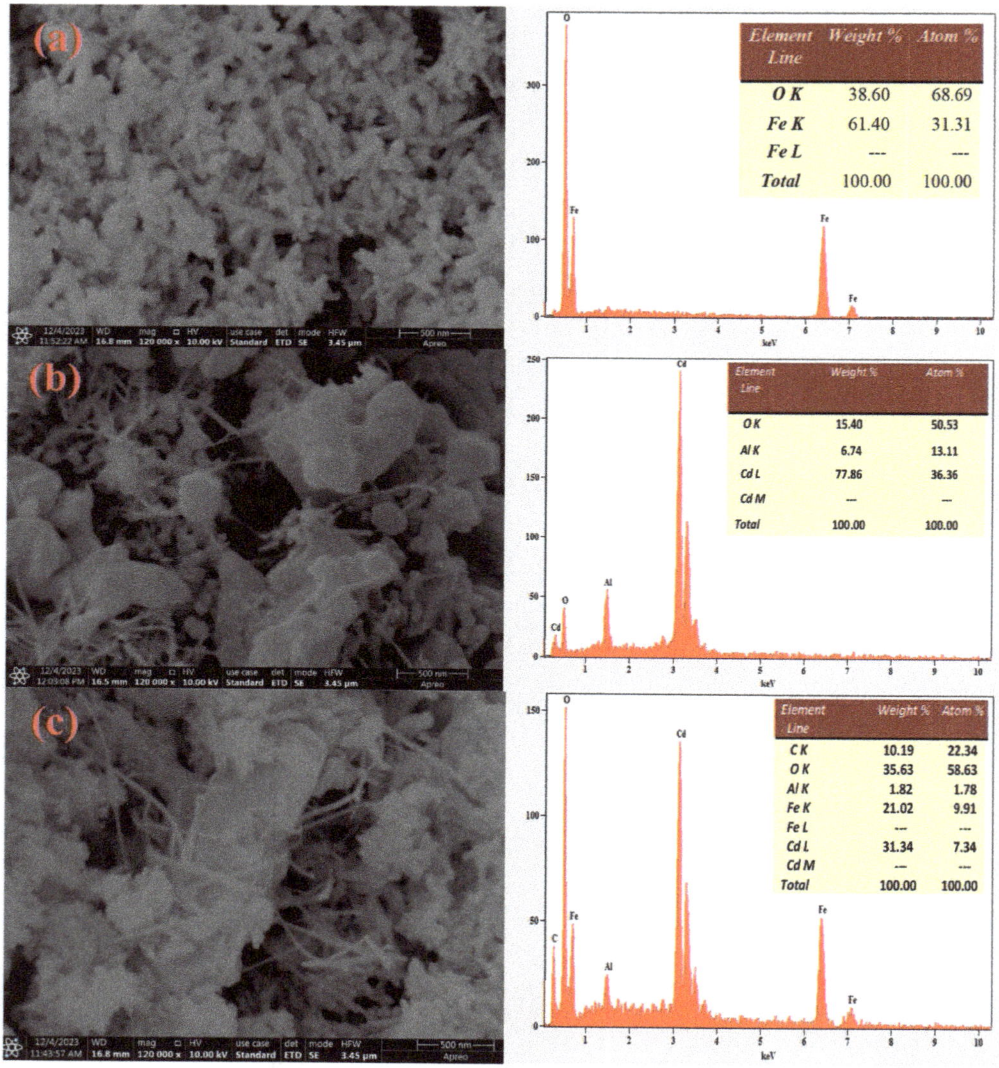

Figure 5. SEM images and EDX analysis for (**a**) Fe_3O_4, (**b**) CdO, and (**c**) Fe_3O_4-CdO NPs samples.

To further study the microstructure of the Fe_3O_4, CdO, and Fe_3O_4-CdO NMs, we performed transmission electron microscopy (TEM) analysis. Figure 6a–c presents TEM images of the Fe_3O_4, CdO, and Fe_3O_4-CdO NMs, confirming the formation of needle-shaped Fe_3O_4 NPs, oval-shaped CdO NPs and Fe_3O_4-CdO NC. Needle-shaped Fe_3O_4 NPs have an average size of 70 nm, and CdO NPs have an average size of 41 nm. Figure 6d shows the HRTEM image of Fe_3O_4-CdO NC and observed the lattice spacing of 0.54 nm and 0.69 nm corresponding to the (311) plane of Fe_3O_4 and (111) plane of CdO, respectively, confirming the formation and heterojunction of Fe_3O_4-CdO NC. The distinct lattice fringes confirm the highly crystalline nature of Fe_3O_4, CdO, and Fe_3O_4-CdO NMs.

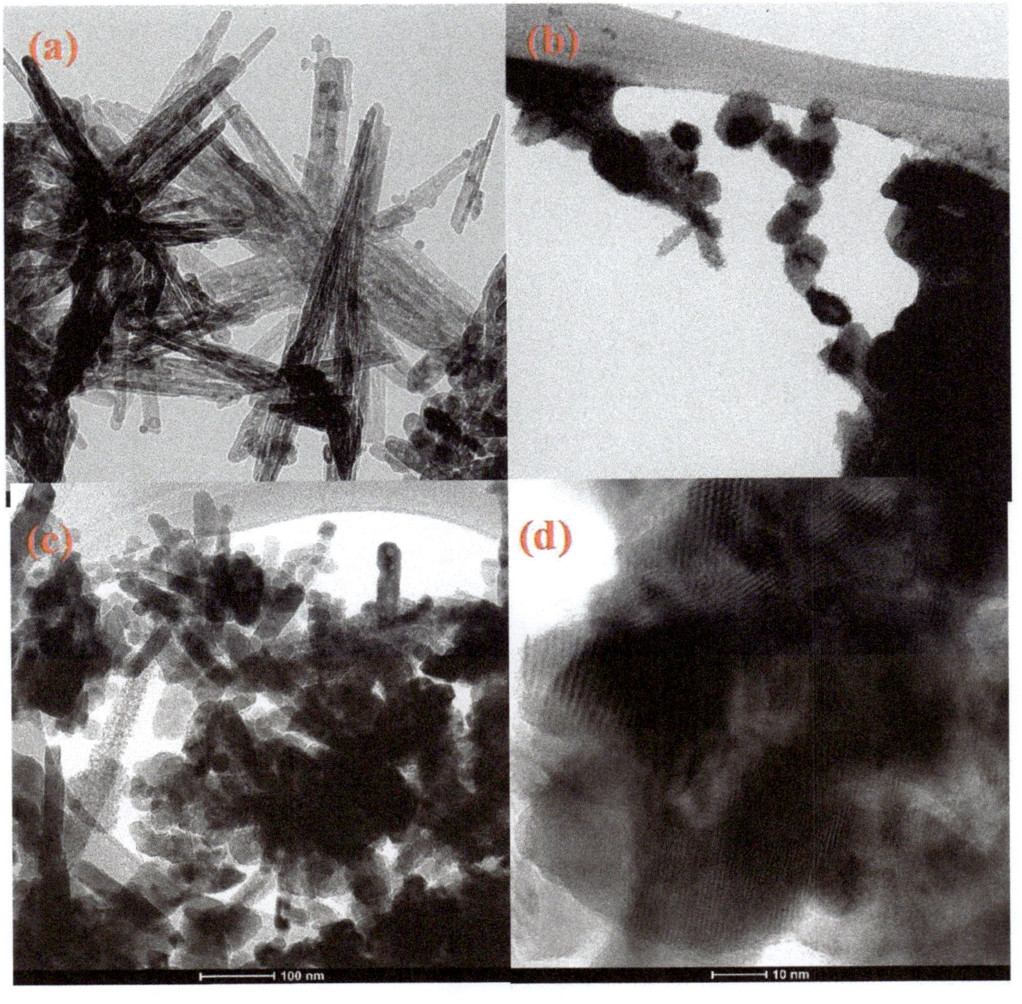

Figure 6. TEM images of (**a**) Fe_3O_4, (**b**) CdO, (**c**) Fe_3O_4-CdO NPs samples, and (**d**) HRTEM lattice fringe image.

2.6. XPS Analysis

XPS analysis of Fe_3O_4-CdO NC offers deep insight into the elemental composition and valence state of the synthesized material. The XPS spectrum of Fe_3O_4-CdO NC shown in Figure 7a indicates the presence of major Fe, Cd, and O components. As seen in Figure 7b, the 711.2 eV and 724.9 eV peaks can be fitted with $Fe2p_{3/2}$ and $Fe2p_{1/2}$ configurations, respectively [56]. The absence of satellite peaks also corroborates the assignment of the final product to Fe_3O_4 rather than Fe_2O_3. This is a vital characteristic used to distinguish between Fe_3O_4 (magnetite) and γ-Fe_2O_3 (maghemite), since the two have the same crystalline structure but differ only in the valence state of iron ions. In the Cd3d core-level scan, $Cd3d_{5/2}$ and $Cd3d_{3/2}$ are located at 405.7 eV and 412.5 eV, respectively [57]. The O1s peaks shown in Figure 7d are centered at 530 eV (low binding energy peak) and 532 eV (higher binding peak). For the O1s spectrum, the lower binding energy peak centered at 530 eV corresponds to the O^{2-} bonded with iron. The peak at 532 eV, which corresponds to oxide formation, dominates in the spectra and points towards the existence of O^{2-} ions in the pure and Fe_3O_4-CdO lattice. XPS analysis confirmed the strong chemical interaction between Fe_3O_4 and CdO in the Fe_3O_4-CdO NC. Thus, the Fe_3O_4 NPs are embedded into the composite matrix through chemical conjugation with CdO rather than simple physical adsorption.

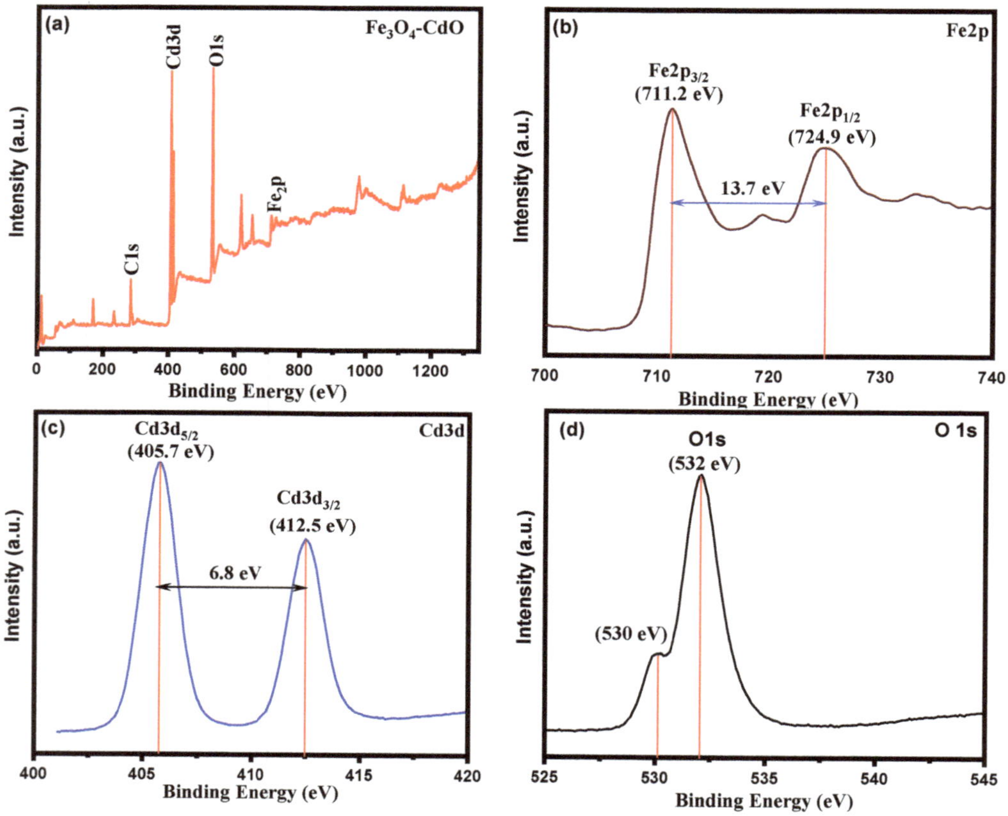

Figure 7. (**a**) Full scan XPS spectrum of Fe_3O_4-CdO NPs and core-level scan of (**b**) Fe2p, (**c**) Cd3d, and (**d**) O1s.

2.7. Photocatalytic Activity of Fe_3O_4-CdO

The potential effectiveness of the Fe_3O_4-CdO NC in mitigating the photodegradation of MB dye molecules under visible-light irradiation was investigated. Metal-doped NMs are frequently employed to enhance the photocatalytic degradation of dye molecules [58].

In addition, it is worth noting that heterogeneous photocatalysis has been used to detoxify air contaminants [59]. The photocatalytic activity of manufactured NPs is generally associated with recombining electron–hole pairs in the reaction medium. Furthermore, determining these parameters, including size, surface area, and radiation source, is crucial to understanding the overall impact. This has been discussed extensively in previous studies [60–63].

Nevertheless, bare cadmium oxide NPs (CdO NPs) exhibit a considerable bandgap energy, rendering them unsuitable for utilization as a viable catalyst for visible light reactions [64]. Hence, the surfaces of the Fe_3O_4-CdO NC, doped with CdO, exhibit promising potential as a novel heterogeneous catalyst under visible-light irradiation. During this investigation, the researchers used a prominent emission of λ_{max} at 365 nm as a visible-light source. This study employed a tungsten lamp with a power rating of 60 W to generate a continuous spectrum of light ranging from 400 to 800 nm. The purpose of utilizing this light source was to investigate the phenomenon of photocatalysis. The photodegradation of MB molecules was visually detected through the physical manifestation of a color change, specifically from blue to colorless.

Furthermore, it was observed that there was a gradual reduction in the peak intensity at λ_{max}, namely at 670 nm, as the time intervals increased. The highest level of degradation, amounting to 92.0%, was achieved following a 120 min exposure to MB dye molecules. This was determined by analyzing the observed minimum or constant peak intensity, as shown in Figure 8a. The percentage degradation over time at various intervals was determined based on the collected data, as depicted in Figure 8b.

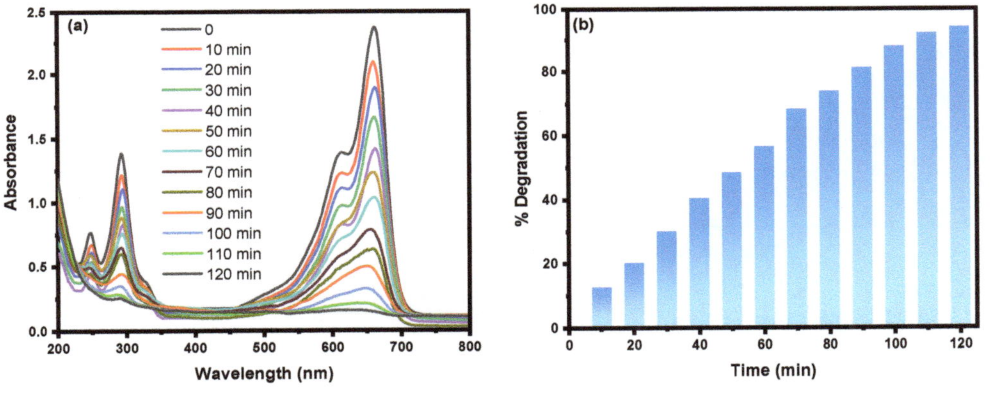

Figure 8. (**a**) UV-vis spectra of MB dye under visible light and (**b**) degradation percentage of MB dye Fe_3O_4-CdO CSNP.

Nevertheless, it has been observed that in CdO NPs (NPs), recombining electron–hole pairs occur within nanoseconds, resulting in restricted photocatalytic effectiveness. The enhanced electron transport observed in the CdO-supported Fe_3O_4 nanocatalyst can be attributed to the disparity in work functions between Fe_3O_4 and CdO. Notably, the augmentation of electron transfer is accomplished by doping CdO onto the surface of Fe_3O_4, whereby metal NPs function as electron sinks or become trapped on the surface of NPs. Moreover, within a photolytic reaction involving an enriched reaction medium, oxidative species (ROS) lead to a notable augmentation in the number of surface traps. Specifically, the electrons are acquired from the catalytic surface of Fe_3O_4 and are employed through Cd in the degradation of MB dye molecules [65].

2.8. Effect Photocatalyst Dosage

The photocatalytic properties of the Fe_3O_4-CdO NC are determined by the process of photocatalytic degradation of Dye and MB molecules. Furthermore, this research investigates the influence of several concentrations of photocatalysts, spanning from 10 mg to 40 mg, on a solution comprising 30 ppm of MB molecules at a pH level of eight while being exposed to visible light irradiation. Typically, dye molecule photodegradation is characterized by the alteration or disappearance of color when subjected to visible-light irradiation. This process involves the transformation of dye carbon molecules into biodegradable chemical species that are less harmful, such as CO_2, O_2^-, or H_2O molecules [66]. Moreover, to understand the photocatalytic process involving MB molecules comprehensively, we employed the pseudo-first-order kinetics approach based on the Hinshelwood kinetic model [67]. In the present study, the photodegradation kinetic curve of $\ln C_o/C_t$ as a function of time (expressed in minutes) was obtained by employing different catalytic dosages ranging from 10 mg to 40 mg. These data are illustrated in Figure 9a.

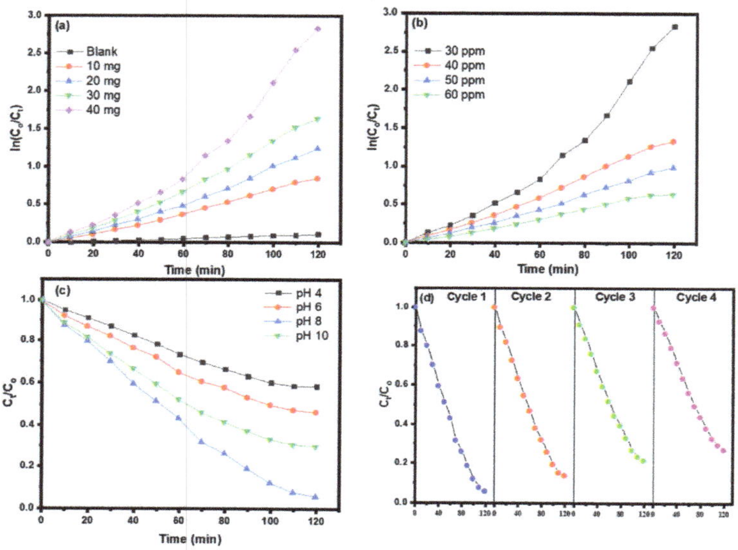

Figure 9. Effect of (**a**) photocatalyst dosage, (**b**) MB dye concentration, (**c**) varying pH and (**d**) reusability test.

Photocatalytic experiments of pure Fe_3O_4 and CdO NPs were performed as shown in Figure S2 and the degradation efficiency of Fe_3O_4-CdO NC was significantly better compared to that of pure Fe_3O_4 and CdO NPs. The photodegradation of MB dye molecules was more significant when the catalytic dose was increased to a numerical value of 40 mg. The degradation experiment was also performed in the absence of Fe_3O_4-CdO NC and negligible dye degradation was observed, as shown in Figure 9a. The catalytic dosage was determined to be optimal for the photodegradation of MB molecules. The maximum number of active sites is anticipated to be available, resulting in the highest catalytic efficiency, as indicated by the rate constant, k, which was calculated to be 0.02345 min^{-1} (Table S1).

Figure S1a,b show the photocatalytic degradation curves of MB over pure Fe_3O_4 and CdO NPs. A good linear correlation, $\ln(C_o/C_t)$ versus time, is obtained in Figure S1b, and the rate constant of this catalytic reaction was found to be 0.0175 and 0.0113/min for pure Fe_3O_4 and CdO NPs, respectively. By comparing both cases, the Kapp value was found to be the maximum for Fe_3O_4-CdO NC because an increase in the surface area and a reduction in the particle size of Fe_3O_4-CdO NC caused an increase in the photoactivity. No reactivity modifications were observed at the end of the reaction, which demonstrated the stability of the photocatalyst. CdO deposition influenced the photocatalytic activity of Fe_3O_4 NPs by capturing the photoinduced electrons and inhibiting the electron–hole (e^-–h^+) recombination. This photo-induced reaction occurs at the surface of a Fe_3O_4 catalyst, which was stimulated by the absorption of a photon with sufficient energy (equal to or higher than the bandgap energy of the Fe_3O_4). The absorption leads to a charge separation due to the promotion of an electron from the conduction band of the Fe_3O_4 catalyst to the conduction band of CdO and a hole from the valence band (VB) of CdO to the VB of the Fe_3O_4 photocatalyst, and this directs the efficient separation of the photo-generated electrons and holes. The optical absorption capability or optical bandgap, minimal surface-to-volume ratio, and photo-generated charge carriers are attributed to the improved catalytic properties of Fe_3O_4-CdO NC. The increase in the photodegradation process observed when utilizing Fe_3O_4-CdO NC as photocatalysts for the degradation of MB dye molecules can be attributed to the achievement of an appropriate surface-to-volume ratio, which is applicable within a small bandgap. This characteristic allows for effective photo-generated precursor generation of charge carriers. Furthermore, during this investigation, we noted that a catalytic dose of 40 mg exhibits optimal catalytic activity, resulting in enhanced mobility and the highest production of photo-carriers from the bulk to the surface. This, in turn, facilitates the maximum adsorption of dye molecules onto the active sites of the Fe_3O_4-CdO NC, leading to a comparatively higher catalytic efficiency.

2.9. *Effect of MB Dye Concentrations*

The relationship between the initial concentration of MB dye and the efficiency of photodegradation utilizing Fe_3O_4-CdO NC as a photocatalyst was determined by analyzing the kinetics based on the pseudo-first-order kinetic model. The experimental procedure involved measuring the plot of $\ln C_o/C_t$ versus time (in minutes) under the condition of a consistent catalytic dose of 40 mg, while the dye concentrations varied between 30 and 60 ppm. The pH was kept at 8 during the 120 min irradiation period. This information is illustrated in Figure 9b. The degradation efficiency for different dye concentrations at a consistent catalytic dose was determined using the experimental data. The optimal photodegradation of MB dye was seen when utilizing 40 mg of Fe_3O_4-CdO NC with a dye concentration of 30 ppm. Nevertheless, when the catalytic concentration remained constant at 40 mg, and the concentration of MB fluctuated from 30 to 60 ppm, the likelihood of binding contacts between the nanocatalyst and dye molecules on their active sites reduced.

Moreover, this phenomenon can be comprehended better by maintaining a consistent catalytic concentration, whereby the highest number of accessible sites are utilized at relatively lower concentrations of dye. Moreover, a higher concentration of dye molecules has been found to result in an inhibitory effect due to catalytic poisoning and sedimentation, as evidenced by the numerical values obtained for the percentage degradation efficiencies and rate constant ($k = 0.02345\ min^{-1}$). During this investigation, we determined that the most favorable concentration of dye was 30 ppm when combined with a catalytic quantity of 40 milligrams (mg) under visible-light illumination.

2.10. *Effect of pH*

The investigation focused on examining the influence of pH on the photodegradation of MB molecules while using the Fe_3O_4-CdO NC as a catalyst. The pH range explored in this study spanned from pH 4 to pH 10. The data that were acquired were analyzed using the kinetic pseudo-first-order approach by plotting C_t/C_o versus time (in minutes), as depicted in Figure 9c. The experiment was conducted using visible light, with an optimal dye concentration of 30 ppm and an optimal concentration of the catalytic Fe_3O_4-CdO NC of 40 mg. The study examines the relationship between degradation efficiency and MB dye molecules. It was shown that the highest degradation efficiency (%) of MB molecules occurred at pH 8, followed by a decrease in photodegradation efficiency at pH 10. Conversely, minimal dye degradation was detected in an acidic pH environment, namely at pH 4. In this study, we examined the degradation of MB dye molecules through photochemical processes, specifically by investigating the degradation rates at different pH levels. The observed degradation sequence was as follows: pH 8 < pH 10 < pH 6 < pH 4. An increase in pH levels was reported to accelerate the photodegradation of MB molecules on the active surface of the Fe_3O_4-CdO NC. This rise in degradation efficacy was detected within the pH range of 4 to 8. However, a drop in the percentage of degradation efficacy was observed at pH levels of 10. Moreover, the Fe_3O_4-CdO NC, which serves as a nanocatalyst, exhibits a dual polarity and can dissolve in acidic and alkaline environments due to its amphoteric properties. The pH dependence of nanocatalysts is generally attributed to their surface properties. The degradation efficiency of dye molecules using nanocatalysts is constrained while altering pH values, contingent upon these catalysts' active surface charge characteristics. Nevertheless, the surface charge of the Fe_3O_4-CdO NC experiences an augmentation when the pH values rise. This is attributed to the adsorption of OH^- ions, which leads to an intensified generation of hydroxyl radicals. Nevertheless, when the pH is reduced, specifically at numerical values of pH 8, the MB dye molecules exhibit a preference for occupying the nanocatalyst Fe_3O_4-CdO NC. Furthermore, it has been shown that when the pH values increase, there is a corresponding increase in the generation of a more significant number of hydroxyl radicals ($^{\bullet}OH$). This increase in hydroxyl radicals leads to enhanced degradation of MB. However, it is essential to note that the highest degradation of MB occurs at a pH of 8. However, as the pH values increase, such as at pH 10, the electrostatic repulsion between the MB anion and the positive surface charge of the Fe_3O_4-CdO NC becomes stronger. On the other hand, when the pH values are relatively higher, there is a potential decrease in the breakdown of MB molecules because of the limited diffusion of $OH^{\bullet}$ formed at the surface towards the double layer, leading to low concentrations of MB. This decline occurs in contrast to the direct charge transfer at the optimal pH of 8. The decolorization efficacy was found to be at its maximum at a pH of 8; this was attributed to the electrostatic repulsion and direct charge transfer mechanisms. However, the decolorization abilities were significantly reduced at pH level 10.

2.11. Recyclability Tests

The recyclability assessment of the produced Fe_3O_4-CdO NC as a nano photocatalyst was conducted subsequent to its demonstrated efficacy in the degradation of MB dye molecules. In this study, we devised a recyclability test to assess the stability and reusability of a manufactured Fe_3O_4-CdO NC under visible-light conditions. The objective was to evaluate its effectiveness as a nanocatalyst with regard to reducing MB dye molecules. Consequently, the photodegradation of MB molecules was terminated, leading to total degradation within 120 min in each cycle. Following each round of exposure, the nanocatalyst was retrieved using centrifugation, subjected to a washing process including 100% alcohol and deionized water, and subsequently dried at a temperature of 70 °C in an oven for one night. The nanocatalyst that had been gathered was introduced into a recently created reaction solution containing MB dye to advance to the subsequent cycle. The degradation of MB molecules was seen to occur in the presence of visible light after four consecutive cycles. The degradation rates ranged from 78% to 92% within 120 min of exposure. This degradation process was facilitated by using Fe_3O_4-CdO NC as nanocatalysts, as depicted in Figure 9d. The results obtained regarding the recyclability of the Fe_3O_4-CdO NC highlight its potential as a nanocatalyst for the photodegradation of MB dye molecules. Nevertheless, it is anticipated that the efficiency of the Fe_3O_4-CdO NC as a nanocatalyst will diminish after four consecutive cycles due to the inevitable loss of the sample (Fe_3O_4-CdO NC) during each cycle's collection and centrifugation procedures. In general, the findings of this study indicate that the Fe_3O_4-CdO NC exhibits recyclability and can serve as a catalyst in the photodegradation of MB molecules. Furthermore, it demonstrates a photocatalytic window that can effectively degrade various thiazine dye molecules. To test the stability of the prepared material, the XRD analysis of the spent catalyst was performed, and the results are shown in Figure S2. The XRD pattern of the spent catalyst shows no significant change in phase after four cycles of reuse. Moreover, we compared our findings with those of earlier studies, as illustrated in Table 1.

Table 1. Literature survey concerned Fe_3O_4-CdO nNC in photodegradation of organic dyes and pharmaceuticals.

Catalyst	Dye	Degradation Time (min)	Degradation (%)	Ref.
Fe_3O_4	MB	180	71	[68]
Fe_3O_4	Congo red	60	77	[69]
CdO	Congo Red	120	100	[70]
	Malachite Green		100	
	Crystal Violet	240	95	
CdO/TiO_2	Reactive Orange 4	60	90.9	[71]
$CdO-TiO_2$	Imazapyr	180	96	[32]
Fe_3O_4/ZnO	Methyl Orange	60	93.6	[40]
Fe_3O_4/ZnO	MB	120	88.5	[72]
$Fe_3O_4@ZnO$	Rhodamine B	60	99.3	[60]
Fe_3O_4/ZnO	MB	60	63.02	[73]
Fe_3O_4-CdO	MB	120	92	Present work

2.12. Photocatalytic Degradation Mechanism

The proposed mechanism for the catalytic photo-assisted degradation of MB dye molecules entails the generation of reactive oxidative species, such as superoxide radical anion ($O_2^{-\bullet}$), hydroxyl radical ($^{\bullet}OH$), electron (e^-), and holes (h^+), through a photolytic reaction, as depicted in Figure 10. In the context of a photocatalytic reaction, a phenomenon occurs whereby electrons residing in the valence band (VB) undergo excitation upon exposure to light radiation, leading to their transition to the conduction band (CB). Consequently, this process results in the generation of holes inside the valence band. The reduction reaction taking place on the surface of the catalyst is assisted by the excited electrons present in the conduction band (CB). In contrast, the movement of holes from the photocatalytic surface drives the oxidation reaction. The process of photodegradation of dye molecules involves both reduction and oxidation reactions [74]. In this study, we comprehensively analyze the sequential reactions that occur following the photodegradation of MB molecules using Fe_3O_4-CdO NC.

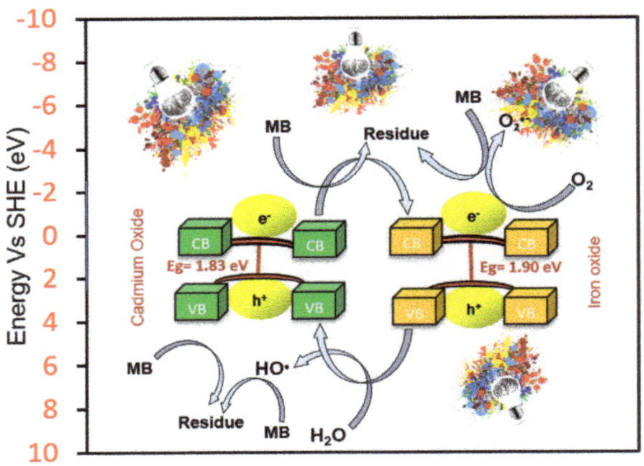

Figure 10. Possible photocatalytic degradation mechanism of MB dye using Fe_3O_4-CdO under visible light.

It has been established that the formation of the stable intermediate(s) during the degradation of MB depends on the nature of the catalyst as well as the oxidizing agents. For example, the formation of different intermediates such as azure B, azure B, azure C, and thionine were reported at λ_{max} 655 nm, 638 nm, 618 nm, and 603 nm, respectively, during the oxidation of MB with MnO_2-SnO_2 [75], zerovalent Fe kmnO_4 [76], MnO_2-cellulose [77], and $KMnO_4$ [78] by using UV-visible and mass spectrometry. Figure 10 shows that the intensity of MB (all absorption peaks) decreases as the reaction time increases, which demonstrates the complete degradation of MB molecule (N-S conjugated system with aromatic heterocyclic system, chromophore, and N-containing methyl groups on the benzene rings, auxochrome) under photocatalytic degradation with Fe_3O_4-CdO NC. The presence of auto-generated reactive radical species ($^{\bullet}OH$ radicals) might be responsible for the complete degradation (decolorization and mineralization) of MB.

Consequently, it was anticipated that the Fe_3O_4-CdO NC, which is a semiconductor photocatalyst supported by CdO, would experience excitation when exposed to visible–visible light radiation at a lower energy level compared to the bandgap energy of CdO. In addition to experiencing excitation, the process of electron (e^-) transfer from the valence band (VB) to the conduction band (CB) results in the creation of holes (h^+) within the VB. During this investigation, it was observed that when the nanocatalyst (namely, the Fe_3O_4-CdO NC) is exposed to visible–visible light, it produces electrons in the conduction band (CB) and holes in the valence band (VB), in accordance with Equation (1). The electron generated in the conduction band (CB) combines with an oxygen molecule to form $O_2^{\bullet -}$. This photo-generated O_2^- then undergoes further reactions with water molecules in a mixed reaction media, resulting in the conversion to $^{\bullet}HO_2$ as described by Equations (2) and (4).

Furthermore, the dissociation of VB results in the release of H^+ ions, which then adsorb onto water molecules, leading to the formation of $\bullet$OH radicals, as depicted in Equation (2). In general, the degradation of MB molecules through the Fe_3O_4-CdO NC is mainly attributed to generating reactive oxygen species (ROS) via photochemical reactions. This process may be described gradually, as shown by Equations (1)–(6), illustrating the photocatalytic reduction process.

$$Fe_2O_3 - Cd + hv \rightarrow Fe_2O_3 - Cd\left(e_{CB}^- + h_{VB}^+\right) \tag{1}$$

$$Fe_2O_3 - Cd\left(e_{CB}^-\right) + O_2 \rightarrow Fe_2O_3 - Cd + O_2^{*-} \tag{2}$$

$$Fe_2O_3 - Cd\left(h_{VB}^+\right) + H_2O \rightarrow Fe_2O_3 - Cd + OH^* \tag{3}$$

$$O_2^{*-} + H^* \rightarrow *_{HO_2} \tag{4}$$

$$OH^*, O_2^{*-}, *_{HO_2} + MB \rightarrow MB^{*+} + H_2O \tag{5}$$

$$OH^*, O_2^{*-}, *_{HO_2} + MB^{*+} \rightarrow MB\ deriv. + H_2O + CO_2 \tag{6}$$

Characterization analysis confirms that the present photocatalyst is a heterojunction material that is formed through the conjugation of two different (Fe_3O_4 and CdO) semiconductors in the photocatalyst. The band-edge positions of these semiconductors in heterojunction were found to be different from each other (Table 2).

Table 2. Band edge position of CdO and Fe_3O_4 in semiconductors in developed heterojunction.

S. No.	Semiconductor	X (eV)	Eg (eV)	E_{VB}	E_{CB}
1.	CdO	5.71	1.83	2.125	2.180
2.	Fe_3O_4	5.73	1.90	0.295	0.280

In order to investigate the position of the conduction band (CB) and valence band (VB) edges in these semiconductors (CdO and Fe_3O_4), the following equations were utilized [34]:

$$E_{CB} = \chi - Ee - 0.5Eg \tag{7}$$

$$E_{VB} = E_{CB} + Eg \tag{8}$$

Here, the χ is electronegativity for CdO and Fe_3O_4 in eV and the symbol "Ee" represents the energy of a free electron on the hydrogen scale, equivalent to 4.5 electron volts (eV). The Eg reveals the energy bandgap of CdO and Fe_3O_4.

From the results shown in Table 2, it can be concluded that in the present photocatalyst, due to the different positions of the band edge of semiconductors, a barrier is formed at the interface of CdO and Fe_3O_4 which may be responsible for the charge separation and long-lasting electron–hole distribution.

In the context of a photocatalytic reaction, the phenomenon occurs whereby e^- residing in the VB of one side (i.e., CdO) undergo excitation upon exposure to light radiation, leading to their transition to the CB. Consequently, this process results in the generation of h^+ inside the VB of CdO. Notably, the heterojunction material (CdO-Fe_3O_4) is made of type 2, which gives a better condition of the junction due to which these two radicals (e^- in CB and h^+ inside VB) move on the other side (i.e., Fe_3O_4). This allows charge separation and ease resistance of e^--h^+ recombination. This leads to the distribution of e^- and h^+ at interface of CdO and Fe_3O_4. Therefore, the electric charge distributes on the entire surface of the catalyst with minimum e^--h^+ recombination and reacts with water molecules to form free radicals.

These photo-generated free radicals such as e^-, h^+, $\bullet O_2^-$, and $\bullet OH$ undergo further reactions with dye molecules in a reaction media, resulting in the degradation of MB molecules. This process may be described gradually, as shown by Figure 10, illustrating the photocatalytic reduction process.

3. Materials and Methods

3.1. Materials

Ferric nitrate nonahydrate ([$Fe(NO_3)_3 \cdot 9H_2O$]), Cadmium acetate-2-hydrate ([$(CH_3COO)_2Cd \cdot 2H_2O$]), Cetyltrimethylammonium bromide ([$(C_{16}H_{33})N(CH_3)_3$]Br), Ammonium hydroxide solution 30–33% NH_3 in H_2O (NH_4OH), and Methylene blue ([$C_{16}H_{18}ClN_3S$]). In addition, all chemicals used in this study were from Sigma Aldrich, Louis, MO, USA. Deionized water was used throughout the experiments.

3.2. Preparation of Fe_3O_4, CdO and Fe_3O_4-CdO NPs

The synthesis of Fe_3O_4 and CdO NPs was conducted in the presence of a template CTAB using a straightforward combined co-precipitation and combustion method. Two different ferric nitrate and cadmium acetate solutions were prepared in deionized water. Subsequently, all solutions were stirred gently for a duration of 1 h individually to ensure thorough homogenization while maintaining a constant temperature of 90 °C. Then, 30 mL of CTAB (0.2 M) was individually introduced into the respective solutions, followed by continuous stirring for a duration of 2 h at a temperature of 90 °C. Afterwards, ammonium hydroxide solution was gradually added to the reaction, followed by the measurement of hydrogen potential, denoted as pH, which was determined to be 12. The combustion process was conducted for all solutions, and the resulting materials were washed several times with water and ethanol, followed by drying at 80 °C for 24 h. The acquired material was calcinated at a temperature of 450 °C for a duration of 4 h.

The Fe_3O_4-CdO NC was synthesized using the one-pot approach, which involved a straightforward co-precipitation and combustion process in the presence of a template CTAB. During the experimental procedure, 5 g of ferric nitrate and cadmium acetate salts was individually dissolved in 50 milliliters of deionized water. Subsequently, the two solutions were combined in a single beaker and stirred for a duration of 1 h to ensure thorough mixing. Throughout this process, the temperature was maintained at 90 degrees Celsius. A volume of 40 mL of CTAB (0.2 M) was introduced into the solution, followed by continuous stirring for a duration of 2 h, ensuring the temperature remained constant. Ammonium hydroxide solution was added to this reaction mixture to achieve a pH 12. The combustion process of the reaction mixture was performed until a fine dark-brown powder was formed. The acquired powdered material was washed with water and ethanol and dried at 80 °C for 24 h. Ultimately, the powdered material was calcined for 4 h at 450 °C. Figure 11 shows a schematic illustration of the preparation of Fe_3O_4, CdO, and Fe_3O_4-CdO NC.

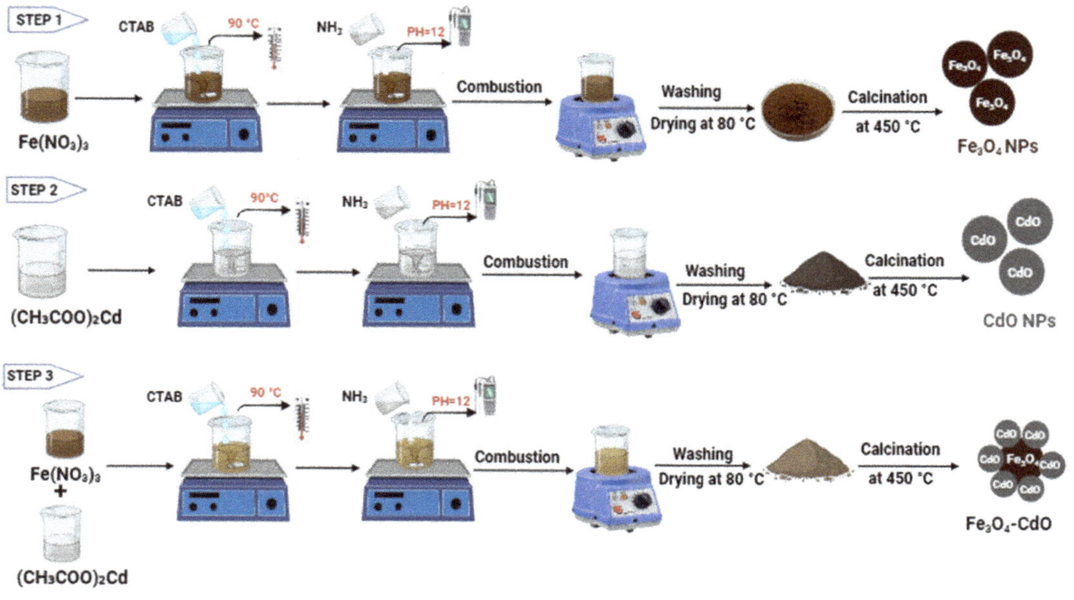

Figure 11. Schematic representation of the preparation of Fe_3O_4 NPs, CdO NPs, and Fe_3O_4-CdO NC.

3.3. *Characterization*

The crystalline nature and the purity of prepared nanomaterials (NMs), CdO, Fe_3O_4, and Fe_3O_4-CdO, were analyzed through X-ray diffraction (XRD). The XRD patterns were recorded using a diffractometer (D8, Advance, Bruker, Germany) with Cu Kα (1.542 Å) radiation source in the 10–80 (2θ diffraction angle) range. The FTIR spectroscopy technique was employed to assess the involvement of functional groups in the reduction and surface capping processes of CdO, Fe_3O_4, and Fe_3O_4-CdO. The measurements were conducted using a Bruker Alpha spectrometer (Billerica, MA, USA), covering a wavelength range of 4000–400 cm^{-1}. The scanning electron microscopy (SEM) (Bruker, Billerica, MA, USA) equipped with energy-dispersive X-ray spectroscopy (EDX) (SEM-EDX) technique was employed to analyze the surface morphology, size, and elemental composition of CdO, Fe_3O_4, and Fe_3O_4-CdO. UV-vis DRS measurements were conducted using a Perkin Elmer spectrophotometer (Waltham, MA, USA) to monitor the optical characteristics of CdO, Fe3O4, and Fe_3O_4-CdO. Transmission electron microscopy (TEM) analysis was performed to determine the size and shape of the prepared NMs. X-ray photoelectron spectroscopy (XPS) (ThermoFisher ESCALAB 250Xi spectrometer, Waltham, MA, USA) was used to examine the chemical states of the materials at their surfaces. The Raman spectroscopy analysis of CdO, Fe_3O_4, and Fe_3O_4-CdO samples was conducted using a Raman microscope

(Bruker, Billerica, MA, USA), operating within the 100 to 4000 cm^{-1} spectral range at ambient temperature.

3.4. Photocatalytic Analysis

The photocatalytic efficiency of CdO, Fe_3O_4 NPs, and Cd-supported Fe_3O_4 NCs was evaluated in the presence of cationic MB dye. The experiments were conducted utilizing visible-light radiation as the illumination source, employing a 500 W Xenon lamp (Shilpent, Nagpur, India) equipped with an ultraviolet cut-off filter. The experimental setup maintained a fixed distance of approximately 15 cm between the visible light source and the dye reaction solution vessel. The experimental setup involved a photochemical reaction in which 40 mg of prepared NPs were dissolved in 50 mL of a 30 ppm MB solution. This solution was placed in a quartz reaction vessel with an outside cooling jacket to maintain a consistent temperature. Before the commencement of the photodegradation method, the reaction mixture underwent a stirring process for 30 min in a lightless environment, with the aim of achieving an equilibrium solution within the photochemical reactor.

Furthermore, the pH of the reaction mixture solution was modified by adding a 0.05 M NaOH solution. At regular time intervals of 10 min, a designated volume of the reaction mixture (3 mL) was extracted. A centrifugation process was employed to achieve the separation of the photocatalyst from the solution. Concurrently, the spectrophotometer was utilized to measure the absorbance of the dye solution across the wavelength spectrum of 200–800 nm. Moreover, several experimental parameters were adjusted for different investigations, such as the concentration of dye, the duration of the irradiation, the pH of the reaction mixture, the reusability of the photocatalyst, and the catalytic load. The determination of the photocatalytic degradation of both dyes was conducted utilizing Equation (9) [79,80].

$$\%\ degradation = \frac{C_o - C_t}{C_o} \times 100 \tag{9}$$

The terms C_o and C_t are commonly used to denote the starting concentration and final concentration of MB after the introduction of the photocatalyst.

4. Conclusions

The synthesis of CdO nanoparticles, magnetic Fe_3O_4, and nanocomposites of Fe_3O_4-CdO was carried out utilizing the solution combustion method, with CTAB serving as a template. The materials show exceptional efficacy in the degradation of dyes by photocatalytic mechanisms. The present work introduces a methodology for producing metal/semiconductor nanocomposites with exceptional stability. The approach involves the creation of CdO nanoparticles on the Fe_3O_4 surface during the production of CdO-supported Fe_3O_4 nanocomposites, utilizing the combustion process. The Fe_3O_4, CdO, and Fe_3O_4-CdO nanomaterials were subjected to various analytical techniques for characterization purposes. These techniques encompassed UV-DRS, XRD, XPS, FTIR, Raman spectroscopy, TEM, SEM, and EDS. The deposition of Cd nanoparticles onto the surface of Fe_3O_4 was confirmed through the application of X-ray diffraction (XRD) analysis, scanning electron microscope-energy dispersive X-ray spectroscopy (SEM-EDX) micrographs. The present study aimed to evaluate the photocatalytic efficiency of Fe_3O_4-CdO nanocomposites through visible-light irradiation by investigating the photocatalytic decolorization of MB under ambient reaction conditions. The experimental results suggest that composites, including Fe_3O_4 and CdO, which were generated under simple reaction conditions, demonstrate a substantial degree of catalytic efficacy. A novel photocatalytic system was devised, utilizing a nanocomposite catalyst comprising Fe_3O_4-CdO. The study provided evidence that the principal cause of degradation is attributed to the involvement of hydroxyl radicals ($^{\bullet}OH$) via photocatalytic processes. Moreover, empirical evidence has substantiated that the composite of Fe_3O_4-CdO demonstrates acceptable stability, enabling it to undergo multiple recycling cycles. The nanocomposites composed of Fe_3O_4 and CdO demonstrate a notable level of photocatalytic efficacy, making them a viable candidate for wastewater treatment.

Supplementary Materials: The following supporting information can be downloaded at: https://www.mdpi.com/article/10.3390/catal14010071/s1, Figure S1: (a) Photocatalytic degradation efficiency and (b) plot of ln (Co/At) versus time of pure Fe_3O_4 and CdO nanoparticles; Figure S2: XRD of fresh and spent CdO-Fe_3O_4 catalyst; Table S1: Rate constants at different catalyst dosages and dye concentrations.

Author Contributions: Conceptualization, M.A.M. and Z.K.; methodology, M.A.M., Z.K. and S.A.A.-T.; software, M.A.M. and R.P.; validation, M.A.M., Z.K., R.P. and S.A.A.-T.; formal analysis, A.A. and S.A.A.-T.; investigation, A.A.; resources, S.A.A.-T. and Z.K.; data curation, A.A. and S.M.; writing—original draft preparation, M.A.M., A.A. and Z.K.; writing—review and editing, M.A.M., Z.K., S.A.A.-T., S.M. and R.P.; visualization, M.A.M., R.P. and Z.K.; supervision, Z.K. and S.A.A.-T.; project administration, S.A.A.-T.; funding acquisition, S.A.A.-T. All authors have read and agreed to the published version of the manuscript.

Funding: This research was funded by Deanship of Scientific Research (DSR), King Abdulaziz University, Jeddah, Saudi Arabia, grant number (KEP-PhD: 90-130-1443).

Data Availability Statement: The data can be found in both the article and the Supplementary Materials.

Acknowledgments: The authors would like to express their gratitude to the Deanship of Scientific Research (DSR), King Abdulaziz University, Jeddah, Saudi Arabia, for providing financial support for this project under grant no. (KEP-PhD: 90-130-1443).

Conflicts of Interest: The authors declare no conflicts of interest.

References

1. Rathi, G.; Siddiqui, S.I.; Pham, Q. Nigella Sativa Seeds Based Antibacterial Composites: A Sustainable Technology for Water Cleansing—A Review. *Sustain. Chem. Pharm.* **2020**, *18*, 100332. [CrossRef]
2. Choudhry, A.; Sharma, A.; Siddiqui, S.I.; Ahamad, I.; Sajid, M.; Khan, T.A.; Chaudhry, S.A. Origanum Vulgare Manganese Ferrite Nanocomposite: An Advanced Multifunctional Hybrid Material for Dye Remediation. *Environ. Res.* **2023**, *220*, 115193. [CrossRef]
3. Balcha, A.; Yadav, O.P.; Dey, T. Photocatalytic Degradation of Methylene Blue Dye By Zinc Oxide Nanoparticles Obtained from Precipitation and Sol-Gel Methods. *Environ. Sci. Pollut. Res.* **2016**, *23*, 25485–25493. [CrossRef]
4. Shehnaz; Prasher, I.B.; Ahmad, N.; Ahmed, M.; Raghuwanshi, S.; Kumar, V.; Siddiqui, S.I.; Oh, S. Live Biomass of Rigidoporus Vinctus: A Sustainable Method for Decoloration and Detoxification of Dyes in Water. *Microorganisms* **2023**, *11*, 1435. [CrossRef]
5. Deng, H.; Lu, J.; Li, G.; Zhang, G.; Wang, X. Adsorption of Methylene Blue on Adsorbent Materials Produced from Cotton Stalk. *Chem. Eng. J.* **2011**, *172*, 326–334. [CrossRef]
6. Nguyen, H.T.; Siddiqui, S.I.; Maeng, S.K.; Oh, S. Biological detoxification of oxytetracycline using Achromobacter-immobilized bioremediation system. *J. Water Process Eng.* **2023**, *52*, 103491. [CrossRef]
7. Fatima, B.; Siddiqui, S.; Ahmed, R.; Chaudhry, S.A. Preparation of functionalized CuO nanoparticles using Brassica rapa leave extract for water purification. *Desalination Water Treat.* **2019**, *164*, 192–205. [CrossRef]
8. Zhao, R.; Li, Y.; Sun, B.; Chao, S.; Li, X.; Wang, C.; Zhu, G. Highly Flexible Magnesium Silicate Nanofibrous Membranes for Effective Removal of Methylene Blue from Aqueous Solution. *Chem. Eng. J.* **2019**, *359*, 1603–1616. [CrossRef]
9. Zhang, T.; Ma, Q.; Zhou, M.; Li, C.; Sun, J.; Shi, W.; Ai, S. Degradation of Methylene Blue by a Heterogeneous Fenton Reaction Catalyzed by $FeCo_2O_4$-N-C Nanocomposites Derived by ZIFs. *Powder Technol.* **2021**, *383*, 212–219. [CrossRef]
10. Ngullie, R.C.; Alaswad, S.O.; Bhuvaneswari, K.; Shanmugam, P.; Pazhanivel, T.; Arunachalam, P. Synthesis and Characterization of E Ffi Cient ZnO/g-C_3N_4 Nanocomposites Photocatalyst for Photocatalytic Degradation of Methylene Blue. *Coatings* **2020**, *10*, 500. [CrossRef]
11. Nada, A.A.; Nasr, M.; Viter, R.; Miele, P.; Roualdes, S.; Bechelany, M. Mesoporous $ZnFe_2O_4$@TiO_2 Nanofibers Prepared by Electrospinning Coupled to PECVD as Highly Performing Photocatalytic Materials. *J. Phys. Chem. C* **2017**, *121*, 24669–24677. [CrossRef]
12. Zheng, Y.; Liu, Y.; Guo, X.; Chen, Z.; Zhang, W.; Wang, Y.; Tang, X.; Zhang, Y.; Zhao, Y. Sulfur-Doped G-C3N4/RGO Porous Nanosheets for Highly Efficient Photocatalytic Degradation of Refractory Contaminants. *J. Mater. Sci. Technol.* **2020**, *41*, 117–126. [CrossRef]
13. Atla, S.B.; Lin, W.R.; Chien, T.C.; Tseng, M.J.; Shu, J.C.; Chen, C.C.; Chen, C.Y. Fabrication of Fe_3O_4/ZnO Magnetite Core Shell and Its Application in Photocatalysis Using Sunlight. *Mater. Chem. Phys.* **2018**, *216*, 380–386. [CrossRef]
14. Xu, P.; Liu, X.; Zhao, Y.; Lan, D.; Shin, I. Study of Graphdiyne Biomimetic Nanomaterials as Fluorescent Sensors of Ciprofloxacin Hydrochloride in Water Environment. *Desalin. Water Treat.* **2023**, *302*, 129–137. [CrossRef]
15. Li, H.; Wu, Y.; Xu, Z.; Wang, Y. In Situ Anchoring Cu Nanoclusters on Cu-MOF: A New Strategy for a Combination of Catalysis and Fluorescence toward the Detection of H_2O_2 and 2,4-DNP. *Chem. Eng. J.* **2024**, *479*, 147508. [CrossRef]

16. Kang, J.; Liu, G.; Hu, Q.; Huang, Y.; Liu, L.; Dong, L. Parallel Nanosheet Arrays for Industrial Oxygen Production. *J. Am. Chem. Soc.* **2023**, *145*, 25143–25149. [CrossRef]
17. Medhi, R.; Marquez, M.D.; Lee, T.R. Visible-Light-Active Doped Metal Oxide Nanoparticles: Review of Their Synthesis, Properties, and Applications. *ACS Appl. Nano Mater.* **2020**, *3*, 6156–6185. [CrossRef]
18. Liu, W.; Huang, F.; Liao, Y.; Zhang, J.; Ren, G.; Zhuang, Z.; Zhen, J.; Lin, Z.; Wang, C. Treatment of CrVI-Containing $Mg(OH)_2$ Nanowaste. *Angew. Chem. (Int. Ed.)* **2008**, *47*, 5619–5622. [CrossRef]
19. Narasimharao, K.; Al-Thabaiti, S.; Rajor, H.K.; Mokhtar, M.; Alsheshri, M.; Alfaifi, S.Y.; Siddiqui, S.I.; Abdulla, N.K. Fe_3O_4@Date Seeds Powder: A Sustainable Nanocomposite Material for Wastewater Treatment. *J. Mater. Res. Technol.* **2022**, *18*, 3581–3597. [CrossRef]
20. Chen, Z.; Ma, T.; Li, Z.; Zhu, W.; Li, L. Enhanced photocatalytic performance of S-scheme $CdMoO_4$/CdO nanosphere photocatalyst. *J. Mater. Sci. Technol.* **2024**, *179*, 198–207. [CrossRef]
21. Mohamed, S.H.; Hadia, N.M.A.; Diab, A.K.; Hakeem, A.M.A. Synthesis, Photoluminescence and Optical Constants Evaluations of Ultralong CdO Nanowires Prepared by Vapor Transport Method. *J. Alloys Compd.* **2014**, *609*, 68–72. [CrossRef]
22. Rajesh, N.; Kannan, J.C.; Leonardi, S.G.; Neri, G.; Krishnakumar, T. Investigation of CdO Nanostructures Synthesized by Microwave Assisted Irradiation Technique for NO_2 Gas Detection. *J. Alloys Compd.* **2014**, *607*, 54–60. [CrossRef]
23. Shahzadi, I.; Aqeel, M.; Haider, A.; Naz, S.; Imran, M.; Nabgan, W.; Al-Shanini, A.; Shahzadi, A.; Alshahrani, T.; Ikram, M. Hydrothermal Synthesis of Fe-Doped Cadmium Oxide Showed Bactericidal Behavior and Highly Efficient Visible Light Photocatalysis. *ACS Omega* **2023**, *8*, 30681–30693. [CrossRef]
24. Nallendran, R.; Selvan, G.; Balu, A.R. Cdo-Fe_3O_4 Nanocomposite with Enhanced Magnetic and Photocatalytic Properties. *Mater. Sci.-Pol.* **2019**, *37*, 100–107. [CrossRef]
25. Yu, H.; Zhao, M.; Xue, C.; Huang, J.; Zhao, N.; Kong, L. All-Solid-State Z-Scheme Nanojunction PW12/Ag/Zno Photocatalyst: Effective Carriers Transfer Promotion and Enhanced Visible Light Driven. *J. Mol. Str.* **2024**, *1300*, 137272. [CrossRef]
26. Gnanasekaran, L.; Hemamalini, R.; Rajendran, S.; Qin, J.; Yola, M.L.; Atar, N.; Gracia, F. Nanosized Fe_3O_4 Incorporated on a TiO_2 Surface for the Enhanced Photocatalytic Degradation of Organic Pollutants. *J. Mol. Liq.* **2019**, *287*, 110967. [CrossRef]
27. Ahmad, J.; Majid, K. Enhanced Visible Light Driven Photocatalytic Activity of CdO–Graphene Oxide Heterostructures for the Degradation of Organic Pollutants. *New J. Chem.* **2018**, *42*, 3246–3259. [CrossRef]
28. Hamad, H.; Elsenety, M.M.; Sadik, W.; Ghaffar, A.; Demerdash, E. The Superior Photocatalytic Performance and DFT Insights of S-scheme CuO@TiO_2 Heterojunction Composites for Simultaneous Degradation of Organics. *Sci. Rep.* **2022**, *21*, 2217. [CrossRef]
29. May, H.; Zhan, W.; Song, S.; Jia, F.; Zhou, J. Visible-Light Photocatalytic Degradation Pathway of Tetracycline Hydrochloride with Cubic Structured ZnO/SnO_2 Heterojunction Nanocatalyst. *Chem. Phys. Lett.* **2019**, *736*, 136806.
30. Sharma, A.; Kanth, S.; Xu, S.; Han, N.; Lei, Z.; Fan, L.; Liu, C.; Zhang, Q. Visible Light Driven G-C_3N_4/Bi_4NbO_8X (X = Cl, Br) Heterojunction Photocatalyst for the Degradation of Organic Pollutants. *J. Alloys Compd.* **2021**, *895*, 162576. [CrossRef]
31. Althabaiti, S.A.; Malik, M.A.; Kumar Khanna, M.; Bawaked, S.M.; Narasimharao, K.; Al-Sheheri, S.Z.; Fatima, B.; Siddiqui, S.I. One-Pot Facile Synthesis of CuO–$CdWO_4$ Nanocomposite for Photocatalytic Hydrogen Production. *Nanomaterials* **2022**, *12*, 4472. [CrossRef]
32. Mohamed, R.M.; Zaki, Z.I. Degradation of Imazapyr Herbicide Using Visible Light-Active CdO–TiO_2 Heterojunction Photocatalyst. *J. Environ. Chem. Eng.* **2021**, *9*, 104732. [CrossRef]
33. Fatima, B.; Alwan, B.A.; Siddiqui, S.I.; Ahmad, R.; Almesfer, M.; Khanna, M.K.; Mishra, R.; Ravi, R.; Oh, S. Facile Synthesis of Cu-Zn Binary Oxide Coupled Cadmium Tungstate (Cu-ZnBO-Cp-CT) with Enhanced Performance of Dye Adsorption. *Water* **2021**, *13*, 3287. [CrossRef]
34. Fatima, B.; Siddiqui, S.I.; Ahmad, R.; Linh, N.T.; Thai, V.N. CuO-ZnO-$CdWO_4$: A Sustainable and Environmentally Benign Photocatalytic System for Water Cleansing. *Environ. Sci. Pollut. Res.* **2021**, *28*, 53793–53803. [CrossRef]
35. Qin, M.-Z.; Fu, W.-X.; Guo, H.; Niu, C.-G.; Huang, D.-W.; Liang, C.; Yang, Y.-Y.; Liu, H.-Y.; Tang, N.; Fan, Q.-Q. 2D/2D Heterojunction Systems for the Removal of Organic Pollutants: A Review. *Adv. Colloid Interface Sci.* **2021**, *297*, 102540. [CrossRef]
36. Ghanbarnezhad, S.; Baghshahi, S.; Nemati, A.; Mahmoodi, M. Preparation, Magnetic Properties, and Photocatalytic Performance under Natural Daylight Irradiation of Fe_3O_4-ZnO Core/Shell Nanoparticles Designed on Reduced GO Platelet. *Mater. Sci. Semicond. Process.* **2017**, *72*, 85–92. [CrossRef]
37. Fu, J.R.; Zheng, J.; Fang, W.J.; Chen, C.; Cheng, C.; Yan, R.W.; Huang, S.G.; Wang, C.C. Synthesis of Porous Magnetic Fe_3O_4/Fe@ZnO Core-Shell Heterostructure with Superior Capability for Water Treatment. *J. Alloys Compd.* **2015**, *650*, 463–469. [CrossRef]
38. Karunakaran, C.; Vinayagamoorthy, P.; Jayabharathi, J. Nonquenching of Charge Carriers by Fe_3O_4 Core in Fe_3O/ZnO Nanosheet Photocatalyst. *Langmuir* **2014**, *30*, 15031–15039. [CrossRef]
39. Qamar, M.T.; Aslam, M.; Ismail, I.M.I.; Salah, N.; Hameed, A. The Assessment of the Photocatalytic Activity of Magnetically Retrievable ZnO Coated γ-Fe_2O_3 in Sunlight Exposure. *Chem. Eng. J.* **2016**, *283*, 656–667. [CrossRef]
40. Xia, J.; Wang, A.; Liu, X.; Su, Z. Preparation and Characterization of Bifunctional, Fe_3O_4/ZnO Nanocomposites and Their Use as Photocatalysts. *Appl. Surf. Sci.* **2011**, *257*, 9724–9732. [CrossRef]
41. Vinosel, V.M.; Asisi, S.A.M.; Pauline, J.S.; Praveena, S.D.P. Enhanced Photocatalytic Activity of Fe_3O_4/SnO_2 Magnetic Nanocomposite for the Degradation of Organic Dye. *J. Mater. Sci. Mater. Electron.* **2019**, *30*, 9663–9677. [CrossRef]

42. Dwivedi, P.; Jatrana, I.; Khan, A.U.; Khan, A.A.; Satiya, H.; Khan, M.; Moon, I.S.; Alam, M. Photoremediation of Methylene Blue by Biosynthesized ZnO/Fe_3O_4 Nanocomposites Using Callistemon Viminalis Leaves Aqueous Extract: A Comparative Study. *Nanotechnol. Rev.* **2021**, *10*, 1912–1925. [CrossRef]
43. Kong, L.; Liu, Y.; Dong, L.; Zhang, L.; Qiao, L.; Wang, W.; You, H. Enhanced red luminescence in $CaAl_{12}O_{19}$:Mn^{4+} via doping Ga3+ for plant growth lighting. *Dalton Trans.* **2020**, *49*, 1947–1954. [CrossRef]
44. Zhuang, L.; Zhang, W.; Zhao, Y.; Shen, H.; Lin, H.; Liang, J. Preparation and Characterization of Fe_3O_4 Particles with Novel Nanosheets Morphology and Magnetochromatic Property by a Modified Solvothermal Method. *Sci. Rep.* **2015**, *5*, 9320. [CrossRef]
45. Somasundaram, G.; Rajan, J.; Sangaiya, P.; Dilip, R. Hydrothermal Synthesis of CdO Nanoparticles for Photocatalytic and Antimicrobial Activities. *Results Mater.* **2019**, *4*, 100044. [CrossRef]
46. Guo, D.; Li, H.; Xu, Z.; Nie, Y. Development of Pyrene-Based Mofs Probe for Water Content and Investigations on Their Mechanochromism and Acidochromism. *J. Alloys Compds.* **2023**, *968*, 172004. [CrossRef]
47. Xitian, H.; Li, Z.; Xu, W.; Yan, P. Review on near-field detection technology in the biomedical field. *Adv. Photonics* **2023**, *2*, 44002.
48. Hai, N.H.; Phu, N.D.; Luong, N.H.; Chau, N.; Chinh, H.D.; Hoang, L.H.; Leslie-Pelecky, D.L. Mechanism for Sustainable Magnetic Nanoparticles under Ambient Conditions. *J. Korean Phys. Soc.* **2008**, *52*, 1327–1331. [CrossRef]
49. Yew, Y.P.; Shameli, K.; Miyake, M.; Bt Ahmad Khairudin, N.B.; Bt Mohamad, S.E.; Hara, H.; Bt Mad Nordin, M.F.; Lee, K.X. An Eco-Friendly Means of Biosynthesis of Superparamagnetic Magnetite Nanoparticles via Marine Polymer. *IEEE Trans. Nanotechnol.* **2017**, *16*, 1047–1052. [CrossRef]
50. Ganesh, V.; Manthrammel, M.A.; Shkir, M.; AlFaify, S. Investigation on Physical Properties of CdO Thin Films Affected by Tb Doping for Optoelectronics. *Appl. Phys. A Mater. Sci. Process.* **2019**, *125*, 1–9. [CrossRef]
51. Anitha, M.; Anitha, N.; Saravanakumar, K.; Kulandaisamy, I.; Amalraj, L. Effect of Zn Doping on Structural, Morphological, Optical and Electrical Properties of Nebulized Spray-Deposited CdO Thin Films. *Appl. Phys. A Mater. Sci. Process.* **2018**, *124*, 1–13. [CrossRef]
52. Kumar, S.; Ojha, A.K. Synthesis, Characterizations and Antimicrobial Activities of Well Dispersed Ultra-Long CdO Nanowires. *AIP Adv.* **2013**, *3*, 052109. [CrossRef]
53. Zargar, R.A. Fabrication and Improved Response of ZnO-CdO Composite Films under Different Laser Irradiation Dose. *Sci. Rep.* **2022**, *12*, 10096. [CrossRef] [PubMed]
54. Xue, Y.; Yang, T.; Liu, X.; Cao, Z.; Gu, J.; Wang, Y. Enabling efficient and economical degradation of PCDD/Fs in MSWIFA via catalysis and dechlorination effect of EMR in synergistic thermal treatment. *Chemosphere* **2023**, *342*, 140164. [CrossRef]
55. Kurnaz Yetim, N.; Kurşun Baysak, F.; Koç, M.M.; Nartop, D. Characterization of Magnetic Fe_3O_4@SiO_2 Nanoparticles with Fluorescent Properties for Potential Multipurpose Imaging and Theranostic Applications. *J. Mater. Sci. Mater. Electron.* **2020**, *31*, 18278–18288. [CrossRef]
56. Ai, Q.; Yuan, Z.; Huang, R.; Yang, C.; Jiang, G.; Xiong, J.; Huang, Z.; Yuan, S. One-Pot Co-Precipitation Synthesis of Fe_3O_4 Nanoparticles Embedded in 3D Carbonaceous Matrix as Anode for Lithium Ion Batteries. *J. Mater. Sci.* **2019**, *54*, 4212–4224. [CrossRef]
57. Li, W.; Li, M.; Xie, S.; Zhai, T.; Yu, M.; Liang, C.; Ouyang, X.; Lu, X.; Li, H.; Tong, Y. Improving the Photoelectrochemical and Photocatalytic Performance of CdO Nanorods with CdS Decoration. *CrystEngComm* **2013**, *15*, 4212–4216. [CrossRef]
58. Alzahrani, S.A.; Al-Thabaiti, S.A.; Al-Arjan, W.S.; Malik, M.A.; Khan, Z. Preparation of Ultra Long α-MnO_2 and Ag@MnO_2 Nanoparticles by Seedless Approach and Their Photocatalytic Performance. *J. Mol. Struct.* **2017**, *1137*, 495–505. [CrossRef]
59. Ibhadon, A.O.; Fitzpatrick, P. Heterogeneous Photocatalysis: Recent Advances and Applications. *Catalysts* **2013**, *3*, 189–218. [CrossRef]
60. Wang, J.; Yang, J.; Li, X.; Wang, D.; Wei, B.; Song, H.; Li, X.; Fu, S. Preparation and Photocatalytic Properties of Magnetically Reusable Fe_3O_4@ZnO Core/Shell Nanoparticles. *Phys. E Low-Dimens. Syst. Nanostruct.* **2016**, *75*, 66–71. [CrossRef]
61. Zhang, Y.; Qiu, L.; Yuan, Y.; Zhu, Y.; Jiang, X.; Xiao, J. Applied Catalysis B: Environmental Magnetic Fe_3O_4@C/Cu and Fe_3O_4@CuO Core–Shell Composites Constructed from MOF-Based Materials and Their Photocatalytic Properties under Visible Light. *Appl. Catal. B Environ.* **2014**, *144*, 863–869. [CrossRef]
62. Yao, K.F.; Peng, Z.; Liao, Z.H.; Chen, J.J. Preparation and Photocatalytic Property of TiO_2-Fe_3O_4 Core–Shell Nanoparticles. *J. Nanosci. Nanotechnol.* **2009**, *9*, 1458–1461. [CrossRef] [PubMed]
63. Dutta, S.; Banerjee, P.; Das, P.; Mukhopadhyay, A. Phytogenic Synthesis of Nanoparticles and Their Application in Photo Catalysis of Dye Rich Effluents. In *Photocatalytic Degradation of Dyes*; Elsevier: Amsterdam, The Netherlands, 2021; pp. 647–694. [CrossRef]
64. Senthil, S.; Srinivasan, S.; Thangeeswari, T.; Ratchagar, V. Enrichment of Optical, Magnetic and Photocatalytic Properties in PVP Capped CdO/SnO_2 Nanocomposites Synthesized by Microwave Irradiation Method. *J. Mater. Sci. Mater. Electron.* **2019**, *30*, 19841–19853. [CrossRef]
65. Liang, H.; Guo, J.; Shi, Y.; Zhao, G.; Sun, S.; Sun, X. Porous Yolk-Shell Fe/Fe_3O_4 Nanoparticles with Controlled Exposure of Highly Active Fe(0) for Cancer Therapy. *Biomaterials* **2021**, *268*, 120530. [CrossRef] [PubMed]
66. Karthik, K.; Dhanuskodi, S.; Gobinath, C.; Prabukumar, S.; Sivaramakrishnan, S. Photocatalytic and Antibacterial Activities of Hydrothermally Prepared CdO Nanoparticles. *J. Mater. Sci. Mater. Electron.* **2017**, *28*, 11420–11429. [CrossRef]
67. Akyüz, D. RGO-TiO_2-CdO-ZnO-Ag Photocatalyst for Enhancing Photocatalytic Degradation of Methylene Blue. *Opt. Mater.* **2021**, *116*, 111090. [CrossRef]

68. Ramesh, A.V.; Rama Devi, D.; Mohan Botsa, S.; Basavaiah, K. Facile Green Synthesis of Fe_3O_4 Nanoparticles Using Aqueous Leaf Extract of Zanthoxylum armatum DC. for Efficient Adsorption of Methylene Blue. *J. Asian Ceram. Soc.* **2018**, *6*, 145–155. [CrossRef]
69. Maruthupandy, M.; Muneeswaran, T.; Vennila, T. Development of Chitosan Decorated Fe_3O_4 Nanospheres for Potential Enhancement of Photocatalytic Degradation of Congo Red Dye Molecules. *Spectrochim. Acta Part A Mol. Biomol. Spectrosc.* **2022**, *267*, 120511. [CrossRef]
70. Tadjarodi, A.; Imani, M.; Kerdari, H.; Bijanzad, K.; Khaledi, D.; Rad, M. Preparation of CdO Rhombus-like Nanostructure and Its Photocatalytic Degradation of Azo Dyes from Aqueous Solution. *Nanomater. Nanotechnol.* **2014**, *4*, 4–16. [CrossRef]
71. Dhatshanamurthi, P.; Subash, B.; Shanthi, M. Investigation on UV-A Light Photocatalytic Degradation of an Azo Dye in the Presence of CdO/TiO_2 Coupled Semiconductor. *Mater. Sci. Semicond. Process.* **2015**, *35*, 22–29. [CrossRef]
72. Elshypany, R.; Selim, H.; Zakaria, K.; Moustafa, A.H.; Sadeek, S.A.; Sharaa, S.I.; Raynaud, P.; Nada, A.A. Elaboration of Fe_3O_4/ZnO Nanocomposite with Highly Performance Photocatalytic Activity for Degradation Methylene Blue under Visible Light Irradiation. *Environ. Technol. Innov.* **2021**, *23*, 101710. [CrossRef]
73. Długosz, O.; Szostak, K.; Krupiński, M.; Banach, M. Synthesis of Fe_3O_4/ZnO Nanoparticles and Their Application for the Photodegradation of Anionic and Cationic Dyes. *Int. J. Environ. Sci. Technol.* **2021**, *18*, 561–574. [CrossRef]
74. Fatima, B.; Siddiqui, S.I.; Rajor, H.K.; Malik, M.A.; Narasimharao, K.; Ahmad, R.; Vikrant, K.; Kim, T.; Kim, K.H. Photocatalytic Removal of Organic Dye Using Green Synthesized Zinc Oxide Coupled Cadmium Tungstate Nanocomposite under Natural Solar Light Irradiation. *Environ. Res.* **2023**, *216*, 114534. [CrossRef]
75. Zaied, M.; Peulon, S.; Bellakhal, N.; Desmazières, B.; Chaussé, A. Studies of N-Demethylation Oxidative and Degradation of Methylene Blue by Thin Layers of Birnessite Electrodeposited onto SnO_2. *Appl. Catal. B Environ.* **2011**, *101*, 441–450. [CrossRef]
76. Wang, X.; Liu, P.; Fu, M.; Ma, J.; Ning, P. Novel Sequential Process for Enhanced Dye Synergistic Degradation Based on Nano Zero-Valent Iron and Potassium Permanganate. *Chemosphere* **2016**, *155*, 39–47. [CrossRef] [PubMed]
77. Wang, Y.; Zhang, X.; He, X.; Zhang, W.; Zhang, X.; Lu, C. In Situ Synthesis of MnO_2 Coated Cellulose Nanofibers Hybrid for Effective Removal of Methylene Blue. *Carbohydr. Polym.* **2014**, *110*, 302–308. [CrossRef]
78. Zaheer, Z.; Bawazir, W.A.; Alwael, H.; Al-Jefri, F.M.; Salem, M. Decolorization and Mineralization of Methylene Blue by Potassium Permanganate. *J. Mol. Liq.* **2023**, *394*, 123794. [CrossRef]
79. Althabaiti, S.A.; Khan, Z.; Malik, M.A.; Bawaked, S.M.; Al-Sheheri, S.Z.; Mokhtar, M.; Siddiqui, S.I.; Narasimharao, K. Biomass-derived carbon deposited TiO2 nanotube photocatalysts for enhanced hydrogen production. *Nanoscale Adv.* **2023**, *5*, 3671–3683. [CrossRef]
80. Zaidi, Z.; Siddiqui, S.I.; Fatima, B.; Chaudhry, S.A. Synthesis of ZnO nanospheres for water treatment through adsorption and photocatalytic degradation: Modelling and process optimization. *Mater. Res. Bull.* **2019**, *120*, 110584. [CrossRef]

catalysts

Article

Electron Regulation in Pt-M (M = Cu, Co, and Ni) Decorated WO_3 Thin Films for Photocatalytic Degradation Performance

Jinkang Pan [1,2], Lihua Zhang [2,3], Cunxia Wang [2,3], Kangjie Gao [2,3] and Pengyu Dong [2,*]

1 School of Chemistry and Chemical Engineering, Yancheng Institute of Technology, Yancheng 224051, China; apanjingkang@163.com
2 Key Laboratory for Advanced Technology in Environmental Protection of Jiangsu Province, Key Laboratory for Ecological-Environment Materials of Jiangsu Province, Yancheng Institute of Technology, Yancheng 224051, China; zlh1670@163.com (L.Z.); wcx1538975755@163.com (C.W.); gkangjie0922@163.com (K.G.)
3 School of Mechanical Engineering, Yancheng Institute of Technology, Yancheng 224051, China
* Correspondence: dongpy11@gmail.com or dongpy11@ycit.edu.cn

Abstract: In this study, Pt-M/WO_3 (M = Cu, Co, and Ni) thin films are effectively synthesized by preparing homogeneous precursor sols, spin-coating, toluene-etching, and calcination. Furthermore, the microstructural, chemical, and electrochemical properties of the WO_3, Pt-Cu/WO_3, Pt-Co/WO_3, and Pt-Ni/WO_3 thin films are also systematically compared. The results demonstrate that when compared to the WO_3 thin film, the photocatalytic capability for methylene blue (MB) solution degradation is greatly increased in the Pt-M/WO_3 thin films. Transfer routes for photogenerated charges and an improved photocatalytic process are suggested based on the experimental results. Due to the large difference in the work function (Φ) between the bimetallic alloy Pt-M and WO_3, a bending of the energy bands at the Pt-M/WO_3 interface is presented. Furthermore, the introduction of transition metals such as Cu, Co, or Ni modifies the electronic structure of Pt-M/WO_3 thin films, facilitating the separation and migration of electrons and holes. Specifically, the photogenerated electrons migrate from the CB of WO_3 to Pt-Co or Pt-Ni nanoparticles in the samples of Pt-Co/WO_3 or Pt-Ni/WO_3 thin films, while the hot electrons from the localized surface plasmon resonance (LSPR) effect of Cu could transfer to the conduction band (CB) of WO_3 and other electrons generated from the photoexcitation of the WO_3 semiconductor itself in the sample of the Pt-Cu/WO_3 thin film. In summary, this work proposes a unique strategy for creating electron regulation in Pt-M decorated WO_3 thin films for photocatalytic application.

Keywords: photocatalytic degradation; WO_3 thin films; Pt-based bimetallic cocatalysts; electron regulation

1. Introduction

For the possibility of using them to degrade organic contaminants, semiconductor-based photocatalysts have been the subject of intense study during the last several years [1–3]. With an energy bandgap (E_g) of around 3.2 eV, titanium dioxide (TiO_2) is a well-known conventional semiconducting photocatalyst [4]. Nevertheless, because of its large bandgap, it hardly responds even weakly to ultraviolet light, which only makes up about 4% of sunlight's total energy. Furthermore, the easy recombination of electron–hole pairs produced by light reduces the efficiency of photocatalysis.

As one of the n-type semiconductors and functional materials, tungsten trioxide (WO_3) has been extensively investigated in recent decades and has a wide range of potential applications [5–7]. Due to its smaller bandgap than the popular TiO_2, which possesses a small bandgap of between 2.4 and 2.8 eV and enables it to absorb more solar energy and produce a larger photocurrent, WO_3 has recently attracted attention [8]. In-depth research has shown that the strongly negative charges on the surfaces of WO_3 make them perfect for use in adsorption applications, especially for cationic dyes such as methylene blue (MB) [9]. Due to its low conduction band (CB) level, WO_3 has a low reduction potential of its electrons, resulting in a lower light energy conversion efficacy than TiO_2 [10]. Researchers have suggested a number of methods for boosting the photocatalytic properties of WO_3, including supplementation with other elements [11], surface modification with metal nanoparticles [12,13], and the creation of hybrid composites with other materials, to bypass this restriction [14,15]. These techniques may boost the stability, photocatalytic activity, and light absorption characteristics of WO_3, as well as its photocatalytic efficiency.

Recently, it has been shown that coating WO_3 semiconductors with noble metals is a successful way to increase their photocatalytic activity. Under UV-light irradiation, WO_3 loaded with Pt nanoparticles displayed photocatalytic activity nearly on par with TiO_2 [16,17]. When compared to Pt monometallic cocatalysts, in a lack of light irradiation, Pt-based bimetallic cocatalysts demonstrated a distinct collaborative impact and substantially enhanced their catalytic efficiency. For instance, our group has demonstrated that the Pt-Ag/WO_3 thin film has a substantially improved photocatalytic performance for MB solution degradation [18]. There has also been research on other Pt-based bimetallic cocatalysts, including Pt-Au [19], Pt-Ru [20], and Pt-Pd [21,22]. Non-precious metals, on the other hand, are being thought of as the second metal to be alloyed with platinum owing to the high price of precious metals. Notably, alloys of metals such as Cu, Co, and Ni with Pt have shown a clear synergistic effect. The interaction between the *d* electron orbitals of these transition metals and Pt leads to electron redistribution, enabling precise control over the electronic structure of Pt. The tunability allows for the modulation of photocatalytic properties and reaction activity, ultimately enhancing the efficiency of photocatalytic reactions. Furthermore, the incorporation of Cu, Co, or Ni within the nanoparticle architecture allows for better utilization of Pt atoms by minimizing the core Pt wastage. This strategy enables a significant reduction in Pt loading without compromising the photocatalytic degradation performance, thus offering a promising avenue for cost reduction. These alloys also have the advantages of being easily accessible and environmentally safe, which reduces the cost of the bimetallic electrocatalysts even more [23–26]. However, there is little research on the bimetallic Pt-M (M represents a non-precious metal element) cocatalyst-decorated WO_3 thin film for photocatalytic degradation.

In this work, the Pt-M/WO_3 (M = Cu, Co, and Ni) thin films were created by dip-coating deposition of sol-gel precursors onto WO_3 as the support material and incorporating Cu, Co, and Ni into the Pt alloy. The electrical alteration of the surface Pt atoms and the uniform distribution of the resulting bimetallic cocatalysts led to considerable synergistic effects and increased catalytic activity. The presence of Pt could facilitate the migration of photoexcited electrons, and the introduction of these transition metals could result in modifications of the electronic structure, band alignment, and surface properties of the Pt-M/WO_3 composite. This might lead to improved charge separation, increased visible-light absorption, and enhanced redox reactions, all contributing to the efficient photocatalytic degradation of MB. We suggest paths for photogenerated charge transfer and photocatalytic processes by a methodical comparative examination of the microstructural, chemical, and photoelectrochemical properties of the WO_3, Pt-Cu/WO_3, Pt-Co/WO_3, and Pt-Ni/WO_3 thin films.

2. Results and Discussion

2.1. XRD and Raman Analysis

X-ray diffraction (XRD) patterns were used to identify the crystal phases of the indium tin oxide (ITO) glass and the thin films of WO_3, Pt-Cu/WO_3, Pt-Co/WO_3, and Pt-Ni/WO_3 while they were being manufactured, as shown in Figure 1a. Except for the ITO glass peaks, the results clearly showed that all of the thin films exhibited polycrystalline structures with peaks that corresponded to the cubic phase of WO_3 (JCPDS 41-0905). Due to the relatively thin film thickness, it was found that the ITO glass peak was detected in all the as-prepared thin films. Additionally, no Pt-Cu, Pt-Co, and Pt-Ni diffraction peaks were seen in the

Pt-Cu/WO_3, Pt-Co/WO_3, and Pt-Ni/WO_3 thin films, respectively, suggesting that the composite thin films contained relatively little Pt-M (M = Cu, Co, and Ni).

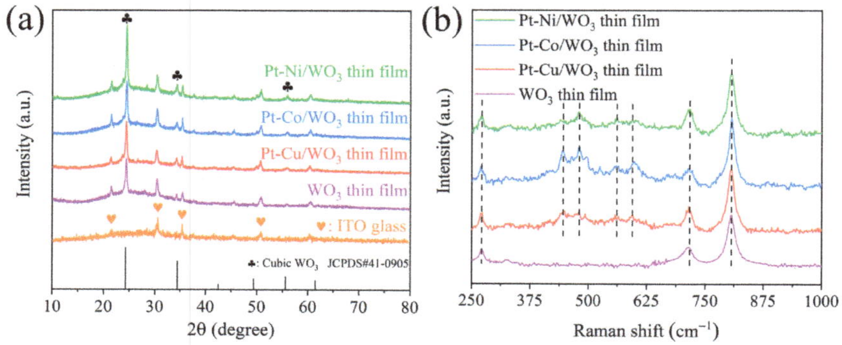

Figure 1. (**a**) XRD patterns (the orange symbols correspond to the diffraction peaks of ITO glass) and (**b**) Raman spectra of the WO_3, Pt-Cu/WO_3, Pt-Co/WO_3, and Pt-Ni/WO_3 thin films.

Figure 1b displays the Raman spectra for all the as-prepared thin films. The peak at 269 cm^{-1} was attributed to the W–O stretching vibrations of the monoclinic WO_3 phase. The major Raman modes seen in all thin films were found at 714 and 806 cm^{-1}, which corresponded to the W–O stretching vibrations [27,28]. In the low wavenumber range, the Raman bands at 445 and 478 cm^{-1} were ascribed to PtO [29] and Pt–CO stretching with on-top configurations [30], respectively. The overlapping band of the stretching vibrations of Pt–OH and Pt–O ($\nu_{Pt\text{–}O(H)}$) was thus allocated to the bands at 560 and 597 cm^{-1}, respectively [31,32]. These findings support the XRD findings that the obtained bimetallic Pt-M/WO_3 thin films have nearly identical Raman bands.

2.2. XPS Analysis

X-ray photoelectron spectroscopy (XPS) analysis was employed to investigate the chemical binding energies of the WO_3, Pt-Cu/WO_3, Pt-Co/WO_3, and Pt-Ni/WO_3 thin films. The full spectra for all as-prepared thin films revealed the presence of C, O, W, Pt, Cu, Co, and Ni elements, as shown in Figure 2a. The W 4f spectra in Figure 2b showed double peaks at approximately 35.1 and 37.3 eV for WO_3 thin films, which were attributed to W $4f_{7/2}$ and W $4f_{5/2}$, respectively [33]. In comparison with the WO_3 thin film, the binding energies of the W $4f_{7/2}$ and W $4f_{5/2}$ peaks for Pt-Cu/WO_3, Pt-Co/WO_3, and Pt-Ni/WO_3 thin films increased by approximately 0.4 eV, indicating a chemical interaction between Pt-M (M = Cu, Co, and Ni) and WO_3 in the thin films, respectively. In accordance with the Raman findings, the XPS analysis of Pt 4f for the Pt-M/WO_3 (M = Cu, Co, and Ni) thin films (Figure 2c) showed two splitting peaks that were assigned to the Pt (II) chemical state (PtO and $Pt(OH)_2$) in all as-prepared thin films [34,35]. When compared to the typical Pt 4f peak, the binding energies of the Pt-Cu, Pt-Co, and Pt-Ni peaks increased [36,37]. In Figure 2d, the Cu $2p_{3/2}$ and Cu $2p_{1/2}$ binding energies of Pt-Cu/WO_3 were determined to be at 932.0 and 952.0 eV, respectively [38,39]. Cu (0) was identified by the greater peak at a lower binding energy of 932.0 eV, whereas Cu (II) and (I) were identified by the weaker peaks at higher binding energies of 934.0 eV and 952.0 eV, respectively [40,41]. Additionally, strong shake-up satellite features at 942.0 eV for Cu $2p_{3/2}$ and 955 eV for Cu $2p_{1/2}$ were observed as a fingerprint of the Cu^{2+} species [42]. The peaks of the Co 2p XPS spectrum in Figure 2e were attributed to Co $2p_{1/2}$ and Co $2p_{3/2}$, which represent the species of Co_2O_3 and CoOOH, respectively [43,44]. In Figure 2f, the high-resolution Ni 2p XPS spectrum shows fitting peaks at 878.0 and 859.1 eV, which belonged to Ni $2p_{1/2}$ and Ni $2p_{3/2}$, respectively, and they were ascribed to NiO and $Ni(OH)_2$ [45,46]. Consequently, these findings support the XPS analysis' finding that the Pt-Cu (Ni, Co) alloy contains oxide species.

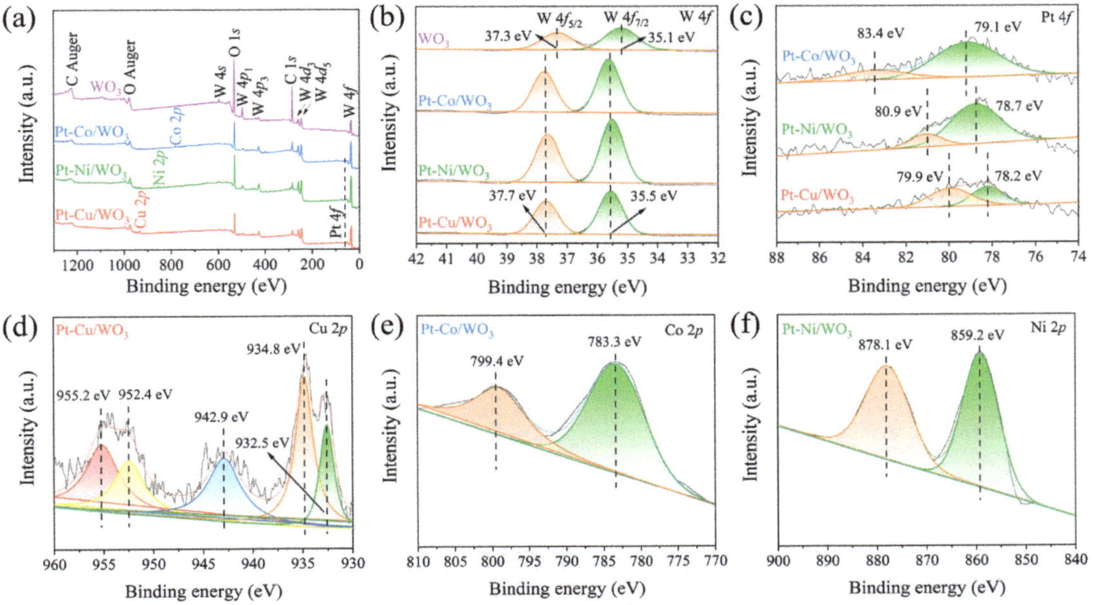

Figure 2. Full XPS spectra (**a**), high-resolution W 4f (**b**), Pt 4f (**c**), Cu 2p (**d**), Co 2p (**e**), and Ni 2p (**f**) XPS spectra of the WO_3, Pt-Cu/WO_3, Pt-Co/WO_3, and Pt-Ni/WO_3 thin films.

2.3. Morphological Features

For the purpose of assessing the surface morphological properties of the thin films in their as-prepared state, scanning electron microscopy (SEM) was used. Figure 3a–d reveals that the WO_3 thin film consisted of compact spherical particles with a size of approximately 50 nm [47]. On the other hand, we demonstrated that the Pt-Cu/WO_3, Pt-Cu/WO_3, and Pt-Ni/WO_3 thin films exhibited uniform and compact blackberry-like morphologies with little roughness. Moreover, SEM images of the cross-section of the films are provided in Figure 3. It was observed that the thickness of the pure WO_3, Pt-Cu/WO_3, Pt-Co/WO_3, and Pt-Ni/WO_3 thin films was about 66, 68, 75, and 72 μm, respectively. In addition, the corresponding transmission electron microscopy (TEM) and high-resolution TEM (HRTEM) images of the Pt-Cu/WO_3 thin film are presented in Figure S1 in the Supplementary Material. The HRTEM image revealed distinct lattice fringes corresponding to d-spacing values of 0.371 nm and 0.220 nm (Figure S1c), which can be attributed to the (100) plane of WO_3 and the (111) plane of the Pt-Cu alloy, respectively [34].

2.4. Optical Absorption Characteristics

Figure 4a shows the optical absorption characteristics of the different photocatalysts, as evaluated by UV-vis diffuse reflectance absorption (DRS) spectra. The inter-band optical transition of the WO_3 thin film is responsible for the prominent peaks seen in all thin films below 400 nm [48]. The Pt-Cu/WO_3, Pt-Co/WO_3, and Pt-Ni/WO_3 thin films showed an increased absorbance in the visible-light range, which could be owing to the Pt-M particulates scattering light [49,50]. The Pt-Cu/WO_3 thin film had a significantly widened and intense peak in the 430–550 nm range, suggesting that localized surface plasmon resonance (LSPR) may be responsible for the photothermal effect being shown by the Pt-Cu nanoparticles [38]. Notably, the Cu particles' plasmon resonance absorbance was not obvious, owing to the reduced Cu content of the Pt-Cu/WO_3 thin film. Furthermore, the WO_3, Pt-Co/WO_3, and Pt-Ni/WO_3 thin films exhibited comparable absorption spectra to the Pt-Cu/WO_3 thin film, likely owing to the weakly expressed texture and microcrystalline structure of the ITO films [51,52]. The following equation was used to obtain the E_g of all the thin films using the value regarding absorption:

$$\alpha h\nu = A(h\nu - E_g)^n \tag{1}$$

The absorption coefficient (α), absorption constant (A), and constant (n) are interdependent and vary with the type of electronic transition involved. For direct transitions, the value of n is 1/2, whereas for indirect transitions, n is 2. The intersection with the tangent in the plots of $(\alpha h\nu)^{1/2}$ against the energy of photons, $h\nu$, was used to determine the E_g values of the thin films since WO_3 showed an indirect bandgap (Figure 4b). According to our research, the WO_3, Pt-Cu/WO_3, Pt-Co/WO_3, and Pt-Ni/WO_3 thin films had E_g values of 2.63, 2.54, 2.60, and 2.57 eV, respectively. Future research is required to ascertain whether alterations in the electronic structure of the WO_3 film result in a reduction in the bandgap energy of the Pt-Cu/WO_3, Pt-Co/WO_3, and Pt-Ni/WO_3 thin films relative to the WO_3 film.

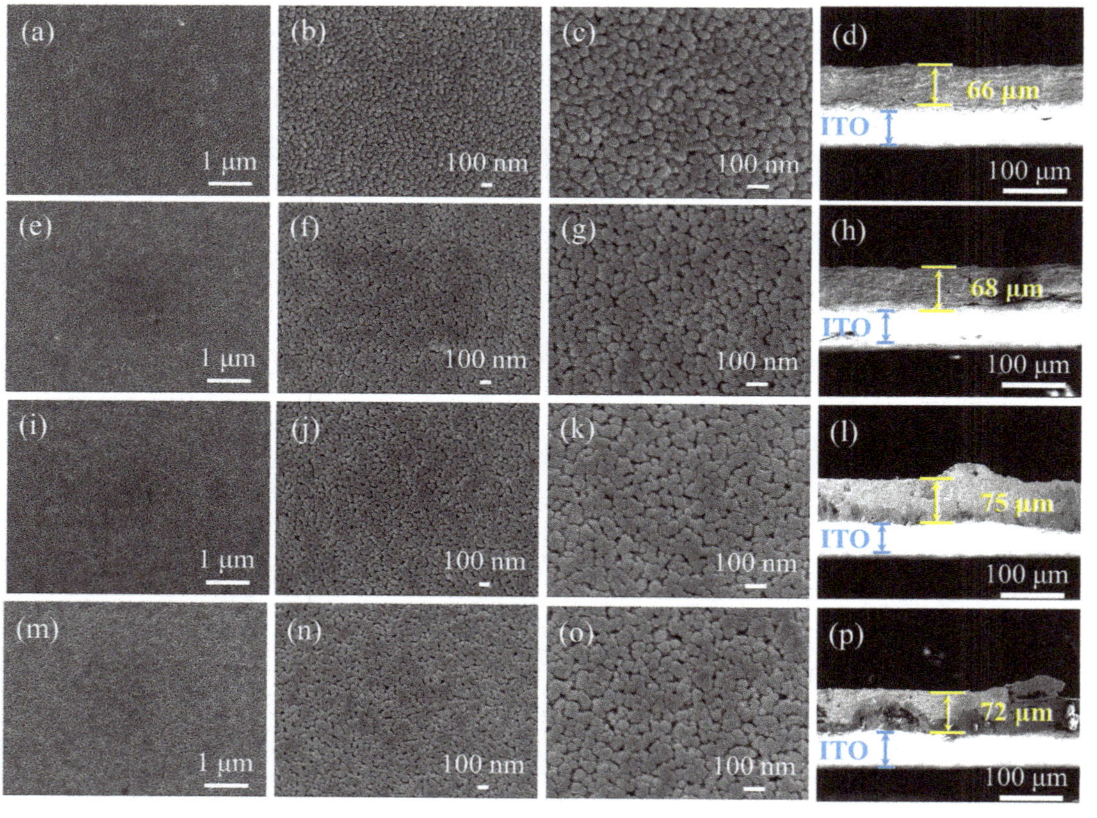

Figure 3. SEM images of (**a**–**d**) WO_3, (**e**–**h**) Pt-Cu/WO_3, (**i**–**l**) Pt-Co/WO_3, and (**m**–**p**) Pt-Ni/WO_3 thin films.

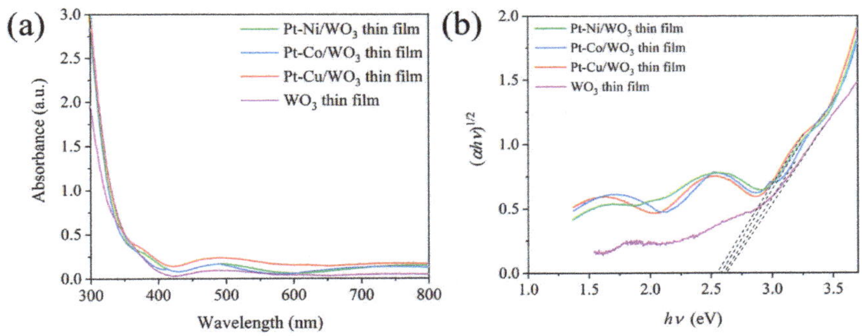

Figure 4. (**a**) UV-vis DRS spectra of the WO_3, Pt-Cu/WO_3, Pt-Co/WO_3, and Pt-Ni/WO_3 thin films (the inset shows the region from 400 to 800 nm). (**b**) Plots of $(\alpha h\nu)^{1/2}$ versus photon energy $h\nu$ of these thin films.

2.5. Photocatalytic Degradation Performance

The adsorption isotherms before the irradiation process of Pt-M/WO_3 were measured and provided in Figure S2. It was observed that adsorption–desorption equipment was achieved within 40 min in the darkness. Due to its low self-degradation under visible light, the pollutant MB was selected to test the photocatalytic efficiency of the synthesized thin films. The efficiency of the photocatalytic degradation of the MB solution measured and plotted in the presence of WO_3, Pt-Cu/WO_3, Pt-Co/WO_3, and Pt-Ni/WO_3 thin films is shown in Figure 5a. With a degradation efficiency of 98.9% after 120 min, the Pt-Cu/WO_3 thin film exhibited the greatest photocatalytic efficiency of all the thin films tested, whereas the WO_3 thin film only managed a 60% degradation efficiency under the same circumstances. Such a high photocatalytic efficiency value for the Pt-Cu/WO_3 thin film exceeded most of the previously reported WO_3-based photocatalysts (Table S1). The Pt-Co/WO_3 and Pt-Ni/WO_3 thin films also exhibited good photocatalytic activity, with degradation efficiencies of approximately 88% and 90%, respectively. Despite the low initial concentration of MB (5 mg L^{-1}), it is crucial to remember that the total degradation of MB under the condition of the Pt-Cu/WO_3 thin film required a lengthy irradiation time of nearly 120 min. This was primarily produced by placing the thin film photocatalyst on a substrate, which resulted in reduced light absorption and pollutant–molecule interactions. As a result, thin-film photocatalysts usually have lower photocatalytic activity than powder photocatalysts. While conventional powder photocatalysts could undergo laborious recycling procedures, thin-film photocatalysts have the advantages of easy reuse and minimal loss during photocatalytic reactions.

By employing the Langmuir–Hinshelwood model, it was found that the degradation of dyestuffs occured through a pseudo-first-order reaction, as long as the initial concentration (C_0) of the dyestuff solution was low. This model assumes that the photocatalytic degradation proceeds via a rapid surface adsorption and reaction step, with the reaction rate being proportional to the concentration of MB. The concentration of the Pt-M/WO_3 catalyst had a minor influence on the reaction rate in this model [53–55]:

$$\ln(C_0/C) = kt \tag{2}$$

where C, C_0, k, and t are the remaining concentration of MB solution, the initial concentration, the rate constant of a first-order reaction, and the irradiation time, respectively. Among them, the Langmuir–Hinshelwood model was utilized in the pseudo-first-order reaction kinetics with a 5 mg L^{-1} initial concentration of MB in our study. The graph of $\ln(C_0/C)$ vs. irradiation time is presented in Figure 5b, and a steeper slope implies a quicker rate of MB solution degradation. Using the Langmuir–Hinshelwood model, the rate constant of a first-order reaction was estimated. The rate constant of a first-order reaction in the

Pt-Cu/WO_3 thin film was discovered to be approximately 6 times larger compared to that found in the WO_3 thin film. Furthermore, the correlation coefficient (R^2) was used to evaluate the relationship between the actual practical photocatalytic reaction data and the linear fitting model [56]. It was observed that all the fabricated thin films exhibited R^2 values exceeding 0.9, suggesting that the experimental results conformed to the first-order kinetics.

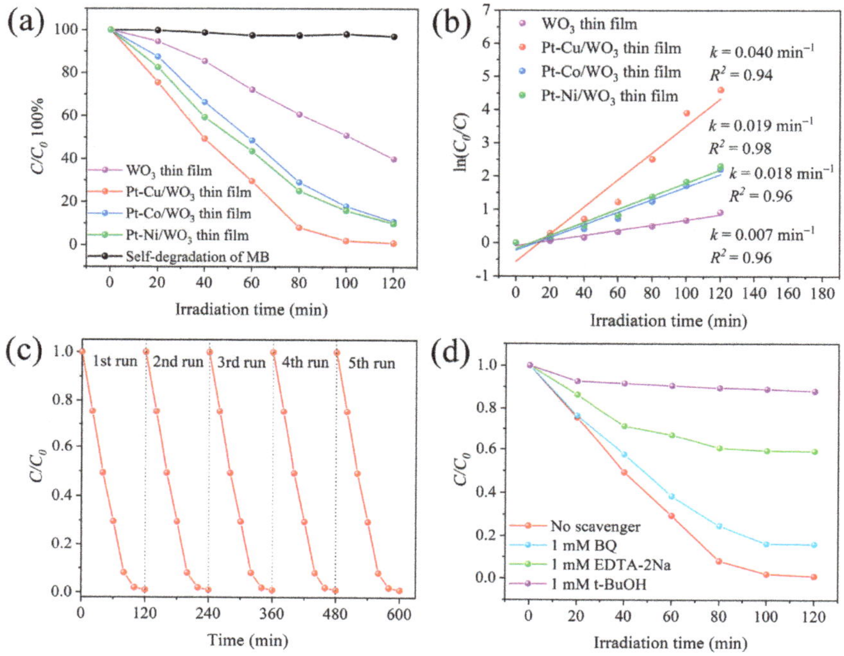

Figure 5. (**a**) Photocatalytic degradation ratio of MB solution in the presence of WO_3, Pt-Cu/WO_3, Pt-Co/WO_3, and Pt-Ni/WO_3 thin films. (**b**) Plots of ln(C_0/C) versus irradiation time. (**c**) Recycling experiments towards MB degradation over the Pt-Cu/ WO_3 thin film. (**d**) Plots of photogenerated carrier-trapping during the photodegradation of MB by the Pt-Cu/WO_3 thin film under visible-light irradiation.

MB was degraded in numerous cycling cycles in order to evaluate the recycling photostability of the Pt-Cu/WO_3 thin-film catalyst, as shown in Figure 5c. The results showed that even after five cycles, the Pt-Cu/WO_3 thin film still maintained an excellent degradation rate under light irradiation. The photocatalytic stability of the Pt-Cu/WO_3 thin film for MB degradation was remarkable, according to our unmistakable data.

It is crucial to carry out free radical and hole-trapping studies to identify these species in order to assure the formation of active species during the photodegradation process in photocatalytic reactions. In our study, we utilized ethylene diamine tetraacetic acid disodium salt (EDTA-2Na), 1,4-benzoquinone (BQ), and *tert*-butyl alcohol (*t*-BuOH) as scavengers to trap the superoxide anion radical ($\bullet O_2^-$), the hole (h^+), and the hydroxy radical (•OH), respectively [57–59]. Figure 5d illustrates how the various scavengers affected the photocatalytic activity of the Pt-Cu/WO_3 thin film for MB photodegradation under visible light. The reaction system with the addition of BQ (1 mM) did not noticeably inhibit the photodegradation efficiency of MB, compared to the Pt-Cu/WO_3 thin-film system without any scavenger. However, the photodegradation efficiency of MB was clearly decreased after adding EDTA-2Na (1 mM) to the system. Additionally, the photocatalytic degradation activity of the Pt-Cu/WO_3 thin film also showed a significant decrease after adding *t*-BuOH (1 mM) to the system. These research results indicate that under visible-light exposure, •OH was the primary active species that facilitated the oxidization of MB, while h^+ was also one of the reactive species. Furthermore, in the Pt-Cu/WO_3 thin-film photocatalytic degradation system, $•O_2^-$ was produced in negligible amounts.

2.6. Photogenerated Chargers Separation and Migration

The photoelectrochemical (PEC) characteristics of the several as-prepared thin films were assessed under AM 1.5 G exposure at 100 mW cm^{-2} in the 0.5 M Na_2SO_4 solution. The linear sweep voltammetry (LSV) curves of the as-prepared thin films are shown in Figure 6a. The photocurrent density of the WO_3 thin film was 0.65 mA cm^{-2} at 1.23 V vs. RHE. In contrast, the photocurrent density of Pt-Cu/WO_3, Pt-Co/WO_3, and Pt-Ni/WO_3 thin films was ca. 5.81 mA cm^{-2}, 5.01 mA cm^{-2}, and 4.74 mA cm^{-2}, respectively, at 1.23 V vs. RHE, which were almost 8.9, 7.7, and 7.3 times higher than that of the WO_3 thin film. These findings imply that the Pt-Cu/WO_3, Pt-Co/WO_3, and Pt-Ni/WO_3 thin films could effectively reduce electron–hole recombination in the PEC reaction and increase effective electron mobility, resulting in improved photocatalytic properties. Furthermore, the slopes of the curves for Pt-Cu/WO_3, Pt-Co/WO_3, and Pt-Ni/WO_3 thin films in the 0.9 to 1.2 V vs. RHE region were significantly steeper than that of the bare WO_3 thin film, indicating that these thin films had higher carrier-filling factors.

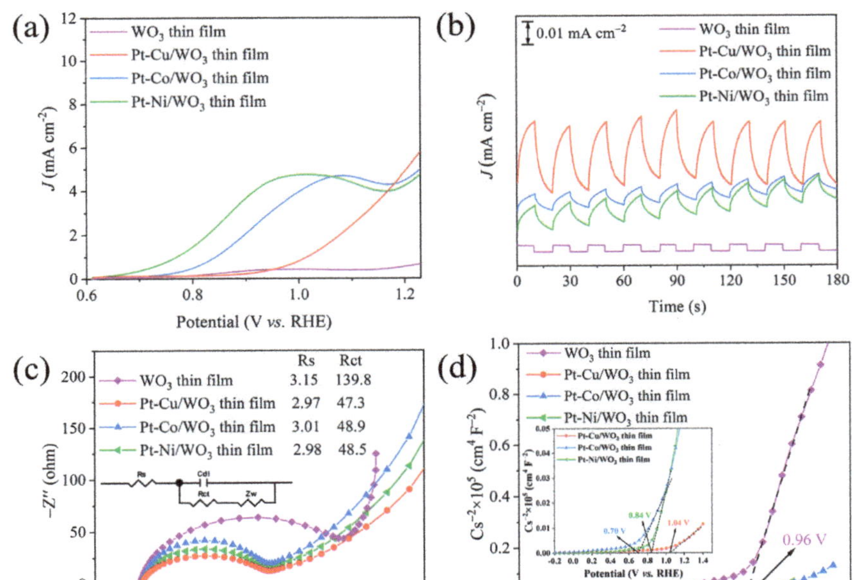

Figure 6. (**a**) Comparison of LSV curves of WO_3, Pt-Cu/WO_3, Pt-Co/WO_3, and Pt-Ni/WO_3 thin films. (**b**) Transient photocurrent responses in several on–off cycles. (**c**) Nyquist plots of EIS of as-prepared thin films with AM 1.5 G and bias of 1.23 V vs. RHE. (**d**) Mott–Schottky plots of all thin films measured in the dark at 1 kHz (the inset is the enlarged part for the three thin films).

The transient photocurrent responses of the different thin films were examined under chopped AM 1.5 G exposure with a bias of 0.2 V in order to examine the effectiveness of charge separation, as shown in Figure 6b. The transient photocurrent density describes the sequence of Pt-Cu/WO_3, Pt-Ni/WO_3, Pt-Co/WO_3, and WO_3 among the as-prepared thin films. The Pt-Cu/WO_3 thin film had the strongest photocurrent response in numerous on–off cycles, which was notable. Additionally, when the light source was turned on and off, distinct transient anodic peaks could be seen in the Pt-Cu/WO_3, Pt-Co/WO_3, and Pt-Ni/WO_3 thin films. This suggests that photogenerated holes from the WO_3 thin film may be stored in the Pt-Cu, Pt-Co, and Pt-Ni layers, preventing the recombination of photogenerated electron–hole pairs. This result suggests that the Pt-Cu, Pt-Co, and Pt-Ni layers could act as efficient hole storage layers for capturing and accumulating photogenerated holes.

Through electrochemical impedance spectroscopy (EIS), a comparison of the semicircular Nyquist plot diameter was performed, and the resultant Nyquist plots were matched to assess the electron transfer resistance (R_{ct}) with an analogous circuit using Zview software [60]. The fitting results showed that the R_{ct} values of Pt-Cu/WO_3 (47.3 Ω), Pt-Co/WO_3 (48.9 Ω), and Pt-Ni/WO_3 (48.5 Ω) were considerably lower than that of WO_3 (139.8 Ω), as presented in Figure 6c. The incorporation of Pt-Cu, Pt-Co, and Pt-Ni alloy nanoparticles significantly improved the charge transfer because they were more efficient at capturing electrons from the CB of WO_3, as shown by the fact that the Pt-Cu/WO_3, Pt-Co/WO_3, and Pt-Ni/WO_3 thin films exhibited a smaller semicircle than the WO_3 thin-film electrodes.

To determine the semiconductor type and the flat-band potentials (E_{fb}), Mott–Schottky plots were employed, and the findings are depicted in Figure 6d. The calculated E_{fb} values of WO_3, Pt-Cu/WO_3, Pt-Co/WO_3, and Pt-Ni/WO_3 thin films were 0.96, 1.04, 0.70, and 0.84 V vs. RHE, respectively, and were similar to each other. The positive slopes of all four curves implied that the thin films were composed of n-type semiconductors. It should be noted that the slope of the Pt-Cu/WO_3 thin-film curve was lower than that of the WO_3, Pt-Co/WO_3, and Pt-Ni/WO_3 thin-film curves, which implies that the concentration of carriers in the former was higher and the transfer rate of photogenerated electrons was faster. This is important to note because there is an inverse correlation between the slopes and the charge carrier density [61]. These results showed that charge transfer was better-facilitated by Pt-Cu/WO_3, Pt-Co/WO_3, and Pt-Ni/WO_3 thin films than by WO_3, and that the photocatalytic activity was significantly increased.

Based on the information that is currently accessible, the energy-level diagram of WO_3, Pt-Cu/WO_3, Pt-Co/WO_3, and Pt-Ni/WO_3 thin films was provided. Figure 7 illustrates the energy-level diagram obtained from the Mott–Schottky plot, where the positive slope (Figure 6d) suggests that the material exhibited n-type semiconductor properties, and the conduction band minimum (E_{CBM}) was approximately equal to the E_{fb}. Additionally, the bandgaps were determined as 2.63, 2.54, 2.60, and 2.57 eV, respectively, by extrapolating the intercept on the x-axis in the linear region of the Tauc plot (Figure 4b). Consequently, the valence band positions (E_{VB}) of WO_3, Pt-Cu/WO_3, Pt-Co/WO_3, and Pt-Ni/WO_3 thin films were inferred to be 3.59, 3.58, 3.30, and 3.41 V, respectively.

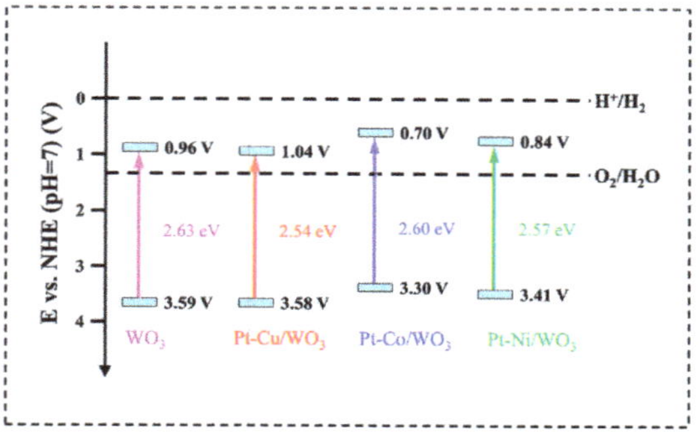

Figure 7. Schematic energy-level diagram of WO_3, Pt-Cu/WO_3, Pt-Co/WO_3, and Pt-Ni/WO_3 thin films.

2.7. Photocatalytic Mechanism Analysis

Understanding the increased photocatalytic activity requires research into the improved photocatalytic mechanism. We have presented a thorough description based on the band structure and work functions in this study. The work functions (Φ) of the WO_3 thin film, metallic Pt, metallic Cu, metallic Co, and metallic Ni were reported to be 5.7 eV, 5.65 eV, 4.65 eV, 5.0 eV, and 5.15 eV, respectively [62–65]. Moreover, the work function of the bimetallic alloy Pt-M (M = Cu, Co, and Ni) can be calculated as the sum of the work functions of the two metals [66], indicating that the work function of the alloy is at an intermediate level between two monometallic Pt and Cu (Co or Ni) (4.65 eV < $\Phi_{Pt\text{-}Cu}$ < 5.65 eV, 5.0 eV < $\Phi_{Pt\text{-}Co}$ < 5.65 eV, and 5.15 eV < $\Phi_{Pt\text{-}Cu}$ < 5.65 eV). When Pt-M (M = Cu, Co, and Ni) came into contact with WO_3, the charges transferred from Pt-M (M = Cu, Co, and Ni) to the CB of WO_3 to achieve Fermi-level equilibrium. Consequently, the accumulation of excess electrons on the surface area of WO_3 and the excess of positive charge by Pt-M (M = Cu, Co, and Ni) led to a bending of the energy bands at the Pt-M/WO_3 interface (M = Cu, Co, and Ni). As demonstrated in Scheme 1a–c, the distinction between the WO_3 and bimetallic work functions ($\Phi_{WO_3} - \Phi_{Pt\text{-}M}$) at the Pt-Cu/$WO_3$, Pt-Co/$WO_3$, or Pt-Ni/$WO_3$ juncture was merely approximately 0.05–1.05 eV, 0.05–0.7 eV, and 0.05–0.55 eV, respectively. Under visible-light illumination, when the WO_3 thin-film catalyst was exposed, the photoinduced electrons (e^-) in the valence band (VB) of WO_3 were promoted to the CB, and at the same time, an equal number of photoinduced h^+ were produced in the VB. The introduction of transition metals such as Cu, Co, or Ni modified the electronic structure of Pt-M/WO_3 thin films, facilitating the separation and migration of electrons and holes. Specifically, the photogenerated electrons migrated from the CB of WO_3 to Pt-Co or Pt-Ni nanoparticles in the samples of Pt-Co/WO_3 or Pt-Ni/WO_3 thin films, while the hot electrons from the LSPR effect of Cu could transfer to the CB of WO_3 and other electrons generated from the photoexcitation of the WO_3 semiconductor itself in the sample of Pt-Cu/WO_3 thin films. Additionally, in the space charge region, the rapidly occurring transport of the photoinduced electrons from WO_3 to Pt-M (M = Cu, Co, and Ni) nanoparticles was rendered achievable by the downward band bending, lowering the recombining of the photoexcited charges and holes. The low CB energy level (1.04 V, 0.70 V, and 0.84 V) of Pt-Cu/WO_3, Pt-Co/WO_3, and Pt-Ni/WO_3 thin films was more positive compared to the reduction potential of oxygen for one electron [16], indicating that WO_3 is more capable of accepting electrons. Consequently, the accumulated electrons on Pt-Cu/WO_3, Pt-Co/WO_3, or Pt-Ni/WO_3 nanoparticles may transfer to oxygen molecules adsorbed on the surface, forming H_2O_2 through the multi-electron reduction of molecular oxygen [8]. In the Pt-Cu/WO_3, Pt-Co/WO_3, and Pt-Ni/WO_3 thin films, an electron donation from Cu (Co and Ni) to Pt may occur since Pt (2.28) exhibited a greater electronegativity compared to Cu (1.93), Co (1.88), and Ni (1.91), causing an electron-rich state in Pt atoms. In other words, the efficient transfer of two electrons for O_2 reduction in the Pt-Cu/WO_3, Pt-Co/WO_3, and Pt-Ni/WO_3 thin films is believed to be promoted by the photoexcited electrons from the CB of the WO_3 being transported first to Cu (Co or Ni), and subsequently to Pt, as demonstrated in Scheme 1d. Additionally, this mechanism caused O_2 to undergo a double-electron reduction on the Pt surface rather than on Cu (Co or Ni), thus preventing contact between O_2 and Cu (Co or Ni) and suppressing the oxidation of Cu (Co or Ni) during photodegradation in the presence of the atmosphere. The presence of these alloys increased the yield of H_2O_2 on the Pt-Cu/WO_3, Pt-Co/WO_3, and Pt-Ni/WO_3 thin-film catalysts, similar to those observed in Au-Ag/TiO_2 catalysts [67]. Additionally, surface-bound H_2O or OH^- readily reacted with the departing h^+ to form •OH due to its strong oxidizing potential (+3 V vs. NHE) [68], in agreement with the results of investigations using photoinduced carrier-trapping (Figure 5d). Furthermore, it has been reported that H_2O_2 can degrade dye molecules [69]. Thus, under the influence of reactive oxidative species such •OH and H_2O_2, thin-film photocatalysts made of Pt-Cu/WO_3, Pt-Co/WO_3, and Pt-Ni/WO_3 can efficiently destroy the adsorbed MB molecules during visible-light exposure.

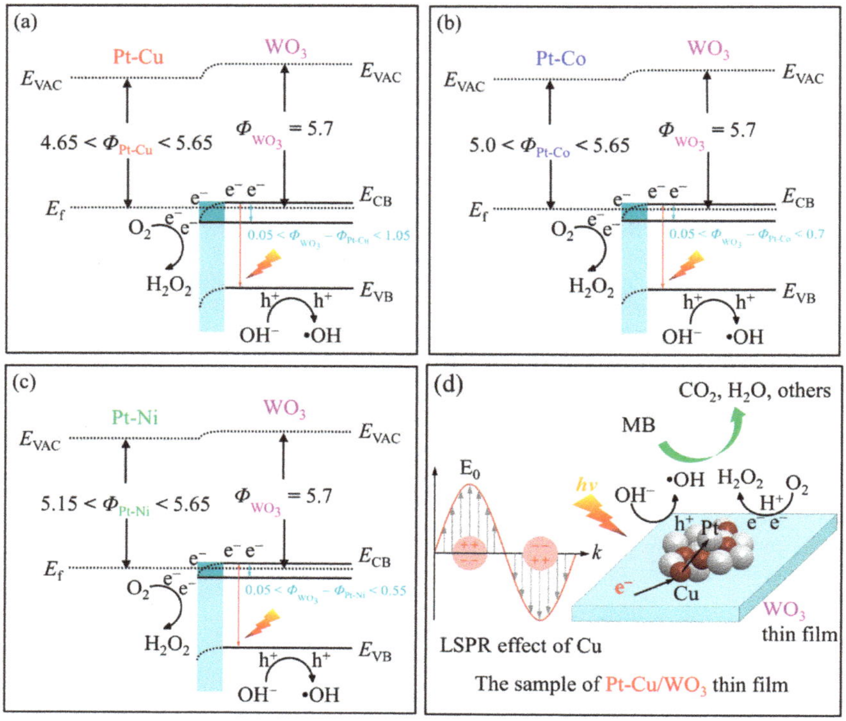

Scheme 1. Schematic energy-band diagrams for (**a**) Pt-Cu/WO_3, (**b**) Pt-Co/WO_3, and (**c**) Pt-Ni/WO_3 thin films (E_{VAC}, E_{CB}, E_{VB}, E_f, $\Phi_{Pt\text{-}M}$, and Φ_{WO_3} denote vacuum level, Fermi level, CB level, VB level, the work function of bimetallic cocatalysts, and the work function of WO_3, respectively (in eV)). (**d**) The photocatalytic mechanism for the degradation of MB on the Pt-Cu/WO_3 thin-film catalyst.

3. Materials and Methods

3.1. Preparation of Pt-M/WO_3 (M = Cu, Co, and Ni) Thin Films

First, the colloidal poly(methyl methacrylate) (PMMA) aqueous suspension, which was previously prepared and described in the Supplementary Material of [70], was used to dissolve 0.3460 g of ammonium paratungstate ($(NH_4)_6H_2W_{12}O_{40}$) in order to create the WO_3 precursor suspension. The resulting mixture was then stirred at ambient temperatures for a few minutes in order to create a homogeneous suspension. Secondly, the aforementioned precursor solution (2 mL) was then combined with H_2PtCl_6 (0.5 mL 10 mM) and $CuCl_2$ (or $CoCl_2{\cdot}6H_2O$ or $Ni(NO_3)_2{\cdot}6H_2O$) (0.5 mL 10 mM) in *N*, *N'*-dimethylformamide (DMF) solution. The combination was then agitated and blended to create a homogeneous suspension. Next, the thin film was then dried in air for around 10 min at 60 °C after 3 mL of the suspension had been spin-coated for 50 s at 3000 rpm onto a 3 × 3 cm ITO substrate. The aforementioned procedure was repeated 5 times, then toluene was used to etch the thin films for 5 min, removing majority of the organic template. Finally, in order to thoroughly break down the PMMA residue, they were then annealed at 450 °C (heating rate 2 °C min^{-1}) in a stream of air (80 cm^3 min^{-1}) for 5 h. The procedure for growing is depicted in Scheme 2 as a diagrammatic illustration.

For the sake of comparison, without the addition of H_2PtCl_6 and $CuCl_2$ ($CoCl_2{\cdot}6H_2O$ or $Ni(NO_3)_2{\cdot}6H_2O$) to the DMF solution, the WO_3 thin film was created under identical circumstances.

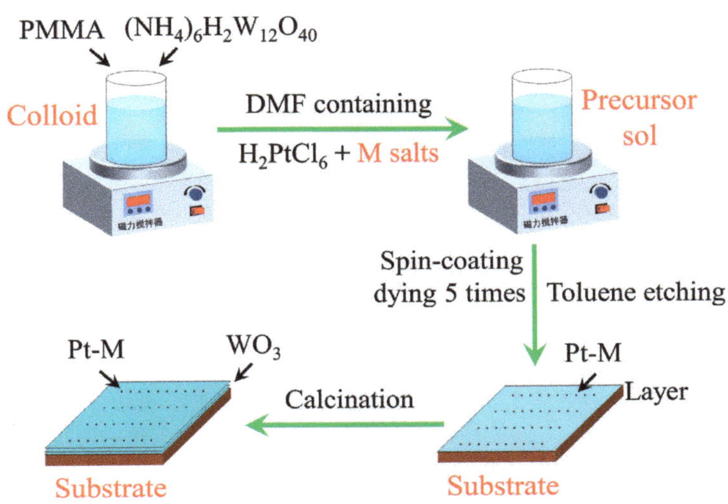

Scheme 2. Schematic illustration of the preparation process of the Pt-M/WO_3 (M = Cu, Co, and Ni) thin films.

3.2. Characterization

The D/max-2400 diffractometer (Rigaku, Tokyo, Japan) was used to conduct XRD tests utilizing Cu-Kα radiation. Raman spectra were obtained on the Thermo Fischer DXR by using a 532 nm laser as the light source. With the use of SEM (Nova NanoSEM 450, FEI, Hillsboro, OR, USA), the surface morphologies were examined. TEM images were collected with a JEM-2100F electron microscope (JEOL, Tokyo, Japan). In order to explore the surface chemical compositions and states, XPS measurements were performed using Al Ka X-ray (hv = 1486.6 eV) radiation (K-Alpha, Thermo Electron, Waltham, MA, USA). On a UV-vis spectrophotometer (UV-2600, Shimadzu, Kyoto, Japan), diffuse reflectance spectra in the UV-vis range were obtained.

3.3. Photocatalytic Tests

The degradation process of MB in an aqueous solution was used to assess the visible-light photocatalyst according to the procedure for testing photocatalytic materials for the purification of the water solution (GB/T 23762-2009, China). In a standard photocatalytic experiment, the as-prepared thin films employing ITO glass as the support were mounted in a quartz frame and immersed in MB solution (5 mg L^{-1}, 100 mL). To ensure the formation of an equilibrium between adsorption and desorption, the solution was magnetically agitated for 40 min in the dark before irradiation. The visible-light irradiation source, a 350 W Xe lamp with a cutoff filter of 420 nm, was used and placed 10 cm from the reactor. After the light was turned on, 4 mL of a clear MB solution was removed at predetermined intervals (20 min) and analyzed by noting the maximum absorbance of MB at 664 nm in the UV-visible spectrum. C/C_0 was used for determining the proportion of degradation, where C is the concentration of leftover MB solution at each irradiated time interval, and C_0 is the initial concentration.

Utilizing the film in numerous cycles of the photocatalytic degradation of MB allowed the researchers to examine the photocatalytic stability of the ideal Pt-Cu/WO_3 thin film. The photocatalyst was filtered and properly cleaned with deionized water after completing one cycle. The next cycle was then started by adding 5 mg L^{-1} of fresh MB solution to the photocatalyst. Each cycle took 120 min to complete, and there were 5 cycles in a row.

Experiments were performed to explore the photocatalytic process using active species capture. The MB aqueous solution was supplemented with several radical scavengers, including *t*-BuOH (1 mmol L^{-1}), EDTA-2Na (1 mmol L^{-1}), and BQ (1 mmol L^{-1}). The subsequent experimental procedures mirrored the photocatalytic test described above.

3.4. PEC Measurements

PEC measurements were conducted using a three-electrode setup and a quartz cell on the electrochemical workstation (CHI660E, Shanghai Chenhua Instruments Co., Ltd., Shanghai, China). As the electrolyte, a 0.5 M Na_2SO_4 solution was utilized. With a surface area of 2.2 cm^2, the as-obtained thin films were employed as photoanodes. The reference electrode was formed of an Ag/AgCl (saturated KCl) electrode, while the cathode was a Pt sheet. As the lamp housing for the PEC measurement, a simulated sunshine Xe lamp (CEL-HXF300, NbeT Group Corp., Beijing, China) with an AM 1.5 G filter was used. The light power density (i.e., light intensity) was tuned to be 100 mW cm^{-2}, which was validated by an optical power meter (Model 843-R, Newport, Irvine, CA, USA). On the CHI660E electrochemical workstation, the LSV curves with a scan rate of 10 mV s^{-1}, transient photocurrent responses, EIS, and Mott-Schottky plots were performed.

4. Conclusions

In the present study, homogeneous precursor sols, spin-coating, toluene-etching, and calcination procedures were used to effectively synthesize WO_3, Pt-Cu/WO_3, Pt-Co/WO_3, and Pt-Ni/WO_3 thin films. In conclusion, the resultant thin films had comparable morphologies and were mostly made of low-roughness, blackberry-like particles. The XPS investigation showed a modification in the electrical structure of Pt when alloyed with Cu (Co, Ni). The Pt-Cu/WO_3, Pt-Co/WO_3, and Pt-Ni/WO_3 thin films considerably enhanced the charge transfer and photocatalytic activity compared to the WO_3 thin film, according to the EIS and Mott–Schottky plot analyses. Capture experiments of active species demonstrated that •OH is essential for the oxidization of MB. Furthermore, the introduction of transition metals such as Cu, Co, or Ni modified the electronic structure of Pt-M/WO_3 thin films, facilitating the separation and migration of electrons and holes. Specifically, the photogenerated electrons migrated from the CB of WO_3 to Pt-Co or Pt-Ni nanoparticles in the samples of Pt-Co/WO_3 or Pt-Ni/WO_3 thin films, while the hot electrons from the LSPR effect of Cu could transfer to the CB of WO_3 and other electrons generated from the photoexcitation of the WO_3 semiconductor itself in the sample of Pt-Cu/WO_3 thin films.

Supplementary Materials: The following supporting information can be downloaded at: https://www.mdpi.com/article/10.3390/catal13071044/s1. Synthesis of polymethylmethacrylate (PMMA) spheres; Supplementary Figures and Table [71–79]. Figure S1. (a,b) TEM and (c) HRTEM images of Pt-Cu/WO_3 thin film. Figure S2. The adsorption isotherms of MB in the presence of as-prepared samples under dark conditions. Table S1. Comparison between this research and some commonly used WO_3-based photocatalysts for the removal of organic pollutants.

Author Contributions: Conceptualization, P.D.; methodology, J.P.; software, L.Z.; validation, C.W.; formal analysis, K.G.; investigation, J.P.; resources, P.D.; data curation, J.P.; writing—original draft preparation, J.P.; writing—review and editing, P.D.; visualization, J.P.; supervision, P.D.; project administration, P.D.; funding acquisition, P.D. All authors have read and agreed to the published version of the manuscript.

Funding: This work has been funded by the National Natural Science Foundation of China (Grant No. 21403184), the Qinglan Project of Jiangsu Province, the Natural Science Foundation of the Jiangsu Higher Education Institutions of China (Grant No. 22KJA430008), and the Scientific Research and Practical Innovation Project for Graduate Students in Jiangsu Province (No. KYCX22_XY002).

Data Availability Statement: Data are available upon request from the authors.

Conflicts of Interest: The authors declare no conflict of interest.

References

1. Chen, C.; Ma, W.; Zhao, J. Semiconductor-mediated photodegradation of pollutants under visible-light irradiation. *Chem. Soc. Rev.* **2010**, *39*, 4206–4219. [CrossRef] [PubMed]
2. Dong, P.; Dong, F.; Fiorenza, R. Photocatalysts for air purification: Design, synthesis, and mechanism investigations. *Front. Chem.* **2022**, *10*, 870550. [CrossRef] [PubMed]
3. Suhaimi, N.A.A.; Kong, C.P.; Shahri, N.N.M.; Nur, M.; Hobley, J.; Usman, A. Dynamics of diffusion- and Immobilization-limited photocatalytic degradation of dyes by metal oxide nanoparticles in binary or ternary solutions. *Catalysts* **2022**, *12*, 1254. [CrossRef]
4. Piątkowska, A.; Janus, M.; Szymański, K.; Mozia, S. C-, N- and S-doped TiO_2 photocatalysts: A review. *Catalysts* **2021**, *11*, 144. [CrossRef]
5. Dong, P.; Hou, G.; Xi, X.; Shao, R.; Dong, F. WO_3-based photocatalysts: Morphology control, activity enhancement and multifunctional applications. *Environ. Sci. Nano* **2017**, *4*, 539–557. [CrossRef]
6. Ni, Z.; Wang, Q.; Guo, Y.; Liu, H.; Zhang, Q. Research progress of tungsten oxide-based catalysts in photocatalytic reactions. *Catalysts* **2023**, *13*, 579. [CrossRef]
7. Wang, Z.; Zhu, C.; Ni, Z.; Hojo, H.; Einaga, H. Enhanced photocatalytic benzene oxidation to phenol over monoclinic WO_3 nanorods under visible light. *ACS Catal.* **2022**, *12*, 14976–14989. [CrossRef]
8. Sun, S.; Wang, W.; Zeng, S.; Shang, M.; Zhang, L. Preparation of ordered mesoporous Ag/WO_3 and its highly efficient degradation of acetaldehyde under visible-light irradiation. *J. Hazard. Mater.* **2010**, *178*, 427–433. [CrossRef]
9. Tahir, M.B.; Nabi, G.; Rafique, M.; Khalid, N.R. Nanostructured-based WO_3 photocatalysts: Recent development, activity enhancement, perspectives and applications for wastewater treatment. *Int. J. Environ. Sci. Technol.* **2017**, *14*, 2519–2542. [CrossRef]
10. Zhao, Z.G.; Miyauchi, M. Nanoporous-walled tungsten oxide nanotubes as highly active visible-light-driven photocatalysts. *Angew. Chem. Int. Ed.* **2008**, *47*, 7051–7055. [CrossRef]
11. Takeuchi, M.; Shimizu, Y.; Yamagawa, H.; Nakamuro, T.; Anpo, M. Preparation of the visible light responsive N^{3-}-doped WO_3 photocatalyst by a thermal decomposition of ammonium paratungstate. *Appl. Catal. B* **2011**, *110*, 1–5. [CrossRef]
12. Adhikari, R.; Gyawali, G.; Sekino, T.; Wohn Lee, S. Microwave assisted hydrothermal synthesis of Ag/AgCl/WO_3 photocatalyst and its photocatalytic activity under simulated solar light. *J. Solid State Chem.* **2013**, *197*, 560–565. [CrossRef]
13. Poudel, M.B.; Lohani, P.C.; Kim, A.A. Synthesis of silver nanoparticles decorated tungsten oxide nanorods as high-performance supercapacitor electrode. *Chem. Phys. Lett.* **2022**, *804*, 139884. [CrossRef]
14. Theerthagiri, J.; Senthil, R.A.; Malathi, A.; Selvi, A.; Madhavan, J.; Ashokkumar, M. Synthesis and characterization of a CuS–WO_3 composite photocatalyst for enhanced visible light photocatalytic activity. *RSC Adv.* **2015**, *5*, 52718–52725. [CrossRef]
15. Poudel, M.B.; Awasthi, G.P.; Kim, H.J. Novel insight into the adsorption of Cr(VI) and Pb(II) ions by MOF derived Co–Al layered double hydroxide @hematite nanorods on 3D porous carbon nanofiber network. *Chem. Eng. J.* **2021**, *417*, 129312. [CrossRef]
16. Abe, R.; Takami, H.; Murakami, N.; Ohtani, B. Pristine simple oxides as visible light driven photocatalysts: Highly efficient decomposition of organic compounds over platinum-loaded tungsten oxide. *J. Am. Chem. Soc.* **2008**, *130*, 7780–7781. [CrossRef]
17. Dong, P.; Xu, N.; Xu, Y.; Wang, X. A study of Pt/WO_3-carrier catalysts for photocatalytic purification of NO gas. *Catal. Commun.* **2016**, *84*, 142–146. [CrossRef]
18. Dong, P.; Yang, B.; Liu, C.; Xu, F.; Xi, X.; Hou, G.; Shao, R. Highly enhanced photocatalytic activity of WO_3 thin films loaded with Pt–Ag bimetallic alloy nanoparticles. *RSC Adv.* **2017**, *7*, 947–956. [CrossRef]
19. Kim, Y.; Hong, J.W.; Lee, Y.W.; Kim, M.; Kim, D.; Yun, W.S.; Han, S.W. Synthesis of AuPt heteronanostructures with enhanced electrocatalytic activity toward oxygen reduction. *Angew. Chem. Int. Ed.* **2010**, *49*, 10197–10201. [CrossRef]
20. Zhao, S.; Yin, H.; Du, L.; Yin, G.; Tang, Z.; Liu, S. Three dimensional N-doped graphene/PtRu nanoparticle hybrids as high performance anode for direct methanol fuel cells. *J. Mater. Chem. A* **2014**, *2*, 3719–3724. [CrossRef]
21. Guo, S.; Dong, S.; Wang, E. Three-dimensional Pt-on-Pd bimetallic nanodendrites supported on graphene nanosheet: Facile synthesis and used as an advanced nanoelectrocatalyst for methanol oxidation. *ACS Nano* **2010**, *4*, 547–555. [CrossRef] [PubMed]
22. Yin, A.X.; Min, X.Q.; Zhu, W.; Wu, H.S.; Zhang, Y.W.; Yan, C.H. Multiply twinned Pt–Pd nanoicosahedrons as highly active electrocatalysts for methanol oxidation. *Chem. Commun.* **2012**, *48*, 543–545. [CrossRef] [PubMed]
23. Papadimitriou, S.; Armyanov, S.; Valova, E.; Hubin, A.; Steenhaut, O.; Pavlidou, E.; Kokkinidis, G.; Sotiropoulos, S. Methanol oxidation at Pt–Cu, Pt–Ni, and Pt–Co electrode coatings prepared by a galvanic replacement process. *J. Phys. Chem. C* **2010**, *114*, 5217–5223. [CrossRef]
24. Xia, B.Y.; Wu, H.B.; Li, N.; Yan, Y.; Lou, X.W.; Wang, X. One-pot synthesis of Pt–Co alloy nanowire assemblies with tunable composition and enhanced electrocatalytic properties. *Angew. Chem. Int. Ed.* **2015**, *54*, 3797–3801. [CrossRef] [PubMed]
25. Yin, Z.; Zhou, W.; Gao, Y.; Ma, D.; Kiely, C.J.; Bao, X. Supported Pd–Cu bimetallic nanoparticles that have high activity for the electrochemical oxidation of methanol. *Chem. Eur. J.* **2012**, *18*, 4887–4893. [CrossRef] [PubMed]
26. Habibi, B.; Delnavaz, N. Carbon–ceramic supported bimetallic Pt–Ni nanoparticles as an electrocatalyst for oxidation of formic acid. *Int. J. Hydrogen Energy* **2011**, *36*, 9581–9590. [CrossRef]
27. Luo, J.Y.; Deng, S.Z.; Tao, Y.T.; Zhao, F.L.; Zhu, L.F.; Gong, L.; Chen, J.; Xu, N.S. Evidence of localized water molecules and their role in the gasochromic effect of WO_3 nanowire films. *J. Phys. Chem. C* **2009**, *113*, 15877–15881. [CrossRef]
28. Jian, L.Y.; Xue, X.C.; Wei, D.L.; Deng, W.Y.; Wei, L.; Hao, Y.W.; Lian, F.Z.; Qing, G.Z. Variable-temperature raman spectroscopic study of the hydrogen sensing mechanism in Pt–WO_3 nanowire film. *Appl. Phys. Lett.* **2013**, *102*, 113104. [CrossRef]

29. McBride, J.R.; Graham, G.W.; Peters, C.R.; Weber, W.H. Growth and characterization of reactively sputtered thin–film platinum oxides. *J. Appl. Phys.* **1991**, *69*, 1596–1604. [CrossRef]
30. Liang, M.-M.; Wang, Y.-H.; Shao, R.; Yang, W.-M.; Zhang, H.; Zhang, H.; Yang, Z.-L.; Li, J.-F.; Tian, Z.-Q. In situ electrochemical surface-enhanced Raman spectroscopy study of CO electrooxidation on PtFe nanocatalysts. *Electrochem. Commun.* **2017**, *81*, 38–42. [CrossRef]
31. Sugimura, F.; Sakai, N.; Nakamura, T.; Nakamura, M.; Ikeda, K.; Sakai, T.; Hoshi, N. In-situ observation of Pt oxides on the low index planes of Pt using surface enhanced raman spectroscopy. *Phys. Chem. Chem. Phys.* **2017**, *19*, 27570–27579. [CrossRef] [PubMed]
32. Huang, Y.F.; Kooyman, P.J.; Koper, M.T. Intermediate stages of electrochemical oxidation of single-crystalline platinum revealed by in situ Raman spectroscopy. *Nat. Commun.* **2016**, *7*, 12440. [CrossRef]
33. Li, L.; Xiao, S.; Li, R.; Cao, Y.; Chen, Y.; Li, Z.; Li, G.; Li, H. Nanotube array-like WO_3 photoanode with dual-layer oxygen-evolution cocatalysts for photoelectrocatalytic overall water splitting. *ACS Appl. Energy Mater.* **2018**, *1*, 6871–6880. [CrossRef]
34. Lei, F.; Li, Z.; Zhang, L.; Wang, Y.; Xu, S.; Lin, S. Facile synthesis of Pt–Cu (Ni, Co)/GNs–CD and their enhanced electro-catalytic activity for methanol oxidation. *J. Electrochem. Soc.* **2016**, *163*, F913–F918. [CrossRef]
35. Wang, H.; Wang, R.; Li, H.; Wang, Q.; Kang, J.; Lei, Z. Facile synthesis of carbon-supported pseudo-core@shell PdCu@Pt nanoparticles for direct methanol fuel cells. *Int. J. Hydrogen Energy* **2011**, *36*, 839–848. [CrossRef]
36. Han, C.; Lu, Y.; Zhang, J.; Ge, L.; Li, Y.; Chen, C.; Xin, Y.; Wu, L.; Fang, S. Novel PtCo alloy nanoparticle decorated 2D g-C_3N_4 nanosheets with enhanced photocatalytic activity for H_2 evolution under visible light irradiation. *J. Mater. Chem. A* **2015**, *3*, 23274–23282. [CrossRef]
37. Li, X.; Liu, Y.; Zhu, J.; Tsiakaras, P.; Shen, P.K. Enhanced oxygen reduction and methanol oxidation reaction over self-assembled Pt–M (M = Co, Ni) nanoflowers. *J. Colloid Interface Sci.* **2022**, *607*, 1411–1423. [CrossRef]
38. Qi, Y.; Yang, Z.; Jiang, Y.; Han, H.; Wu, T.; Wu, L.; Liu, J.; Wang, Z.; Wang, F. Platinum–copper bimetallic nanoparticles supported on TiO_2 as catalysts for photo–thermal catalytic toluene combustion. *ACS Appl. Nano Mater.* **2022**, *5*, 1845–1854. [CrossRef]
39. Peng, X.; Zhao, Y.; Chen, D.; Fan, Y.; Wang, X.; Wang, W.; Tian, J. One-pot synthesis of reduced graphene oxide supported $PtCu_y$ catalysts with enhanced electro-catalytic activity for the methanol oxidation reaction. *Electrochim. Acta* **2014**, *136*, 292–300. [CrossRef]
40. Georgieva, J.; Valova, E.; Mintsouli, I.; Sotiropoulos, S.; Armyanov, S.; Kakaroglou, A.; Hubin, A.; Steenhaut, O.; Dille, J. Carbon-supported Pt(Cu) electrocatalysts for methanol oxidation prepared by Cu electroless deposition and its galvanic replacement by Pt. *J. Appl. Electrochem.* **2013**, *44*, 215–224. [CrossRef]
41. Ghodbane, O.; Roué, L.; Bélanger, D. Study of the electroless deposition of Pd on Cu-modified graphite electrodes by metal exchange reaction. *Chem. Mater.* **2008**, *20*, 3495–3504. [CrossRef]
42. Khalakhan, I.; Vorokhta, M.; Xie, X.; Piliai, L.; Matolínová, I. On the interpretation of X-ray photoelectron spectra of Pt-Cu bimetallic alloys. *J. Electron Spectrosc. Relat. Phenom.* **2021**, *246*, 147027. [CrossRef]
43. Mathe, N.R.; Scriba, M.R.; Coville, N.J. Methanol oxidation reaction activity of microwave-irradiated and heat-treated Pt/Co and Pt/Ni nano-electrocatalysts. *Int. J. Hydrogen Energy* **2014**, *39*, 18871–18881. [CrossRef]
44. Biesinger, M.C.; Payne, B.P.; Grosvenor, A.P.; Lau, L.W.M.; Gerson, A.R.; Smart, R.S.C. Resolving surface chemical states in XPS analysis of first row transition metals, oxides and hydroxides: Cr, Mn, Fe, Co and Ni. *Appl. Surf. Sci.* **2011**, *257*, 2717–2730. [CrossRef]
45. Park, K.W.; Choi, J.H.; Kwon, B.K.; Lee, S.A.; Sung, Y.E. Chemical and electronic effects of Ni in Pt/Ni and Pt/Ru/Ni alloy nanoparticles in methanol electrooxidation. *J. Phys. Chem. B* **2002**, *106*, 1869–1877. [CrossRef]
46. Grosvenor, A.P.; Biesinger, M.C.; Smart, R.S.C.; McIntyre, N.S. New interpretations of XPS spectra of nickel metal and oxides. *Surf. Sci.* **2006**, *600*, 1771–1779. [CrossRef]
47. Desseigne, M.; Dirany, N.; Chevallier, V.; Arab, M. Shape dependence of photosensitive properties of WO_3 oxide for photocatalysis under solar light irradiation. *Appl. Surf. Sci.* **2019**, *483*, 313–323. [CrossRef]
48. Miyauchi, M. Photocatalysis and photoinduced hydrophilicity of WO_3 thin films with underlying Pt nanoparticles. *Phys. Chem. Chem. Phys.* **2008**, *10*, 6258–6265. [CrossRef]
49. Shiraishi, Y.; Sugano, Y.; Tanaka, S.; Hirai, T. One-pot synthesis of benzimidazoles by simultaneous photocatalytic and catalytic reactions on Pt@TiO_2 nanoparticles. *Angew. Chem. Int. Ed.* **2010**, *49*, 1656–1660. [CrossRef]
50. Shiraishi, Y.; Takeda, Y.; Sugano, Y.; Ichikawa, S.; Tanaka, S.; Hirai, T. Highly efficient photocatalytic dehalogenation of organic halides on TiO_2 loaded with bimetallic Pd–Pt alloy nanoparticles. *Chem. Commun.* **2011**, *47*, 7863–7865. [CrossRef]
51. Gulen, M.; Yildirim, G.; Bal, S.; Varilci, A.; Belenli, I.; Oz, M. Role of annealing temperature on microstructural and electro-optical properties of ITO films produced by sputtering. *J. Mater. Sci. Mater. Electron.* **2012**, *24*, 467–474. [CrossRef]
52. Kotlyar, K.P.; Kudryashov, D.A.; Soshnikov, I.P.; Uvarov, A.V.; Reznik, R.R. Temperature annealing effect on ITO film. *J. Phys. Conf. Ser.* **2018**, *1124*, 041035. [CrossRef]
53. Wang, X.H.; Li, J.G.; Kamiyama, H.; Moriyoshi, Y.; Ishigaki, T. Wavelength-sensitive photocatalytic degradation of methyl orange in aqueous suspension over Iron(III)-doped TiO_2 nanopowders under UV and visible light irradiation. *J. Phys. Chem. B* **2006**, *110*, 6804–6809. [CrossRef] [PubMed]
54. Haleem, A.; Shafiq, A.; Chen, S.Q.; Nazar, M. A comprehensive review on adsorption, photocatalytic and chemical degradation of dyes and nitro-compounds over different kinds of porous and composite materials. *Molecules* **2023**, *28*, 1081. [CrossRef]

55. Begum, R.; Najeeb, J.; Sattar, A.; Naseem, K.; Irfan, A.; Al-Sehemi, A.G.; Farooqi, Z.H. Chemical reduction of methylene blue in the presence of nanocatalysts: A critical review. *Rev. Chem. Eng.* **2020**, *36*, 749–770. [CrossRef]
56. Dong, P.; Wang, Y.; Li, H.; Li, H.; Ma, X.; Han, L. Shape-controllable synthesis and morphology-dependent photocatalytic properties of Ag_3PO_4 crystals. *J. Mater. Chem. A* **2013**, *1*, 6955–6963. [CrossRef]
57. Chen, C.; Wang, Q.; Lei, P.; Song, W.; Ma, W.; Zhao, J. Photodegradation of dye pollutants catalyzed by porous $K_3PW_{12}O_{40}$ under visible Irradiation. *Environ. Sci. Technol.* **2006**, *40*, 3965–3970. [CrossRef]
58. Pan, C.; Zhu, Y. New type of $BiPO_4$ oxy-acid salt photocatalyst with high photocatalytic activity on degradation of dye. *Environ. Sci. Technol.* **2010**, *44*, 5570–5574. [CrossRef]
59. Lee, H.; Choi, W. Photocatalytic oxidation of arsenite in TiO_2 suspension: kinetics and mechanisms. *Environ. Sci. Technol.* **2002**, *36*, 3872–3878. [CrossRef]
60. Poudel, M.B.; Kim, A.A.; Lohani, P.C.; Yoo, D.J.; Kim, H.J. Assembling zinc cobalt hydroxide/ternary sulfides heterostructure and iron oxide nanorods on three-dimensional hollow porous carbon nanofiber as high energy density hybrid supercapacitor. *J. Energy Storage* **2023**, *60*, 106713. [CrossRef]
61. Klahr, B.; Gimenez, S.; Fabregat-Santiago, F.; Bisquert, J.; Hamann, T.W. Photoelectrochemical and impedance spectroscopic investigation of water oxidation with "Co–Pi"-coated hematite electrodes. *J. Am. Chem. Soc.* **2012**, *134*, 16693–16700. [CrossRef] [PubMed]
62. Vida, G.; Josepovits, V.K.; Gyor, M.; Deak, P. Characterization of tungsten surfaces by simultaneous work function and secondary electron emission measurements. *Microsc. Microanal.* **2003**, *9*, 337–342. [CrossRef] [PubMed]
63. Yella, A.; Gautam, U.K.; Mugnaioli, E.; Panthöfer, M.; Bando, Y.; Golberg, D.; Kolb, U.; Tremel, W. Asymmetric tungsten oxide nanobrushes via oriented attachment and ostwald ripening. *CrystEngComm* **2011**, *13*, 4074–4081. [CrossRef]
64. Eastman, D.E. Photoelectric work functions of transition, rare-earth, and noble metals. *Phys. Rev. B* **1970**, *2*, 1–2. [CrossRef]
65. Michaelson, H.B. The work function of the elements and its periodicity. *J. Appl. Phys.* **1977**, *48*, 4729–4733. [CrossRef]
66. Ishii, R.; Matsumura, K.; Sakai, A.; Sakata, T. Work function of binary alloys. *Appl. Surf. Sci.* **2001**, *169*, 658–661. [CrossRef]
67. Tsukamoto, D.; Shiro, A.; Shiraishi, Y.; Sugano, Y.; Ichikawa, S.; Tanaka, S.; Hirai, T. Photocatalytic H_2O_2 production from ethanol/O_2 system using TiO_2 loaded with Au–Ag bimetallic alloy nanoparticles. *ACS Catal.* **2012**, *2*, 599–603. [CrossRef]
68. Arai, T.; Yanagida, M.; Konishi, Y.; Iwasaki, Y.; Sugihara, H.; Sayama, K. Efficient complete oxidation of acetaldehyde into CO_2 over $CuBi_2O_4/WO_3$ composite photocatalyst under visible and UV light irradiation. *J. Phys. Chem. C* **2007**, *111*, 7574–7577. [CrossRef]
69. Weng, B.; Wu, J.; Zhang, N.; Xu, Y.J. Observing the role of graphene in boosting the two-electron reduction of oxygen in graphene–WO_3 nanorod photocatalysts. *Langmuir* **2014**, *30*, 5574–5584. [CrossRef]
70. D'Arienzo, M.; Armelao, L.; Mari, C.M.; Polizzi, S.; Ruffo, R.; Scotti, R.; Morazzoni, F. Macroporous WO_3 thin films active in NH_3 sensing: Role of the hosted Cr isolated centers and Pt nanoclusters. *J. Am. Chem. Soc.* **2011**, *133*, 5296–5304. [CrossRef]
71. Rick, C.; Schroden, M.A.-D.; Blanford, F.C.; Stein, A. Optical properties of inverse opal photonic crystals. *Chem. Mater.* **2002**, *14*, 3305–3315.
72. Liu, C.; Lü, H.; Yu, C.; Ding, B.; Ye, R.; Ji, Y.; Dai, B.; Liu, W. Novel $FeWO_4/WO_3$ nanoplate with p–n heterostructure and its enhanced mechanism for organic pollutants removal under visible-light illumination. *J. Environ. Chem. Eng.* **2020**, *8*, 104044. [CrossRef]
73. Ibrahim, Y.O.; Gondal, M.A.; Alaswad, A.; Moqbel, R.A.; Hassan, M.; Cevik, E.; Qahtan, T.F.; Dastageer, M.A.; Bozkurt, A. Laser-induced anchoring of WO_3 nanoparticles on reduced graphene oxide sheets for photocatalytic water decontamination and energy storage. *Ceram. Int.* **2020**, *46*, 444–451. [CrossRef]
74. Liu, Y.; Li, M.; Zhang, Q.; Qin, P.; Wang, X.; He, G.; Li, L. One-step synthesis of a WO_3-CuS nanosheet heterojunction with enhanced photocatalytic performance for methylene blue degradation and Cr(VI) reduction. *J. Chem. Technol. Biotechnol.* **2019**, *95*, 665–674. [CrossRef]
75. Sharma, S.; Basu, S. Highly reusable visible light active hierarchical porous WO_3/SiO_2 monolith in centimeter length scale for enhanced photocatalytic degradation of toxic pollutants. *Sep. Purif. Technol.* **2020**, *231*, 115916. [CrossRef]
76. Singh, J.; Arora, A.; Basu, S. Synthesis of coral like WO_3/g-C_3N_4 nanocomposites for the removal of hazardous dyes under visible light. *J. Alloys Compd.* **2019**, *808*, 151734. [CrossRef]
77. Zhang, J.; Fu, X.; Hao, H.; Gan, W. Facile synthesis 3D flower-like Ag@WO_3 nanostructures and applications in solar-light photocatalysis. *J. Alloys Compd.* **2018**, *757*, 134–141. [CrossRef]
78. Xie, R.; Fang, K.; Liu, Y.; Chen, W.; Fan, J.; Wang, X.; Ren, Y.; Song, Y. Z-scheme In_2O_3/WO_3 heterogeneous photocatalysts with enhanced visible-light-driven photocatalytic activity toward degradation of organic dyes. *J. Mater. Sci.* **2020**, *55*, 11919–11937. [CrossRef]
79. Prabhu, S.; Cindrella, L.; Kwon, O.J.; Mohanraju, K. Photoelectrochemical and photocatalytic activity of TiO_2-WO_3 heterostructures boosted by mutual interaction. *Mater. Sci. Semicond. Process.* **2018**, *88*, 10–19. [CrossRef]

catalysts

MDPI

Article

One-Pot Fabrication of 2D/2D $CdIn_2S_4/In_2S_3$ Heterojunction for Boosting Photocatalytic Cr(VI) Reduction

Jiawei Hu, Jiaxin Wu, Siyuan Zhang, Wenxuan Chen, Wen Xiao, Haijun Hou, Xiaowang Lu *, Chao Liu * and Qinfang Zhang

School of Materials Science and Engineering, Yancheng Institute of Technology, Yancheng 224051, China; qfangzhang@gmail.com (Q.Z.)
* Correspondence: luxiaowang@ycit.edu.cn (X.L.); cliu@ycit.edu.cn (C.L.)

Abstract: The development of efficient heterojunction photocatalysts with a facilitated charge carrier separation rate and improved light-harvesting capacity is still a challenging issue for effectively solving environmental pollution. Herein, a one-step refluxing process was employed to construct 2D/2D $CdIn_2S_4/In_2S_3$ (CISI) heterojunction photocatalysts with an intimate interface between these two components. The crystal structure, morphology, light-harvesting capacity, and Cr(VI) photoreduction performance were systematically investigated and discussed in detail. The tight interface formed between $CdIn_2S_4$ (CIS) and In_2S_3 (IS) could effectively facilitate the charge carrier separation and transfer. Thus, the resulting CISI composites exhibited a high efficiency for Cr(VI) photoreduction under visible light, with the optimal sample of 0.5 CISI. The charge transfer kinetics were deeply investigated by multiple techniques. Based on the characterization results, a possible mechanism for Cr(VI) photoreduction was proposed.

Keywords: $CdIn_2S_4$; In_2S_3; photocatalysis; heterojunction; Cr(VI) photoreduction

1. Introduction

Over the past decades, the presence of the toxic heavy metal hexavalent chromium (Cr(VI)) in liquid industrial waste has had a significantly negative impact on the ecological system and human health on account of its acute toxicity, good solubility, and high carcinogenicity [1,2]. Conversely, trivalent chromium (Cr(III)), with a lower toxicity, is considered to be the essential microelement for the metabolism of various organisms. Furthermore, Cr(III) favors the appearance of sedimentation under alkaline and neutral conditions. Thus, the reduction of Cr(VI) to (Cr(III) is regarded as an effective pathway for the removal of Cr pollution. Until now, semiconductor-based photocatalytic technology has been extensively applied in the elimination of pollutants due to its low cost, environmental friendliness, and extraordinary efficacy [3–7]. For the photoreduction of Cr(VI) to Cr(III), constructing high-efficiency photocatalysts with a high charge carrier separation rate and a light harvesting capacity is imperative to meet the demand of industrial development.

Recently, diverse semiconductor-based photocatalysts, such as metal sulfides, TiO_2 [8,9], layered double hydroxides (LDHs) [10], WO_3 [11], and $BiPO_4$ [12], have been developed for Cr(VI) photoreduction. Among these photocatalysts, metal sulfides have attracted ever-increasing attention in various photocatalytic fields on account of their simple synthesis process, low cost, visible-light harvesting capacity, and unique photoelectric properties [13,14]. In particular, the ternary semiconductor $CdIn_2S_4$ (CIS) has been extensively investigated due to its easy accessibility, structural stability, and suitable energy band structure [15,16]. Unfortunately, like most single photocatalysts, the rapid charge carrier recombination rate of CIS restrains its potential application in achieving the desirable photocatalytic efficiency [17].

Generally, photocatalytic efficiency is profoundly dependent on the charge carrier separation rate. To meet the practical requirements, some modified strategies have been used to suppress the charge recombination in order to boost the photocatalytic efficiency of single photocatalysts. These modification methods mainly include heterojunction formation [18–20], morphology design [21–23], elemental doping [24,25], cocatalyst deposition [26,27], and defect engineering [28,29]. CIS-based heterostructured photocatalysts, in particular, could significantly accelerate the charge separation and thus improve the photocatalytic activity of CIS, including CIS/$ZnIn_2S_4$ for water splitting and CO_2 reduction [28], carbon quantum dots/CIS for photodegradation [30], TiO_2/CIS for photocatalytic hydrogen evolution [31], etc. However, it is still a significant challenge to construct an effective heterojunction with efficient charge carrier separation and highly photocatalytic activity.

For this purpose, herein, a one-step fluxing method was developed to successfully construct 2D/2D $CdIn_2S_4$/In_2S_3 (CISI) heterostructured photocatalysts for visible light-driven Cr(VI) photoreduction. 2D CIS and 2D IS were chosen as primary components to fabricate a 2D/2D CISI heterojunction for Cr(VI) photoreduction based on the following reasons: (1) The suitable energy band structures of CIS and IS show visible light responses and sufficiently strong redox potentials. (2) The one-step synthesis method can be regarded as an advantageous approach to prepare heterojunction photocatalysts compared to other synthesis preparation methods. (3) The constructed 2D/2D heterojunction created via the one-step synthesis method has the maximum amount of contact interfaces between two components, resulting in the efficient separation of photogenerated charge carriers. The resulting 2D/2D CISI heterostructured photocatalysts show a high efficiency for Cr(VI) photoreduction under visible light. A possible mechanism for Cr(VI) photoreduction is also speculated.

2. Results

2.1. Synthesis of In_2S_3 (IS)

To prepare In_2S_3 (IS), 4 mmol of $InCl_3{\cdot}4H_2O$ and 6 mmol of thioacetamide (TAA) were dissolved in 100 mL of deionized water. After stirring for 30 min, the mixed solution was transferred to a three-necked flask (250 mL) and then heated at 90 °C for 5 h in an oil bath. After cooling down to room temperature, the obtained product was filtrated and washed with deionized water several times and dried at 60 °C overnight.

2.2. Synthesis of $CdIn_2S_4$ (CIS)

In a typical process, 1.2 mmol of $Cd(CH_3COO)_2{\cdot}2H_2O$, 2.4 mmol of $InCl_3{\cdot}4H_2O$, and 6.4 mmol of thioacetamide (TAA) were added into 200 mL of deionized water. After constant stirring for 30 min, the uniform solution was transferred to a 250 mL three-necked flask for heating at 100 °C for 12 h in an oil bath. Afterwards, the obtained precipitate was centrifuged, washed with water several times, and dried in an oven at 60 °C overnight.

2.3. Synthesis of $CdIn_2S_4$/In_2S_3 (CISI) Composites

The $CdIn_2S_4$/In_2S_3 (x-CISI) composites (x = 0.25, 0.5, and 0.75) were synthesized by the similar method to that of CIS, where x represented the molar ratio of In_2S_3 in CISI composites. An amount of $2 - 2x$ mmol of $Cd(CH_3COO)_2{\cdot}2H_2O$, 4 mmol of $InCl_3{\cdot}4H_2O$, and $12 - 3x$ mmol thioacetamide (TAA) were dissolved in 150 mL deionized water under constant stirring for 30 min. Subsequently, the mixed solution was transferred to a 250 mL three-necked flask and then heated at 100 °C for 12 h in an oil bath. After cooling to an ambient temperature, the precipitates were collected by centrifugation, washed with deionized water several times, and dried in an oven at 60 °C overnight.

2.4. Characterization

To observe morphology, scanning electron microscopy (SEM) images were recorded in Nova Nano SEM 450 (Hillsboro, OR, USA). Additionally, transmission electron microscopy (TEM, JEOL JEM-2100F, Tokyo, Japan) with an accelerating voltage of 200 kV was performed to confirm microstructure and elemental distribution. The powder X-ray diffractometer

(XRD, PANalytical B.V., Almelo, Netherlands) with Cu Kα radiation (λ = 0.15418 nm) was used to verify the crystal structure of samples. To determine light absorption capacity, UV–vis diffuse reflectance spectra (DRS) were measured on a UV–vis spectrophotometer (Shimadzu UV-3600plus, Kyoto, Japan) using $BaSO_4$ as a reference. X-ray photoelectron spectrometer (Thermo Scientific, Escalab 250 Xi, Waltham, MA, USA) with Al Kα X-ray source (hν = 1486.6 eV) was employed to confirm the atomic states of photocatalysts using a hemispherical electron analyzer (pass energy of 20 eV). The adventitious C 1 s peak at 284.8 eV acted as the corrected binding energy. Photoluminescence (PL) spectra were recorded on luminescence spectrometer (F-4600, Tokyo, Japan) using the excitation wavelength of 400 nm. Time-resolved photoluminescence (TR-PL) decay spectra were obtained on a fluorescence spectrometer (FLS1000, Edinburgh, UK) using the excitation wavelength of 330 nm.

2.5. Photo-Electrochemical Measurements

The photo-electrochemical data were recorded in an electrochemical workstation (CHI-660E, Shanghai Chenhua, China) with a three-electrode cell. The platinum (Pt) wire and Ag/AgCl were used as the counter electrode and the reference electrode, respectively. The working electrode was gained by the following procedure: An amount of 1.0 mg photocatalyst was mixed with 1 mL Nafion (0.5%). After ultrasonic treatment, the 40 μL uniform slurry was sprayed onto fluorine-doped tin oxide (FTO) glass (1×1 cm^2) and then naturally dried. Photocurrent was measured under a 300 W Xenon arc lamp (λ > 420 nm) in Na_2SO_4 aqueous solution (0.2 M, pH = 7.0). Linear sweep voltammetry (LSV) and Mott–Schottky (M-S) measurements were performed in Na_2SO_4 aqueous solution (0.2 M, pH = 7.0). M-S measurements were performed at a frequency of 1000 Hz. Electrochemical impedance spectra (EIS) were conducted in a mixed solution of KCl (1 mol/L), $K_3[Fe(CN)_6]$ (0.02 mol/L), and $K_4[Fe(CN)_6]$ (0.02 mol/L) at the open circuit potential.

2.6. Photocatalytic Cr(VI) Reduction

The photocatalytic activity of samples were evaluated by Cr(VI) reduction under a 300 W Xenon arc lamp (λ > 420 nm). For a typical process, 10.0 mg photocatalyst and 1 mL citric acid (100 mg/L) were added into Cr(VI) solution (100 mL, 20 mg/L), and then vigorously stirred for 30 min in the dark to reach adsorption–desorption balance. After visible light irradiation, the above suspension was sampled with an interval of 15 min. According to the 1,5-diphenylcarbazide method, the absorbance of Cr(VI) solution was determined on a Shimadzu UV-2250 spectrophotometer at the wavelength of approximately 540 nm. The efficiency for Cr(VI) photoreduction was confirmed by the value of C_t/C_0, where C_t and C_0 are the concentration at time t and the initial concentration, respectively.

3. Discussion

The morphologies of the as-prepared samples were characterized via scanning electron microscopy (SEM) (Figure 1). Clearly, the irregular particles can be observed over CIS and IS due to the agglomeration of their nanosheets (Figure 1a,b), which is confirmed by the below TEM images. Furthermore, the particular morphologies of the CISI composites are similar to that of CIS and IS (Figure 1c–e), indicating the maintenance of the overall structure for the CISI composites.

In the transmission electron microscopy (TEM) image in Figure 2a, CIS nanosheets are visible with some overlayers. Additionally, the sample of IS has the nanosheet-contained structure with the agglomeration phenomenon (Figure 2b). After the combination of CIS with IS, the similar nanosheet-contained structure also achieved over 0.5 CISI compared to IS (Figure 2c). As displayed in Figure 2d,e, the lattice distances of CIS and IS are estimated to be 0.38 nm and 0.62 nm, corresponding to the (220) plane of CIS and the (111) plane of IS, respectively, based on the crystallographic symmetry [32,33]. These two lattice fringes with the spacing value of 0.38 nm and 0.62 nm appear in 0.5 CISI, indicating the successful combination between CIS and IS (Figure 2f). Meanwhile, the formation of an intimate interface between CIS and IS (red line) shows the successful construction of an effective heterojunction, which benefits the efficient separation of the photogenerated charge carriers [34,35]. The scanning transmission electron microscopy high-angle annular dark-field (STEM-HAADF) (Figure 2g) and the corresponding elemental distribution images (Figure 2h–j) of 0.5 CISI show the uniform distribution of Cd, In, and S, further confirming a close interface between CIS and IS.

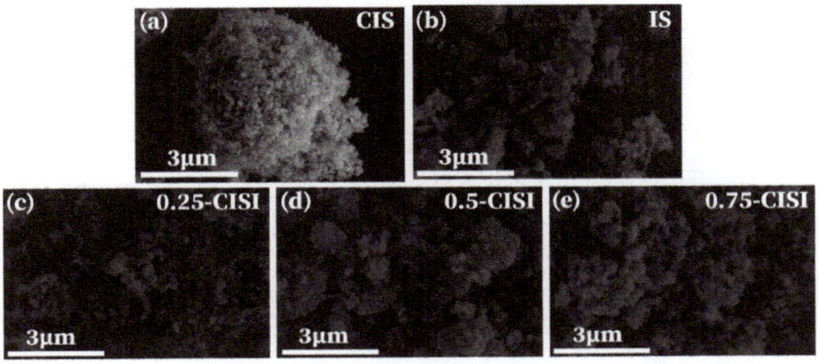

Figure 1. This SEM images of (**a**) CIS, (**b**) IS, (**c**) 0.25 CISI, (**d**) 0.5 CISI, and (**e**) 0.75 CISI.

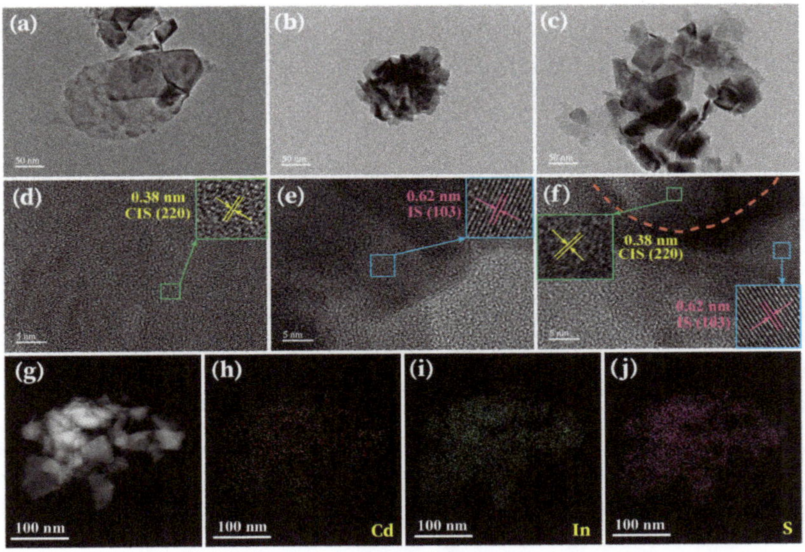

Figure 2. TEM images of (**a**) CIS, (**b**) IS, and (**c**) 0.5 CISI. HRTEM images of (**d**) CIS, (**e**) IS, and (**f**) 0.5 CISI. (**g**) STEM-HAADF image and (**h–j**) the distribution of Cd/In/S on 0.5 CISI.

The crystal structure of the as-prepared samples was monitored by XRD measurements. As shown in Figure 3a, the characteristic peaks of CIS and IS are well matched with their respective standard crystal structures of PDF#27-0060 and PDF#25-0390. Notably, the characteristic peaks of CIS are similar to those of IS except for the differences at 27.3 and 33.0°. After coupling CIS with IS, the characteristic peaks of 0.25 CISI, 0.5 CISI, and 0.75 CISI are almost unchanged related to CIS and IS originally. For the enlarged XRD patterns (Figure 3b), the characteristic peaks of 0.25 CISI and 0.5 CISI are derived from CIS, while the characteristic signals of 0.75 CISI are stemmed from IS. Thus, it can be reasonably concluded that the CISI composites have been successfully synthesized with the components of CIS and IS.

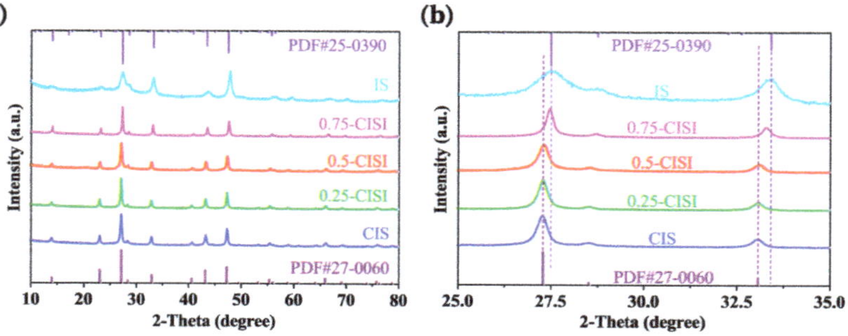

Figure 3. (**a**) XRD patterns and (**b**) enlarged XRD patterns of as-prepared samples.

The optical properties of CIS, IS, and CISI composites were investigated (Figure 4a). The absorption band edges of IS and CIS are at 561 and 616 nm, respectively, showing good visible light harvesting capacity. After coupling CIS with IS, the light absorption capacities of the CISI composites are improved compared to the single IS. Meanwhile, based on the Kubelka–Munk function [36,37], the bandgap values of CIS and IS are calculated to be 2.47 and 2.23 eV, respectively.

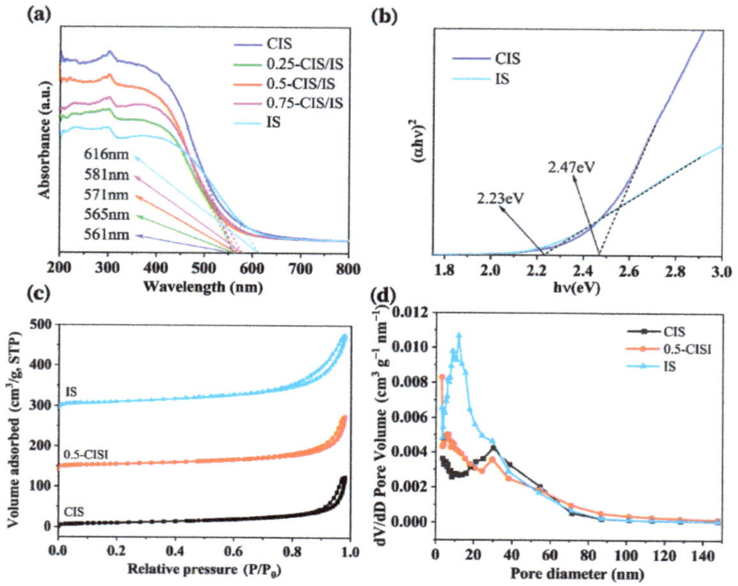

Figure 4. (**a**) UV–vis diffuse reflectance spectra (DRS) of samples. (**b**) Band gap energies of CIS and IS. (**c**) The N_2 absorption–desorption isotherms of CIS, IS, and 0.5 CISI. (**d**) Pore-size distribution plots of CIS, IS, and 0.5 CISI.

As depicted in Figure 4c, all the N_2 adsorption–desorption isotherms can be classified as type IV isotherms with an H3 hysteresis loop at P/P_0 between 0.4 and 1.0, suggesting the mesoporous materials for CIS, IS, and 0.5 CISI. The pore-size distribution curves also confirm the above mesoporous materials (Figure 4d) [38]. The Brunauer-Emmett-Teller (BET) surface area values (S_{BET}) of CIS, IS, and 0.5 CISI are measured to be 37.3, 61.9, and

44.1 m^2/g, respectively. Specifically, after combining IS nanoflakes with CIS nanosheets, the S_{BET} of 0.5 CISI is reduced due to their nanosheets stacking between CIS and IS [39]. Generally, a higher surface-to-volume ratio of the photocatalyst can offer more active sites. Thus, the relatively high BET surface areas of all samples benefit Cr(VI) photoreduction.

The X-ray photoelectron spectroscopy (XPS) technique was employed to determine the elemental composition and atomic states of CIS, IS, and 0.5 CISI. As shown in Figure 5a, the XPS survey spectra of CIS and IS demonstrated that they mainly contain elements of Cd, In, and S, while the sample of IS has no signals of Cd. As for the XPS of the Cd 3d spectrum of CIS (Figure 5b), two strong peaks appear at 411.8 and 405.1 eV, corresponding to Cd 3d3/2 and Cd 3d5/2, respectively [29]. However, the peak positions in 0.5 CISI are positively shifted to 412.2 and 405.3 eV. This indicates that the elemental valences of Cd in CIS and 0.5 CISI are +2 [40]. The XPS In 3d spectrum of CIS shows two peaks centered at 452.3 eV and 444.7 eV (Figure 5c), which are assigned to In 3d3/2 and In 3d5/2, respectively [41]. These two characteristic peaks also appear in pure IS. After coupling CIS with IS, there is an obviously positive shift for XPS iIn the 3d spectrum of 0.5 CISI compared to CIS and IS. As shown in Figure 5d, the S 2p XPS spectra of CIS and IS possess similar signals at ~162.6 and ~161.4 eV, which belong to S 2p1/2 and S 2p3/2, respectively [42]. Clearly, these two peaks of XPS S 2p in 0.5 CISI also exhibit a positive shift related to CIS and IS. Combined with the XPS results of Cd 3d and In 3d, we reasonably conclude that there is a strong interaction existing in the CIS/IS heterojunction, which is conducive to the efficient charge carrier transfer [43].

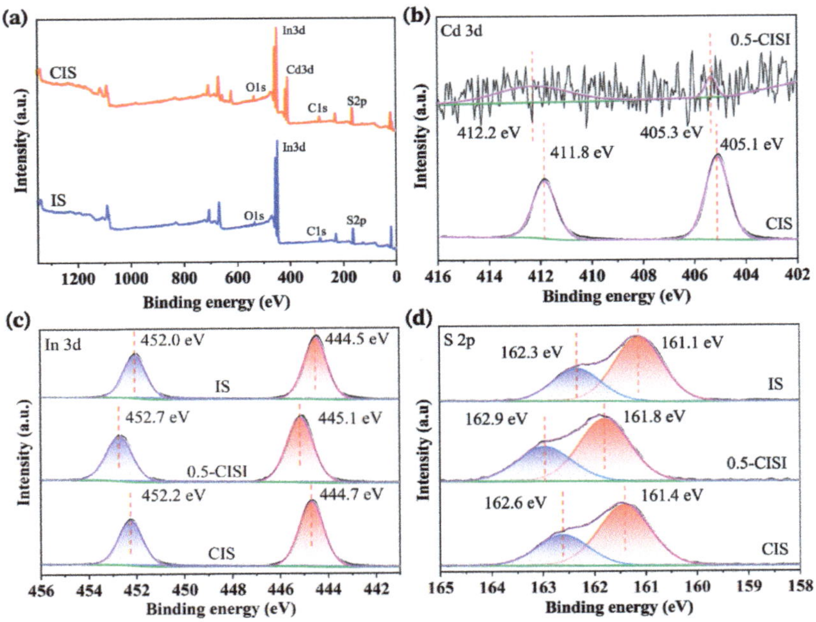

Figure 5. High-resolution XPS spectra of CIS, IS, and 0.5 CISI: (**a**) survey, (**b**) Cd 3d, (**c**) In 3d, and (**d**) S 2p.

The steady-state photoluminescence (PL) spectra of the as-prepared samples were employed to investigate the charge separation behavior (Figure 6a) [44]. The PL peak intensities of 0.25 CISI, 0.5 CISI and 0.75 CISI are weaker than those of CIS and IS, demonstrating that the heterojunction formation between CIS and IS can effectively promote the charge separation. Additionally, the highest PL intensity being that of the 0.5 CISI sample suggests that the proper mass ratio between CIS and IS can achieve the effective heterojunction for efficient charge carrier separation, resulting in the improved photocatalytic activity for Cr(VI) reduction.

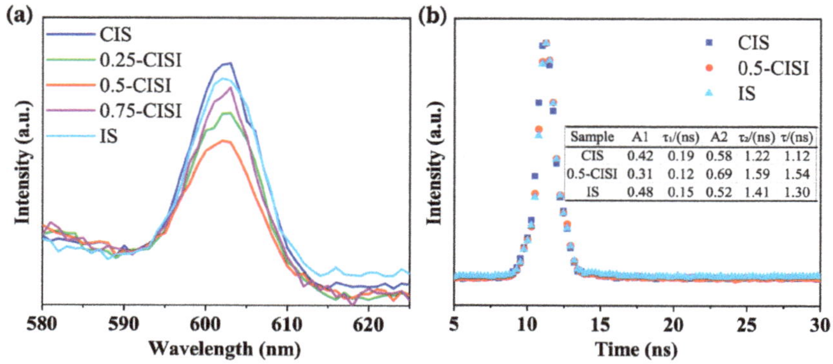

Figure 6. (**a**) PL spectra with the excitation wavelength of 400 nm and (**b**) TR-PL decay spectra of as-prepared samples.

The time-resolved PL (TR-PL) decay curves in Figure 6b also support the above results. The fluorescence lifetime is closely related to the charge carrier separation rate [45,46]. Generally, a longer fluorescence lifetime means a higher charge separation efficiency. Based on our previous work [37], the average lifetimes of CIS, IS, and 0.5 CISI are calculated to be 1.12, 1.30, and 1.54 ns, respectively. Obviously, the average lifetime of 0.5 CISI is higher than that of CIS and IS, reflecting that the heterojunction formation between CIS and IS can prolong the lifetime of photogenerated charge carriers and thus boost activity for Cr(VI) photoreduction.

The electrochemical impedance spectra (EIS) and the transient photocurrent responses (TPRs) were performed to reveal the charge separation and migration behavior of the samples [47–49]. As for the EIS measurements, the smaller arc radius is closely related to the lower interfacial resistance, resulting in a higher charge migration efficiency in the Nyquist plots. As shown in Figure 7a, the arc radii of 0.25 CISI, 0.5 CISI, and 0.75 CISI are smaller than those of CIS and IS, demonstrating that the formation of the heterojunction can effectively reduce the interfacial resistance and thus promote the charge carrier separation. The optimal sample, 0.5 CISI, has the smallest arc radius, which means the highest migration rate of photogenerated charge carriers in all CISI composites.

The TPR spectra of samples are displayed in Figure 7b for a total of nine on–off photoperiod tests. Pure CIS and IS show relatively low photocurrent intensities due to their rapid recombination rate of photogenerated charge carriers, leading to the low availability of photogenerated electrons [50]. When CIS was coupled with IS to construct heterostructured photocatalysts, all CISI composites exhibited the increased TPR intensity, suggesting that the recombination rate of electrons and holes is effectively suppressed [51,52]. Among all composites, the sample of 0.5 CISI shows the highest TPR intensity, which means an excellent separation efficiency of photogenerated charge carriers for potentially achieving the highest photocatalytic activity.

The linear sweep voltammetry (LSV) plots of the samples are presented in Figure 7c in order to investigate the hydrogen evolution kinetics over various samples. Among all samples, 0.5 CISI possesses the lowest onset potential and the highest cathodic current density, which is advantageous for electrocatalytic hydrogen evolution reaction. Thus, the formation of the heterojunction is beneficial to the decreased onset potential and enhanced kinetics for hydrogen evolution, indirectly manifesting the increased reduction capacity.

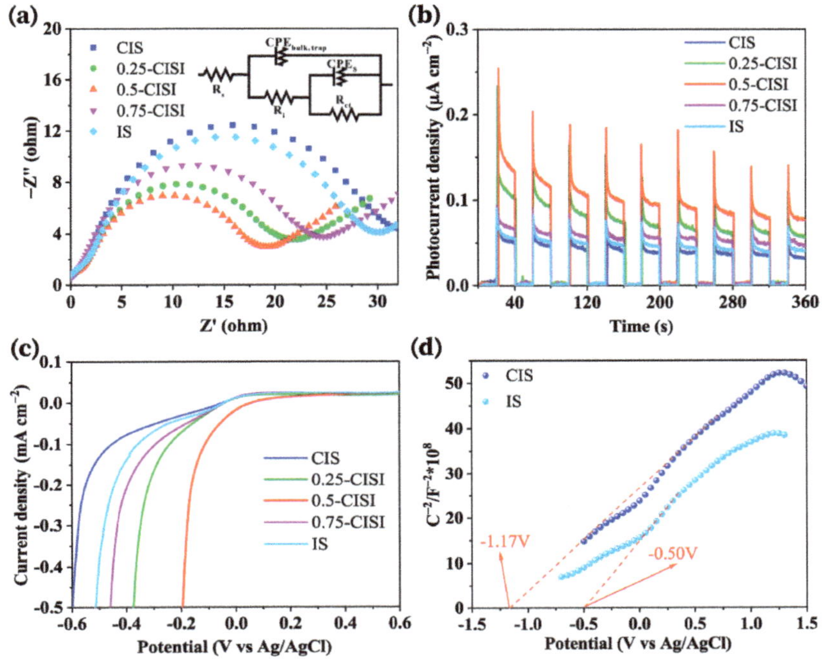

Figure 7. (**a**) Electrochemical impedance spectra (EIS) (inset is the equivalent circuit to fit the plots, where R_s represents the overall series resistance of the circuit, R_i represents the internal resistance of the electrode along with a constant phase element ($CPE_{bulk,\ trap}$), and R_{ct} denotes the charge transfer resistance from the surface states to the solution with the corresponding CPE_s). (**b**) Transient photocurrent responses (TPRs). (**c**) Linear sweep voltammetry (LSV) plots. (**d**) Mott–Schottky curves of as-prepared samples measured at a frequency of 1000 Hz.

The matched band structure of photocatalysts determines their Cr(VI) photoreduction performance. Mott–Schottky (M-S) plots of CIS and IS are measured at a frequency of 1000 Hz to determine their band positions, as displayed in Figure 7d. Based on the X-axis tangents intercept, the flat-band potentials (E_{fb}) of CIS and IS can be estimated to be −1.17 and −0.50 V (vs. Ag/AgCl at pH = 7.0), respectively. The positive slopes over CIS and IS are their n-type semiconductors, where the conduction band potential (E_{CB}) is generally 0~0.2 V more negative than its E_{fb} [53]. Considering the voltage discrepancy of 0.1 V between E_{CB} and E_{fb}, the E_{CB} of CIS and IS are calculated as −1.27 V and −0.60 V (vs. Ag/AgCl at pH = 7.0), respectively [54]. Based on the formula, $E_{NHE} = E_{Ag/AgCl} + 0.059 * pH - E^{\circ}_{Ag/AgCl}$ ($E^{\circ}_{Ag/AgCl} = 0.197$ V, pH = 7.0), the E_{CB} of CIS and IS are equal to −1.05 eV and −0.38 eV versus the normal hydrogen electrode (NHE), respectively. Considering the bandgap values of 2.47 eV for CIS and 2.23 eV for IS, the valence band (VB) potentials (E_{VB}) of CIS and IS are estimated to be +1.42 eV and +1.85 eV (vs. NHE), respectively.

The photocatalytic activity of the as-prepared samples was evaluated by Cr(VI) reduction under visible light irradiation. As shown in Figure 8a, the samples of CIS and IS show relatively low photocatalytic activity due to their rapid recombination rate of the photogenerated charge carrier. After coupling CIS with IS, the CISI composites exhibit a higher Cr(VI) photoreduction activity than CIS and IS. Then, the enhanced photo-activity of the CISI composites is assigned to the efficient charge carrier separation derived from the heterojunction formation between CIS and IS. Correspondingly, the reaction rate constants of all samples are measured with the following order: CIS (0.03554 min^{-1}) < IS (0.05578 min^{-1}) < 0.75 CISI (0.09358 min^{-1}) < 0.25 CISI (0.13313 min^{-1}) < 0.5 CISI (0.15561 min^{-1}) (Figure 8b,c). Notably, the optimal sample of 0.5 CISI has the highest Cr(VI) photoreduction efficiency, indicating that the proper mass ratio between CIS and IS is necessary to achieve the effective heterojunction for boosting photocatalytic activity. To monitor the photocatalytic process of Cr(VI) reduction, the UV-vis spectral changes over 0.5 CISI are displayed in Figure 8d toward the Cr(VI) solution with an absorption characteristic peak at approximately 540 nm. As the irradiation time increases, the peak intensity is gradually decreased and Cr(VI) is almost completely removed within 45 min. In addition, the overall Cr(VI) photoreduction efficiency over 0.5 CISI is higher than that of most reported photocatalysts, as listed in Table 1.

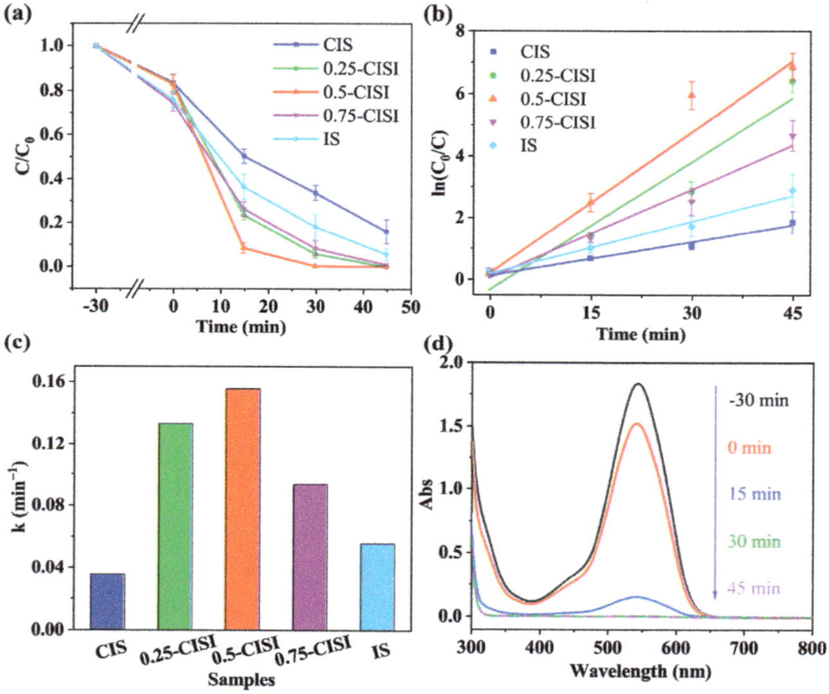

Figure 8. (**a**) Photocatalytic Cr(VI) reduction curves, (**b**) the kinetic curves, and (**c**) the kinetic value of samples. (**d**) UV-vis spectral changes for Cr(VI) photo-reduction over 0.5 CISI.

Table 1. Comparison of photocatalytic Cr(VI) reduction over different catalysts.

Catalyst	Catalyst (mg)	Cr(VI) (mg/L)	Solution (mL)	Time (min)	Refs.
Cu(X)–$CdIn_2S_4$	20	20	40	110	[55]
$CdIn_2S_4/In(OH)_3$/NiCr-LDH	30	50	50	120	[56]
$CdIn_2S_4$@UiO-66-NH_2	50	20	100	120	[57]
Bi_2S_3-In_2S_3	20	70	100	140	[58]
In_2S_3/TiO_2	50	20	100	80	[59]
$Mo_2C/MoS_2/In_2S_3$	50	40	50	90	[60]
In_2S_3/G	15	10	30	50	[61]
In_2S_3/BiOBr	200	50	500	180	[62]
This work	10	20	100	45	-

The cycling experiments for the Cr(VI) photoreduction were conducted to evaluate the photocatalytic stability of the catalyst. As shown in Figure 9a, after five cycles under the same condition, the sample of 0.5 CISI still exhibits the highly photocatalytic efficiency. As shown in Figure 9b, after Cr(VI) photoreduction cycling experiments, the XRD pattern of 0.5 CISI shows no obvious change but does show the slightly decreased intensity compared to the fresh 0.5 CISI sample, indicating the highly structural stability. Additionally, SEM images of 0.5 CISI before (Figure 1d) and after (Figure 9c) Cr(VI) photoreduction are similar to each other. These results verify the excellent structural stability of 0.5 CISI, which demonstrates good reusability in practical application.

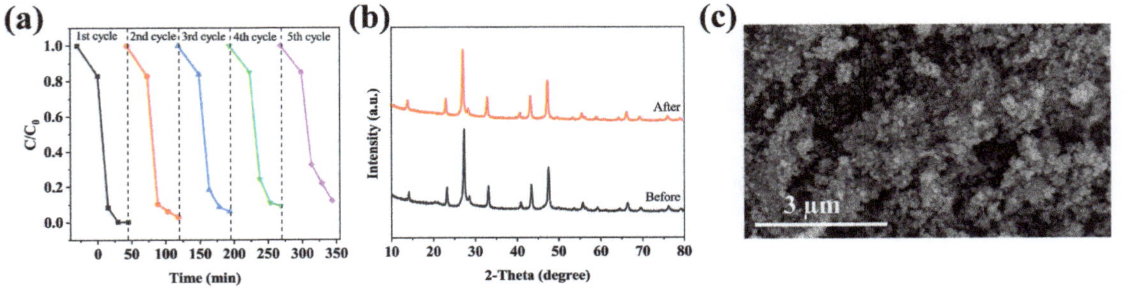

Figure 9. (**a**) Cyclic performance of 0.5 CISI for the reduction of aqueous Cr(VI) under visible light irradiation. (**b**) XRD patterns of 0.5 CISI before and after Cr(VI) photoreduction for five cycles. (**c**) SEM image of 0.5 CISI after Cr(VI) photoreduction for five cycles.

In light of the above results and discussion, a possible mechanism for Cr(VI) photoreduction was proposed for 0.5 CISI (Figure 10). A one-step synthesis approach is beneficial to the formation of an intimate interface between CIS and IS in the 0.5 CISI composite. Under visible light, the components of CIS and IS in 0.5 CISI are excited to produce electron and hole pairs. Considering the potential differences, the photogenerated electrons are quickly migrated from CB of CIS to CB of IS across the intimate coupling interface. Meanwhile, an opposite pathway for the transport of photogenerated holes appears from VB of IS to VB of CIS. Subsequently, parts of the powerful electrons in CB of IS contribute to the formation of $^{\bullet}O_2^-$. Consequently, these generated $^{\bullet}O_2^-$ and e^- radicals mainly participate in Cr(VI) photoreduction.

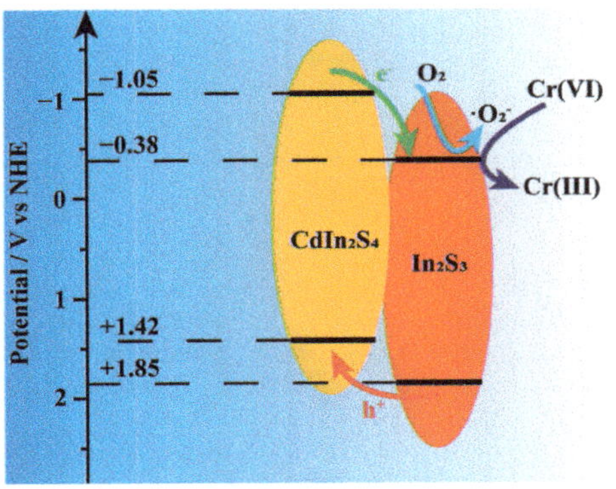

Figure 10. Photocatalytic mechanism for Cr(VI) reduction on 0.5 CISI.

4. Conclusions

In this work, $CdIn_2S_4/In_2S_3$ (CISI) heterostructured photocatalysts were constructed by a one-step refluxing process, showing an intimate interface between CIS and IS. This formed tight interface between two components could effectively promote the separation and migration of photogenerated electrons and holes. Compared to CIS and IS, the resulting

CISI composites exhibited the enhanced activity for visible-light-driven Cr(VI) photoreduction due to the heterojunction formation between CIS and IS. The optimal sample of 0.5 CISI exhibited the highest Cr(VI) photoreduction efficiency and a reaction rate constant of 0.15561 min^{-1} for the complete removal Cr(VI) within 45 min. This implies that the proper mass ratio between CIS and IS could achieve the effective heterojunction for boosting photocatalytic activity. The results of the PL, TR-PL, EIS, and TPR curves significantly verified that the heterojunction formation between CIS and IS could effectively promote the separation and migration of photogenerated electrons and holes, leading to highly photocatalytic activity. Based on the characterization results, a possible photocatalytic mechanism was speculated for Cr(VI) reduction.

Author Contributions: Conceptualization, J.H. and C.L.; methodology, J.W.; software, J.H.; validation, X.L., C.L. and Q.Z.; formal analysis, J.H.; investigation, S.Z.; resources, C.L.; data curation, J.H.; writing—original draft preparation, W.X.; writing—review and editing, C.L.; visualization, W.C.; supervision, X.L.; project administration, H.H.; funding acquisition, C.L. All authors have read and agreed to the published version of the manuscript.

Funding: This research was funded by the National Natural Science Foundation of China, grant numbers No. 51902282 and 12274361. This work was also supported by the Qinglan Project of Jiangsu of China, the Natural Science Foundation of Jiangsu Province (BK20211361), and the College Natural Science Research Project of Jiangsu Province (20KJA430004).

Data Availability Statement: Data are available on request from the authors.

Conflicts of Interest: The authors declare no conflict of interest.

References

1. Wang, C.-C.; Ren, X.; Wang, P.; Chang, C. The state of the art review on photocatalytic Cr(VI) reduction over MOFs-based photocatalysts: From batch experiment to continuous operation. *Chemosphere* **2022**, *303*, 134949. [CrossRef] [PubMed]
2. Jafarzadeh, M. Recent progress in the development of MOF-based photocatalysts for the photoreduction of Cr(VI). *ACS Appl. Mater. Interfaces* **2022**, *14*, 24993–25024. [CrossRef] [PubMed]
3. Li, S.; Cai, M.; Wang, C.; Liu, Y. Ta_3N_5/CdS core-shell S-scheme heterojunction nanofibers for efficient photocatalytic removal of antibiotic tetracycline and Cr(VI): Performance and mechanism insights. *Adv. Fiber Mater.* 2023. [CrossRef]
4. Liu, C.; Xiao, W.; Yu, G.; Wang, Q.; Hu, J.; Xu, C.; Du, X.; Xu, J.; Zhang, Q.; Zou, Z. Interfacial engineering of Ti_3C_2 MXene/$CdIn_2S_4$ Schottky heterojunctions for boosting visible-light H_2 evolution and Cr(VI) reduction. *J. Colloid Interface Sci.* **2023**, *640*, 851–863. [CrossRef]
5. Hunge, Y.M.; Yadav, A.; Kang, S.-W.; Kim, H.; Fujishima, A.; Terashima, C. Nanoflakes-like nickel cobaltite as active electrode material for 4-nitrophenol reduction and supercapacitor applications. *J. Hazard. Mater.* **2021**, *419*, 126453. [CrossRef]
6. Hunge, Y.M.; Yadav, A.A.; Kang, S.-W.; Mohite, B. Role of Nanotechnology in Photocatalysis Application. *Recent Pat. Nanotechnol.* **2023**, *17*, 5–7. [CrossRef]
7. Hunge, Y.M.; Yadav, A.A.; Kang, S.-W.; Kim, H. Facile synthesis of multitasking composite of Silver nanoparticle with Zinc oxide for 4-nitrophenol reduction, photocatalytic hydrogen production, and 4-chlorophenol degradation. *J. Alloys Compd.* **2022**, *928*, 167133. [CrossRef]
8. Liu, C.; Zhang, Q.; Hou, W.; Zou, Z. Two-dimensional titanium/niobium metal oxide-based materials for photocatalytic application. *Sol. RRL* **2020**, *4*, 2000070. [CrossRef]
9. Wang, L.; Zhu, B.; Cheng, B.; Zhang, J.; Zhang, L.; Yu, J. In-situ preparation of TiO_2/N-doped graphene hollow sphere photocatalyst with enhanced photocatalytic CO_2 reduction performance. *Chin. J. Catal.* **2021**, *42*, 1648–1658. [CrossRef]
10. Boumeriame, H.; Da Silva, E.S.; Cherevan, A.S.; Chafik, T.; Faria, J.L.; Eder, D. Layered double hydroxide (LDH)-based materials: A mini-review on strategies to improve the performance for photocatalytic water splitting. *J. Energy Chem.* **2022**, *64*, 406–431. [CrossRef]
11. Shandilya, P.; Sambyal, S.; Sharma, R.; Mandyal, P.; Fang, B. Properties, optimized morphologies, and advanced strategies for photocatalytic applications of WO_3 based photocatalysts. *J. Hazard. Mater.* **2022**, *428*, 128218. [CrossRef]
12. Pan, C.; Bian, G.; Zhang, Y.; Lou, Y.; Dong, Y.; Xu, J.; Zhu, Y. Efficient and stable H_2O_2 production from H_2O and O_2 on $BiPO_4$ photocatalyst. *Appl. Catal. B Environ.* **2022**, *316*, 121675. [CrossRef]
13. Zhang, G.; Guan, Z.; Yang, J.; Li, Q.; Zhou, Y.; Zou, Z. Metal sulfides for photocatalytic hydrogen production: Current development and future challenges. *Sol. RRL* **2022**, *6*, 2200587. [CrossRef]
14. Liu, C.; Zhang, Q.; Zou, Z. Recent advances in designing $ZnIn_2S_4$-based heterostructured photocatalysts for hydrogen evolution. *J. Mater. Sci. Technol.* **2023**, *139*, 167–188. [CrossRef]

15. Zhao, S.; Li, K.; Wu, J.; Zhang, J.; Li, X.; Guo, X.; Song, C. Metal-Organic Framework-Derived Tubular In_2O_3-C/$CdIn_2S_4$ Heterojunction for Efficient Solar-Driven CO_2 Conversion. *ACS Appl. Mater. Interfaces* **2022**, *14*, 20375–20384. [CrossRef]
16. Nagappagari, L.R.; Patil, S.S.; Lee, J.; Park, E.; Yu, Y.-T.; Lee, K. Upgraded charge transport in g-C_3N_4 nanosheets by boron doping and their heterojunction with 3D $CdIn_2S_4$ for efficient photodegradation of azo dye. *Mater. Today Chem.* **2022**, *24*, 100857. [CrossRef]
17. Ma, X.; Li, W.; Ren, C.; Dong, M.; Geng, L.; Fan, H.; Li, Y.; Qiu, H.; Wang, T. Construction of novel noble-metal-free MoP/$CdIn_2S_4$ heterojunction photocatalysts: Effective carrier separation, accelerating dynamically H_2 release and increased active sites for enhanced photocatalytic H_2 evolution. *J. Colloid Interface Sci.* **2022**, *628*, 368–377. [CrossRef]
18. Liu, R.Y.; Minh Man, T.; Chang, M.B. Photocatalytic removal of toluene with $CdIn_2S_4$/CNFs catalyst: Effect of ozone addition. *Sustain. Environ. Res.* **2022**, *32*, 10. [CrossRef]
19. Li, C.; Ding, G.; Liu, X.; Huo, P.; Yan, Y.; Yan, Y.; Liao, G. Photocatalysis over NH_2-UiO-66/$CoFe_2O_4$/$CdIn_2S_4$ double p-n junction: Significantly promoting photocatalytic performance by double internal electric fields. *Chem. Eng. J.* **2022**, *435*, 134740. [CrossRef]
20. Huang, J.; Huang, Z.; Xu, S.; Fan, T.; Yi, X.; Chen, Z. Photocatalytic oxidative coupling of benzylamine to schiff base over 0D/2D CdS/$CdIn_2S_4$ heterojunction. *Energy Technol.* **2022**, *10*, 2200362. [CrossRef]
21. Jiang, W.; Yin, X.; Xin, F.; Bi, Y.; Liu, Y.; Li, X. Preparation of $CdIn_2S_4$ microspheres and application for photocatalytic reduction of carbon dioxide. *Appl. Surf. Sci.* **2014**, *288*, 138–142. [CrossRef]
22. Wang, W.; Ng, T.W.; Ho, W.; Huang, J.; Liang, S.; An, T.; Li, G.; Yu, J.C.; Wong, P.K. $CdIn_2S_4$ microsphere as an efficient visible-light-driven photocatalyst for bacterial inactivation: Synthesis, characterizations and photocatalytic inactivation mechanisms. *Appl. Catal. B Environ.* **2013**, *129*, 482–490. [CrossRef]
23. Gao, L.; Chen, Z.; Zheng, H.; Hu, J. A distinct hollow spindle-like $CdIn_2S_4$ photocatalyst for high-efficiency tetracycline removal. *Mater. Today Chem.* **2022**, *24*, 100800. [CrossRef]
24. Liu, C.; Feng, Y.; Han, Z.; Sun, Y.; Wang, X.; Zhang, Q.; Zou, Z. Z-scheme N-doped $K_4Nb_6O_{17}$/g-C_3N_4 heterojunction with superior visible-light-driven photocatalytic activity for organic pollutant removal and hydrogen production. *Chin. J. Catal.* **2021**, *42*, 164–174. [CrossRef]
25. Liu, C.; Zhu, H.; Zhu, Y.; Dong, P.; Hou, H.; Xu, Q.; Chen, X.; Xi, X.; Hou, W. Ordered layered N-doped $KTiNbO_5$/g-C_3N_4 heterojunction with enhanced visible light photocatalytic activity. *Appl. Catal. B Environ.* **2018**, *228*, 54–63. [CrossRef]
26. Manh-Hiep, V.; Nguyen, C.-C.; Sakar, M.; Do, T.-O. Ni supported $CdIn_2S_4$ spongy-like spheres: A noble metal free high-performance sunlight driven photocatalyst for hydrogen production. *Phys. Chem. Chem. Phys.* **2017**, *19*, 29429–29437.
27. Zhang, P.; Yin, L.; Yang, X.; Wang, J.; Chi, M.; Qiu, J. Cotton-derived 3D carbon fiber aerogel to in situ support Bi_2O_3 nanoparticles as a separation-free photocatalyst for antibiotic removal. *Carbon* **2023**, *201*, 110–119. [CrossRef]
28. Zhang, G.; Wang, Z.; He, T.; Wu, J.; Zhang, J.; Wu, J. Rationally design and in-situ fabrication of ultrasmall pomegranate-like $CdIn_2S_4$/$ZnIn_2S_4$ Z-scheme heterojunction with abundant vacancies for improving CO_2 reduction and water splitting. *Chem. Eng. J.* **2022**, *442*, 136309. [CrossRef]
29. Chen, C.; Jiang, T.; Hou, J.; Zhang, T.; Zhang, G.; Zhang, Y.; Wang, X. Oxygen vacancies induced narrow band gap of BiOCl for efficient visible-light catalytic performance from double radicals. *J. Mater. Sci. Technol.* **2022**, *114*, 240–248. [CrossRef]
30. Wang, Q.; Wang, L.; Jiang, Y.; Liu, Y.; Zhang, W.; Zhang, J.; Macauley, A.L.O. Morphology-engineered carbon quantum dots embedded on octahedral $CdIn_2S_4$ for enhanced photocatalytic activity towards pollutant degradation and hydrogen evolution. *Environ. Res.* **2022**, *209*, 112800. [CrossRef]
31. Zhao, X.; Zhou, Y.; Liang, Q.; Zhou, M.; Li, Z.; Xu, S. Coupling MOF-derived titanium oxide with $CdIn_2S_4$ formed 2D/3D core–shell heterojunctions with enhanced photocatalytic performance. *Sep. Purif. Technol.* **2021**, *279*, 119765. [CrossRef]
32. Zhang, Z.; Cao, Y.; Zhang, F.; Li, W.; Li, Y.; Yu, H.; Wang, M.; Yu, H. Tungsten oxide quantum dots deposited onto ultrathin $CdIn_2S_4$ nanosheets for efficient S-scheme photocatalytic CO_2 reduction via cascade charge transfer. *Chem. Eng. J.* **2022**, *428*, 131218. [CrossRef]
33. Huang, H.; Jiang, X.; Li, N.; Chen, D.; Xu, Q.; Li, H.; He, J.; Lu, J. Noble-metal-free ultrathin MXene coupled with In2S3 nanoflakes for ultrafast photocatalytic reduction of hexavalent chromium. *Appl. Catal. B Environ.* **2021**, *284*, 119754. [CrossRef]
34. Liu, C.; Han, Z.; Feng, Y.; Dai, H.; Zhao, Y.; Han, N.; Zhang, Q.; Zou, Z. Ultrathin Z-scheme 2D/2D N-doped $HTiNbO_5$ nanosheets/g-C_3N_4 porous composites for efficient photocatalytic degradation and H_2 generation under visible light. *J. Colloid Interface Sci.* **2021**, *583*, 58–70. [CrossRef] [PubMed]
35. Yu, G.; Hu, F.; Cheng, W.; Han, Z.; Liu, C.; Dai, Y. ZnCuAl-LDH/Bi_2MoO_6 nanocomposites with improved visible light-driven photocatalytic degradation. *Acta Phys.-Chim. Sin.* **2020**, *36*, 1911016.
36. Liu, C.; Zhang, Y.; Wu, J.; Dai, H.; Ma, C.; Zhang, Q.; Zou, Z. Ag-Pd alloy decorated $ZnIn_2S_4$ microspheres with optimal Schottky barrier height for boosting visible-light-driven hydrogen evolution. *J. Mater. Sci. Technol.* **2022**, *114*, 81–89. [CrossRef]
37. Yu, G.; Zhang, Y.; Du, X.; Wu, J.; Liu, C.; Zou, Z. In-situ synthesis of nickel/palladium bimetal/ZnIn2S4 Schottky heterojunction for efficient photocatalytic hydrogen evolution. *J. Colloid Interface Sci.* **2022**, *623*, 205–215. [CrossRef]
38. Zhang, W.; Zhao, S.; Xing, Y.; Qin, H.; Zheng, Q.; Zhang, P.; Zhang, S.; Xu, X. Sandwich-like P-doped h-BN/$ZnIn_2S_4$ nanocomposite with direct Z-scheme heterojunction for efficient photocatalytic H_2 and H_2O_2 evolution. *Chem. Eng. J.* **2022**, *442*, 136151. [CrossRef]

39. Meng, S.; Chen, C.; Gu, X.; Wu, H.; Meng, Q.; Zhang, J.; Chen, S.; Fu, X.; Liu, D.; Lei, W. Efficient photocatalytic H_2 evolution, CO_2 reduction and N_2 fixation coupled with organic synthesis by cocatalyst and vacancies engineering. *Appl. Catal. B Environ.* **2021**, *285*, 119789. [CrossRef]
40. Huang, L.; Li, B.; Su, B.; Xiong, Z.; Zhang, C.; Hou, Y.; Ding, Z.; Wang, S. Fabrication of hierarchical Co_3O_4@$CdIn_2S_4$ p–n heterojunction photocatalysts for improved CO_2 reduction with visible light. *J. Mater. Chem. A* **2020**, *8*, 7177–7183. [CrossRef]
41. Bai, J.; Chen, W.; Shen, R.; Jiang, Z.; Zhang, P.; Liu, W.; Li, X. Regulating interfacial morphology and charge-carrier utilization of Ti_3C_2 modified all-sulfide $CdS/ZnIn_2S_4$ S-scheme heterojunctions for effective photocatalytic H_2 evolution. *J. Mater. Sci. Technol.* **2022**, *112*, 85–95. [CrossRef]
42. Li, C.; Zhao, Y.; Liu, X.; Huo, P.; Yan, Y.; Wang, L.; Liao, G.; Liu, C. Interface engineering of $Co_9S_8/CdIn_2S_4$ ohmic junction for efficient photocatalytic H_2 evolution under visible light. *J. Colloid Interface Sci.* **2021**, *600*, 794–803. [CrossRef] [PubMed]
43. Zuo, G.; Wang, Y.; Teo, W.L.; Xie, A.; Guo, Y.; Dai, Y.; Zhou, W.; Jana, D.; Xian, Q.; Dong, W.; et al. Enhanced photocatalytic water oxidation by hierarchical 2D-Bi_2MoO_6@2D-MXene Schottky junction nanohybrid. *Chem. Eng. J.* **2021**, *403*, 126328. [CrossRef]
44. Wang, C.; Yan, R.; Cai, M.; Liu, Y.; Li, S. A novel organic/inorganic S-scheme heterostructure of TCPP/$Bi_{12}O_{17}Cl_2$ for boosting photodegradation of tetracycline hydrochloride: Kinetic, degradation mechanism, and toxic assessment. *Appl. Surf. Sci.* **2023**, *610*, 155346. [CrossRef]
45. Liu, C.; Gao, X.; Zhan, C.; Cheng, T.; Wang, Y.; Zhang, B.; Dong, P.; Chen, X.; Xi, X.; Zou, Z. Layered BiOCl/$H^+TiNbO_5^-$ heterojunctions for boosting visible-light-driven photocatalytic RhB degradation. *Sustain. Energy Fuels* **2021**, *5*, 4680–4689. [CrossRef]
46. Liu, C.; Xu, Q.; Zhang, Q.; Zhu, Y.; Ji, M.; Tong, Z.; Hou, W.; Zhang, Y.; Xu, J. Layered BiOBr/Ti_3C_2 MXene composite with improved visible-light photocatalytic activity. *J. Mater. Sci.* **2019**, *54*, 2458–2471. [CrossRef]
47. Hou, J.; Yang, M.; Dou, Q.; Chen, Q.; Wang, X.; Hu, C.; Paul, R. Defect engineering in polymeric carbon nitride with accordion structure for efficient photocatalytic CO_2 reduction and H_2 production. *Chem. Eng. J.* **2022**, *450*, 138425. [CrossRef]
48. Li, S.; Wang, C.; Liu, Y.; Liu, Y.; Cai, M.; Zhao, W.; Duan, X. S-scheme MIL-101(Fe) octahedrons modified Bi_2WO_6 microspheres for photocatalytic decontamination of Cr(VI) and tetracycline hydrochloride: Synergistic insights, reaction pathways, and toxicity analysis. *Chem. Eng. J.* **2023**, *455*, 140943. [CrossRef]
49. Zhang, B.; Zhang, L.; Zhang, Y.; Liu, C.; Xia, J.; Li, H. Ionic Liquid-Assisted Synthesis of Ag_3PO_4 Spheres for Boosting Photodegradation Activity under Visible Light. *Catalysts* **2021**, *11*, 788. [CrossRef]
50. Liu, C.; Wu, Q.; Ji, M.; Zhu, H.; Hou, H.; Yang, Q.; Jiang, C.; Wang, J.; Tian, L.; Chen, J.; et al. Constructing Z-scheme charge separation in 2D layered porous BiOBr/graphitic C_3N_4 nanosheets nanojunction with enhanced photocatalytic activity. *J. Alloys Compd.* **2017**, *723*, 1121–1131. [CrossRef]
51. Li, S.; Cai, M.; Liu, Y.; Wang, C.; Yan, R.; Chen, X. Constructing $Cd_{0.5}Zn_{0.5}S/Bi_2WO_6$ S-scheme heterojunction for boosted photocatalytic antibiotic oxidation and Cr(VI) reduction. *Adv. Powder Mater.* **2023**, *2*, 100073. [CrossRef]
52. Li, S.; Cai, M.; Liu, Y.; Wang, C.; Lv, K.; Chen, X. S-scheme photocatalyst TaON/Bi_2WO_6 nanofibers with oxygen vacancies for efficient abatement of antibiotics and Cr(VI): Intermediate eco-toxicity analysis and mechanistic insights. *Chin. J. Catal.* **2022**, *43*, 2652–2664. [CrossRef]
53. Sun, Y.; Zhu, Q.; Bai, B.; Li, Y.; He, C. Novel all-solid-state Z-scheme $SnO_2/Pt/In_2O_3$ photocatalyst with boosted photocatalytic performance on water splitting and 2,4-dichlorophenol degradation under visible light. *Chem. Eng. J.* **2020**, *390*, 124518. [CrossRef]
54. Yang, W.; Ma, G.; Fu, Y.; Peng, K.; Yang, H.; Zhan, X.; Yang, W.; Wang, L.; Hou, H. Rationally designed Ti_3C_2 MXene@$TiO_2/CuInS_2$ Schottky/S-scheme integrated heterojunction for enhanced photocatalytic hydrogen evolution. *Chem. Eng. J.* **2022**, *429*, 132381. [CrossRef]
55. Chen, Z.; Ren, Z.; Xu, J.; He, Y.; Xiao, G. Hydrothermal synthesis of Cu (X)–$CdIn_2S_4$ microspheres and photocatalytic reduction of aqueous Cr (VI) under visible light irradiation. *Mater. Lett.* **2014**, *117*, 17–20. [CrossRef]
56. Fu, R.; Gong, Y.; Li, C.; Niu, L.; Liu, X. $CdIn_2S_4$/In $(OH)_3$/NiCr-LDH Multi-Interface Heterostructure Photocatalyst for Enhanced Photocatalytic H_2 Evolution and Cr (VI) Reduction. *Nanomaterials* **2021**, *11*, 3122. [CrossRef]
57. He, J.; Hu, J.; Hu, Y.; Guo, S.; Huang, Q.; Li, Y.; Zhou, G.; Gui, T.; Hu, N.; Chen, X. Hierarchical S-Scheme Heterostructure of $CdIn_2S_4$@ UiO-66-NH_2 toward Synchronously Boosting Photocatalytic Removal of Cr (VI) and Tetracycline. *Inorg. Chem.* **2022**, *61*, 19961–19973. [CrossRef]
58. Zheng, X.; Liu, T.; Wen, J.; Liu, X. Flower-like Bi_2S_3–In_2S_3 heterojunction for efficient solar light induced photoreduction of Cr (VI). *Chemosphere* **2021**, *278*, 130422. [CrossRef]
59. Zhang, X.; Li, X.; Shao, C.; Li, J.; Zhang, M.; Zhang, P.; Wang, K.; Lu, N.; Liu, Y. One-dimensional hierarchical heterostructures of In_2S_3 nanosheets on electrospun TiO_2 nanofibers with enhanced visible photocatalytic activity. *J. Hazard. Mater.* **2013**, *260*, 892–900. [CrossRef]
60. Zhang, X.; Tian, F.; Qiu, L.; Gao, M.; Yang, W.; Liu, Y.; Yu, Y. Z-Scheme $Mo_2C/MoS_2/In_2S_3$ dual-heterojunctions for the photocatalytic reduction of Cr (vi). *J. Mater. Chem. A* **2021**, *9*, 10297–10303. [CrossRef]

61. Chen, L.; Yang, G.; Wei, X.; Xu, H.; Jin, S. One-pot soft integration fabrication of graphene-induced phase transition to form dimension control contact In_2S_3/G heterojunction hybrids for enhancing visible photocatalytic purification performance. *J. Alloys Compd.* **2022**, *895*, 162589. [CrossRef]
62. Fu, X.; Tao, J.; He, Z.; Gao, Y.; Xia, Y.; Zhao, Z. Synergy of Z-scheme heterostructure with interfacial S−O bonding in In_2S_3/BiOBr for efficient tetracycline hydrochloride degradation and Cr (VI) reduction. *J. Alloys Compd.* **2023**, *936*, 168202. [CrossRef]

catalysts

MDPI

Article

Preparation of Two-Dimensional Layered CeO_2/Bi_2O_3 Composites for Efficient Photocatalytic Desulfurization

Xiaowang Lu [1,2,*], Wenxuan Chen [1], Haijun Hou [1,2], Junchao Qian [3] and Qinfang Zhang [1,*]

1 School of Material Science and Engineering, Yancheng Institute of Technology, Yancheng 224051, China
2 Jiangsu Provincial Key Laboratory of Eco-Environmental Materials, Yancheng Institute of Technology, Yancheng 224051, China
3 School of Material Science and Engineering, Suzhou University of Science and Technology, Suzhou 215009, China
* Correspondence: luxiaowang@ycit.edu.cn (X.L.); qfangzhang@gmail.com (Q.Z.)

Abstract: A two-dimensional layered CeO_2/Bi_2O_3 composite was synthesized by microwave solvothermal method. X-ray diffraction (XRD), Raman spectroscopy, field emission scanning electron microscope (FESEM), transmission electron microscope (TEM), UV-Vis diffuse reflection spectroscopy (DRS), and X-ray photoelectron spectroscopy (XPS) were used to studied crystal structure, morphology, optical performance, elemental composition and the surface electronic state of the samples. The photocatalytic properties of the prepared samples were evaluated by photocatalytic desulfurization under visible light. When the molar ratio of Ce and Bi was 1:2, CeO_2/Bi_2O_3 composite presented the highest photocatalytic desulfurization rate. Transient Photocurrent measurement, electrochemical impedance spectroscopy (EIS) and photoluminescence spectroscopy (PL) showed that CeO_2 and Bi_2O_3 formed a heterojunction, which could promote the separation of photogenerated electrons and holes, improving the photocatalytic activity. Furthermore, it was found that the active species of hydroxyl radical (·OH) played an important role in the photocatalytic degradation of dibenzothiophene (DBT) based on the active species capture experiment. Finally, a plausible mechanism for the photocatalytic oxidative desulfurization of this nanocomposite was proposed.

Keywords: CeO_2; Bi_2O_3; composite; heterojunction; photocatalytic desulfurization

1. Introduction

Nowadays, environmental pollution and climate change caused by fossil energy combustion have become the focus of global attention. In particular, the combustion of sulfides in motor vehicle fuel generates a large amount of sulfur oxides. These sulfur oxides can cause acid fog, acid rain and haze, lead to environmental pollution and threaten human health [1,2]. Therefore, it is necessary to develop deep desulfurization technology to remove the sulfide in fuel oil and decrease the emission of sulfur oxides from the source. At present, hydrodesulfurization (HDS) is the main technology, widely used in modern industry to reduce the sulfur content of fuel. It requires high pressure, high temperature and high hydrogen consumption. Dibenzothiophene (DBT) and its derivatives are the main sulfur species in diesel and gasoline. However, HDS technology experiences difficult in removing them [3,4]. In order to solve these problems, several hydrogen-free desulfurization technologies have been developed, such as oxidative desulfurization (ODS) [5], adsorptive desulfurization (ADS) [6], extractive desulfurization (EDS) [7], chemical desulfurization (CDS) [8] and biological desulfurization (BDS) [9], etc. Among these, the ODS method [10], as a deep desulfurization technology, has attracted wide attention because it can selectively convert organosulfur compound into corresponding sulfones and sulfate ions under moderate reaction conditions. In recent years, the photocatalytic oxidation desulfurization (PODS) method has shown extraordinary potential due to its special advantages, such as carbon neutrality, chemical reactions driven by photons instead of high pressure and high temperature, and the utilization of abundant available solar energy [11,12]. In this technique, oxidizing reagents such as molecular oxygen(O_2) [13], hydrogen peroxide(H_2O_2) [14], ozone (O_3) [15], and air [16] are used to convert organosulfur compounds present in motor fuels into their corresponding sulfones or SO_4^{2-}, which can be removed through adsorption or extraction to achieve deep desulfurization. In addition, the activity of photocatalytic reactions mainly depends on the photocatalyst, so photocatalytic materials need to have stable structures and excellent optical properties Therefore, a series of excellent photocatalysts for photocatalytic desulfurization have been developed. Hussain's group developed a deep aerobic photocatalytic oxidation desulfurization technology and synthesized a series of excellent photocatalysts such as $LaVO_4$ [17,18], Ag_3VO_4 [19], V_2O_5 [20], etc.; Zhou et al. [21] prepared a Ag_2O/Na-g-C_3N_4 heterojunction for photocatalytic desulfurization of thiophene in fuel. Belousov et al. [22] synthetized nanosized $Bi_2W_xMO_{1-x}O_6$ solid solutions to degrade DBT in model fuel; Zhang et al. [23] prepared a CeF_3/g-C_3N_4 heterojunction photocatalyst with up-conversion performance for photocatalytic removal of DBT from model oil.

Currently, a rare earth oxide (CeO_2) is considered as one of the semiconductor materials with various application prospects, especially in the field of photocatalysis, due to its unique 4f orbital and abundant electronic energy levels. However, single CeO_2 still has some disadvantages, such as large band gap, low absorption in visible region and high electron–hole recombination. Nevertheless, these shortcomings can be overcome by coupling other semiconductors, promoting charge separation and increasing carrier lifetime to improve photocatalytic activity [24]. Therefore, a series of Ceria-based photocatalytic composites have been developed for photocatalytic desulfurization. Mousavi-Kamazani et al. [25] synthetized Cu_2O-CeO_2 nanocomposites, which has 84% photocatalytic desulfurization efficiency within 180 min under visible light; Radwan et al. [26] prepared Fe_2O_3-CeO_2 nanocomposites, and research showed that, when the loading amount of Fe_2O_3 was 15%, the composite had the highest photocatalytic desulfurization activity; Chen's group [27] prepared well-aligned a CeO_2/TiO_2 nanotube array photocatalyst and, when used for photocatalytic oxidation of benzo-thiophene (BT) under visible light irradiation, it was found that more than 90% of sulfur compounds in model oil were removed.

The semiconductor oxide Bi_2O_3 has recently attracted the attention of researchers due to its unique optical and electrical properties. It is well known that bismuth oxide has six crystal forms, denoted as α-Bi_2O_3 (monoclinic), β-Bi_2O_3 (tetragonal), δ-Bi_2O_3 (cubic bcc), γ-Bi_2O_3 (orthorhombic), ε-Bi_2O_3 (orthorhombic), and ω-Bi_2O_3 (triclinic), respectively. Among them, α-Bi_2O_3 is considered as a promising photocatalyst for water decomposition and organic photocatalytic degradation under visible light illumination, with good stability, non-toxicity and small band gap [28]. Dong et al. [29] successfully prepared a porous nanosheet structure α-Bi_2O_3 photocatalyst by the biomimetic-synthesis assisted hydrothermal method, which displayed efficient degradation toward different pollutant molecules; Gupta et.al. [30] synthesized α-Bi_2O_3 nanosheets by a simple annealing assisted thermal decomposition method, which exhibited high photocatalytic degradation of Rhodamine B. Therefore, CeO_2 coupled with α-Bi_2O_3 to construct a heterojunction structure is expected to improve photocatalytic activity. In addition, two-dimensional (2D) semiconductor photocatalysts have been widely investigated due to their good light absorption characteristics, shorter electron and hole migration paths [31,32]. Lu et al. [33] synthesized N-doped two-dimensional CeO_2-TiO_2 nanosheets by a biological template method, which exhibited excellent photocatalytic desulfurization activity; Li et al. [34] utilized attapulgite-CeO_2 decorated two-dimensional MoS_2 to prepare nanocomposite, and the degradation rate of DBT can reach 95% under 3 h irradiation.

Herein, in this paper, we prepared a two-dimensional layered CeO_2/Bi_2O_3 composite using a microwave solvothermal method. Then the activity of the photocatalysts were evaluated by removing DBT from the model oil. In addition, the crystal structure, morphology, optical performance, elemental composition and surface electronic state of the prepared samples were investigated by XRD, Raman spectroscopy, FESEM, TEM, DRS and XPS. Meanwhile, the separation efficiency of photogenerated electrons and holes was studied by photocurrent measurement, EIS and PL. Finally, the possible mechanism of photocatalytic desulfurization was discussed.

2. Results and Discussion

2.1. XRD Analysis

Figure 1 shows the XRD patterns of the prepared photocatalysts. The diffraction peaks near 28.6°, 33.1°, 47.5°, 56.3° and 59.1°correspond to (111), (200), (220), (311) and (222) crystal planes of cubic fluorite structure CeO_2 (PDF#34-3094), respectively. Accordingly, the main characteristic peaks at 24.6° 25.8°, 26.9°, 27.4 °, 28.0 °, 33.0°, 33.2°, 34.0°, 35.0°, 37.6°, 40.0°, 42.4°, 45.1°, 46.3°, 48.6°, 52.4°,54.7°, 57.9°, 59.1°, 62.5° and 71.4 are indexed to (−102), (002), (−112), (−121), (012), (−122), (−202), (022), (−212), (−113), (−222), (−123), (023), (041), (−104), (−332), (−241), (024), (150), (104) and (−161) the crystal planes of monoclinic α-Bi_2O_3 (PDF#71-2274), respectively. In CeBi-2 and CeBi-1 composites, the main characteristic peaks are similar to CeO_2, and some low-intensity Bi_2O_3 characteristic peaks appear. However, in CeBi-0.5 composite, the main characteristic peaks of the composite are similar to α-Bi_2O_3, and some low-intensity CeO_2 characteristic peaks appear. It is worth noting that, with the increase of the Bi_2O_3 component, the characteristic diffraction peaks of the monoclinic α-Bi_2O_3 were gradually enhanced. Moreover, compared with pure CeO_2 and Bi_2O_3, some peak positions of the composites are slightly shifted, which is due to the strong interfacial interaction between CeO_2 and Bi_2O_3 [35]. This indicates that a heterostructure may be formed at the interface between CeO_2 and Bi_2O_3.

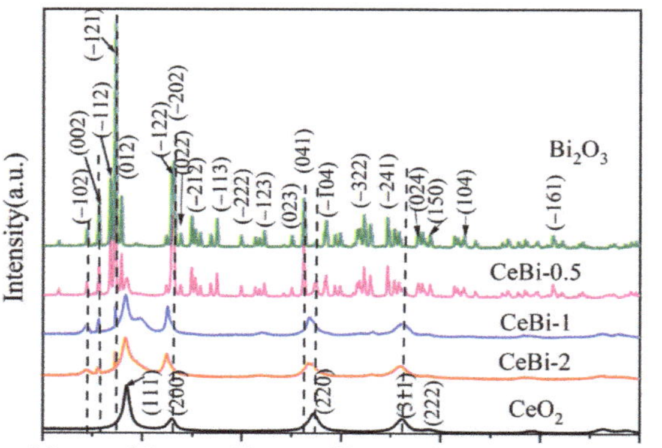

Figure 1. XRD patterns of CeO_2, Bi_2O_3 and CeO_2/Bi_2O_3 composites.

2.2. Raman Spectra Analysis

Figure 2 shows the Raman characteristic peaks of CeO_2, Bi_2O_3 and CeO_2/Bi_2O_3 composites in the wavenumber range of 100–650 cm^{-1}. The peak at 460 cm^{-1} in the pure CeO_2 corresponds to the symmetrical stretching vibration of Ce-O-Ce, which belongs to the triple degenerate F_{2g} vibration mode of CeO_2 in the cubic fluorite phase [36]. However, for pure α-Bi_2O_3, the characteristic peak centered at 117 cm^{-1} is attributed to the vibration modes of bismuth atoms; while the characteristic peaks at 138 cm^{-1} and 150 cm^{-1} correspond to Bi-O stretching vibration modes; the bands at 183 cm^{-1}, 210 cm^{-1}, 284 cm^{-1}, 313 cm^{-1}, 410 cm^{-1}, 445 cm^{-1} and 530 cm^{-1} belong to the oxygen vibration modes [37]. Additionally, in the Raman spectra of CeO_2/Bi_2O_3 composites with different ratios, it can be found that the broad peaks of CeO_2 (445 cm^{-1}) and Bi_2O_3 (460 cm^{-1}) overlap. Meanwhile, it is observed that the characteristic peak intensity of Bi_2O_3 diminishes with the increase in CeO_2 content, all of which indicate that CeO_2/Bi_2O_3 composites are successful prepared.

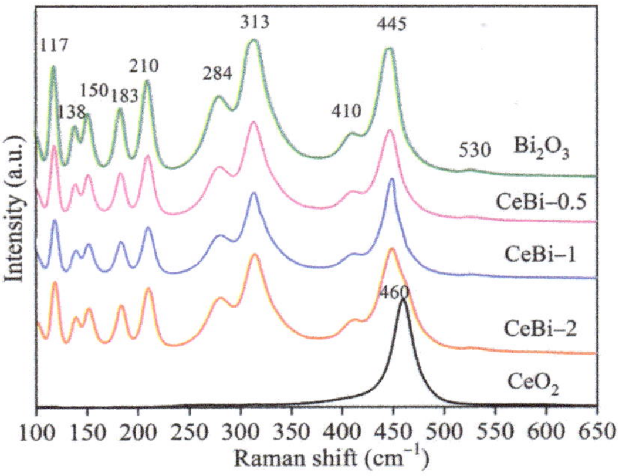

Figure 2. Raman spectra of CeO_2, Bi_2O_3 and CeO_2/Bi_2O_3 composites.

2.3. SEM, TEM, HRTEM and EDS Analysis

The morphology and microstructure of the prepared photocatalyst are studied by FESEM, TEM, HRTEM and EDS. Figure 3a,b show the FESEM images of CeBi-0.5. It can be seen that CeBi-0.5 presents an oval nanosheet structure with uniform shape and size on a large scale. The sizes of the longitudinal axis and horizontal axis are about 500 nm and 1 μm, respectively, and the nanosheets self-assemble into flower-like nanostructures. Figure 3c,d are the TEM and HRTEM images of CeBi-0.5. It is observed that the prepared composite presents a two-dimensional lamellar structure, which is composed of about 5–20 nm nanoparticles. In addition, the lattice fringes with an interplanar spacing of 0.32 nm and 0.27 nm correspond to (111) crystal planes of cubic fluorite CeO_2 and (200) crystal plane of monoclinic α-Bi_2O_3, respectively (Figure 3d). Figure 3e shows the EDS spectrum of CeBi-0.5 composite. Only Bi, Ce and O elements are detected, and no other elements, indicating that the CeO_2 and Bi_2O_3 are successfully combined.

Figure 3. *Cont.*

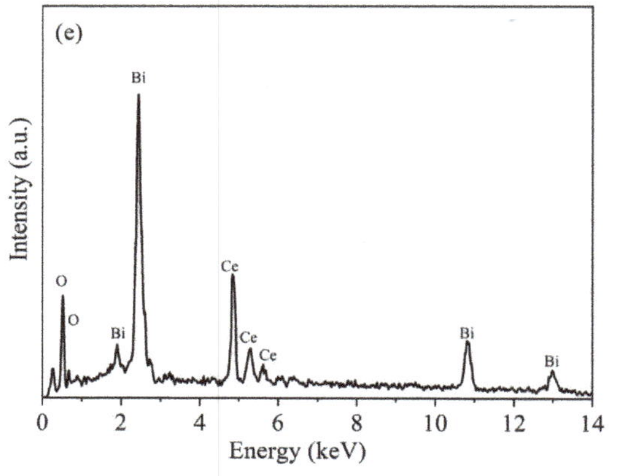

Figure 3. FESEM of images of CeBi-0.5 (**a**,**b**); TEM (**c**) and HRTEM (**d**) images of CeBi-0.5; EDS spectrum of CeBi-0.5 (**e**).

2.4. Nitrogen Adsorption–Desorption Isotherm and Pore Size Distribution

The specific surface area and pore-size distribution of the photocatalysts have a major influence on the enhancement of the photocatalytic performance. Figure 4 shows the N_2 adsorption–desorption isotherm of CeO_2, Bi_2O_3 CeBi-0.5 and the pore-size distribution isotherm of CeBi-0.5. The curve of pure CeO_2 presents an inverse S shape and there is no hysteresis loop, which belongs to a type II adsorption isotherm, indicating that pure CeO_2 has no pores. The Bi_2O_3 and CeBi-0.5 composite exhibit a typical IV like isotherm with H3 hysteresis loop, indicative of the characteristic of mesopores, according to the IUPAC classification [38]. It can be seen that the CeBi-0.5 composite has a relatively narrow pore size distribution, centered at about 5–20 nm (the inset of Figure 4). The BET-specific surface areas of CeO_2, Bi_2O_3 and CeBi-0.5 composite are measured as 4.8 m^2/g, 38.7 m^2/g and 52.5 m^2/g, respectively. In particular, two-dimensional CeBi-0.5 composite exhibit higher surface areas than pure CeO_2 and Bi_2O_3, which may offer a more active site in the process of photocatalytic desulfurization, enhancing the photocatalytic performance.

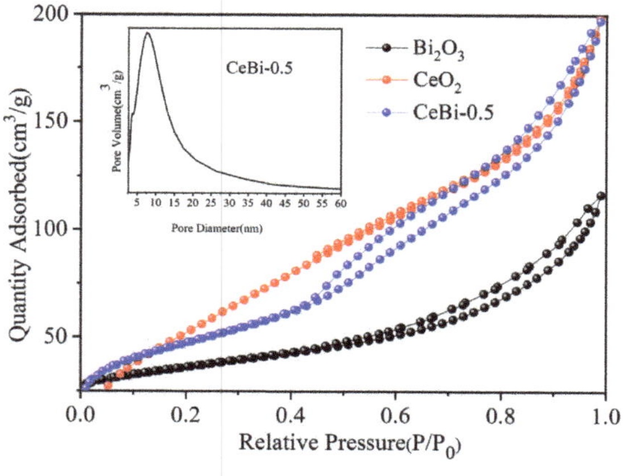

Figure 4. N_2 adsorption–desorption isotherm of CeO_2, Bi_2O_3 CeBi-0.5 and the pore-size distribution isotherm of CeBi-0.5.

2.5. XPS Analysis

The chemical states and bonding environment of the elements on the surface of the composites are analyzed by XPS. Figure 5a shows the high-resolution spectrum of the Ce 3d. The emission peaks of Ce 3d are divided into eight fitting peaks, where the v and u represent spin-orbit coupling 3d $_{5/2}$ and $3d_{3/2}$, respectively. The peaks of v_1 and u_1 are assigned to the Ce IV ($3d^9$ $4f^2$) O ($2p^4$) state; the peaks of v_3 and u_3 are assigned to the Ce IV ($3d^9$ $4f^1$) O ($2p^5$) state and the peaks of v_4 and u_4 are assigned to the Ce IV ($3d^9$ $4f^0$) O ($2p^6$) state. These six fitting peaks belong to the Ce^{4+} species. The other two fitting peaks v_2 and u_2 are related to the Ce III ($3d^9$ $4f^2$) O ($2p^5$) state, which belong to the Ce^{4+} species. Therefore, there are trivalent and tetravalent cerium in CeO_2 and CeBi-0.5 [39]. The presence of trivalent cerium implies the presence of more oxygen vacancies. As shown in Figure 5b, there are characteristic peaks at 159 eV and 164.3 eV, which are attributed to Bi 4 $f_{7/2}$ and Bi 4 $f_{3/2}$ spin orbitals, respectively [40]. This indicates the presence of trivalent bismuth in Bi_2O_3 and CeBi-0.5. The fitting peaks located at 529.7 eV (529.5 eV) and 531.3 eV (531.2 eV) observed in Figure 5c correspond to the lattice oxygen and surface hydroxyl oxygen in CeO_2 (Bi_2O_3), respectively. It is noted that, compared with pure CeO_2 and Bi_2O_3, the binding energies of Ce 3d, Bi 4f and O 1s have slightly shifted, which indicates that there is a strong electron interaction between the CeO_2 and Bi_2O_3 interfaces [41]. Therefore, the heterojunction structure between CeO_2 and Bi_2O_3 may be formed.

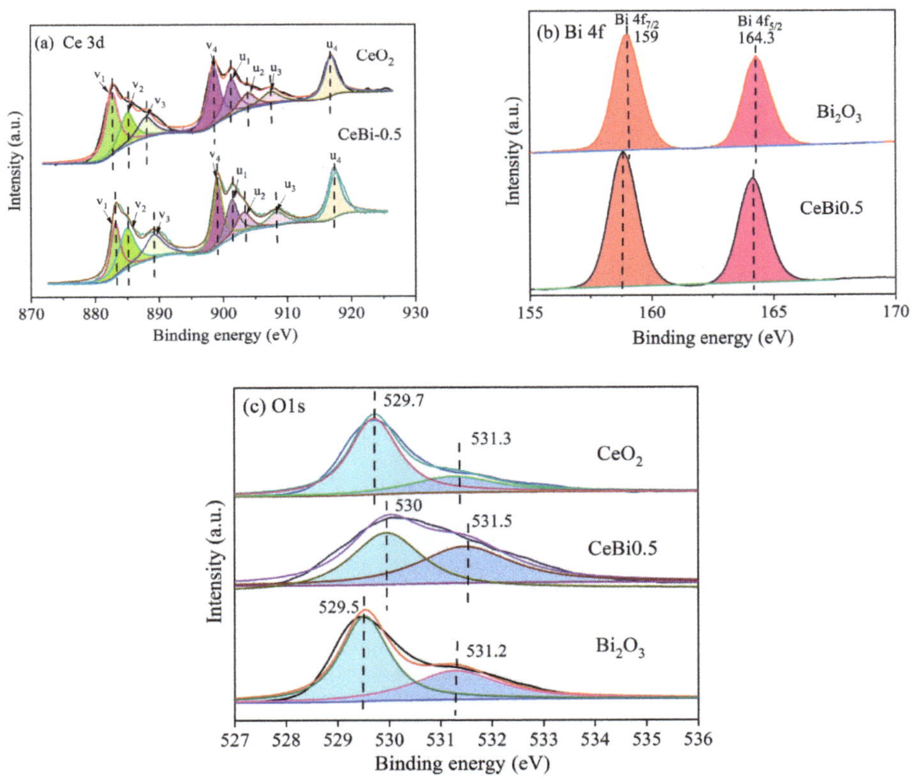

Figure 5. High-resolution XPS spectrum of (**a**) Ce 3d, (**b**) Bi 4f, (**c**) O 1s.

2.6. DRS Analysis

The UV-Vis DRS of CeO_2, Bi_2O_3 and CeO_2/Bi_2O_3 composites are shown in Figure 6a. Compared with the pure CeO_2 and Bi_2O_3, it is observed that the absorption edge of

CeO_2/Bi_2O_3 composites shifted significantly to the visible light region. In addition, the composites have higher absorption than pure CeO_2 in the visible light region, indicating that the heterojunction formed between CeO_2 and Bi_2O_3 could improve the light absorption ability and expand the light response region. The band gap energy of a semiconductor can be calculated by the following formula [42,43]:

$$(\alpha \, h \, v) = \mathrm{A}(h \, v - E_g)^{n/2} \quad (1)$$

where α, h, n, E_g and A are absorption coefficient, Planck constant, light frequency, band gap energy and a constant, respectively, and n is 1 for a direct transition. Plotting $(\alpha h \, v)^2$ versus energy (hv) based on the spectral response in Figure 6a gives the extrapolated intercept corresponding to the E_g value (Figure 6b). The band gap values (E_g) of pure CeO_2 and Bi_2O_3 are about 2.95 eV and 2.85 eV, respectively. Simultaneously, the E_g of the CeO_2/Bi_2O_3 composites decreases gradually with the increase in the Bi_2O_3 molar ratio. Therefore, the prepared composites could make better use of visible light.

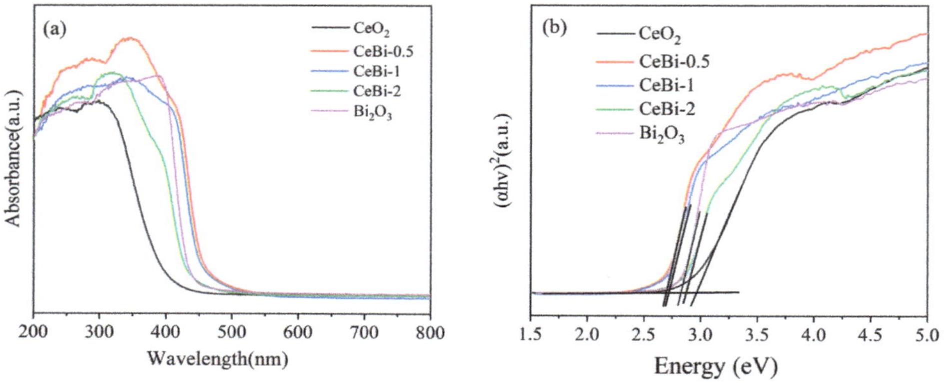

Figure 6. (**a**) UV-Vis DRS of CeO_2, Bi_2O_3 and CeO_2/Bi_2O_3 composites; (**b**) Tauc plots of $(\alpha hv)^2$ vs photon energy (hv).

2.7. Photocatalytic Activity

The removal experiment for DBT in model oil is carried out to study the photocatalytic activity of different photocatalysts under visible light. Figure 7a shows the photocatalytic desulfurization rate of CeO_2, Bi_2O_3 and CeO_2/Bi_2O_3 composites. It was found that all photocatalysts have a much lower adsorption ability for DBT in the dark. In the absence of any photocatalyst, the desulfurization rate of H_2O_2 is about 7.5% in 3 h, indicating that H_2O_2 can oxidize the DBT molecular to some extent, but the oxidation ability of H_2O_2 is extremely limited. When there is a photocatalyst in the reaction system, it can be found that the desulfurization rate is effectively improved. It is worth noting that the Bi_2O_3/CeO_2 composites show a significant desulfurization trend compared with the pure CeO_2 and Bi_2O_3. When the molar ratio of Ce to Bi is 1:2, the composite shows the best photocatalytic efficiency, and its desulfurization rate reaches 90.5 % in 3 h. Meanwhile, the corresponding photocatalytic desulfurization kinetic curves over the prepared photocatalysts are shown in Figure 7b. The reaction data are fitted by a first-order model as depicted by the formula [44]:

$$\ln(C_0/C_t) = kt \quad (2)$$

where k is the pseudo-first-order rate constant, and the relationship between $\ln(C_0/C)$ and catalytic reaction time t is considered as linear. It is worth noting that CeBi-0.5 shows the maximum kinetic constant (k) value of 0.81802 h^{-1}, which is approximately 5.19-flod and 2.45-flod of pure CeO_2 and Bi_2O_3, respectively. This indicates that CeBi-0.5 composite can effectively improve the photocatalytic activity. As shown in Figure 7c, the desulfurization rate of the CeBi-0.5 composite after three photodegradation cycles is 70%, which may be due to the loss of active components and the easy filling of the oxygen vacancy in the photocatalytic process [45]. Consequently, the excellent photocatalytic performance of the composite is attributed to its unique two-dimensional lamellar structure, which is conducive to the enrichment of organic sulfur molecules. Moreover, the formation of heterojunction between CeO_2 and Bi_2O_3 in the composite can promote the effective separation of photogenerated electrons and holes, and can significantly improve the photocatalytic activity.

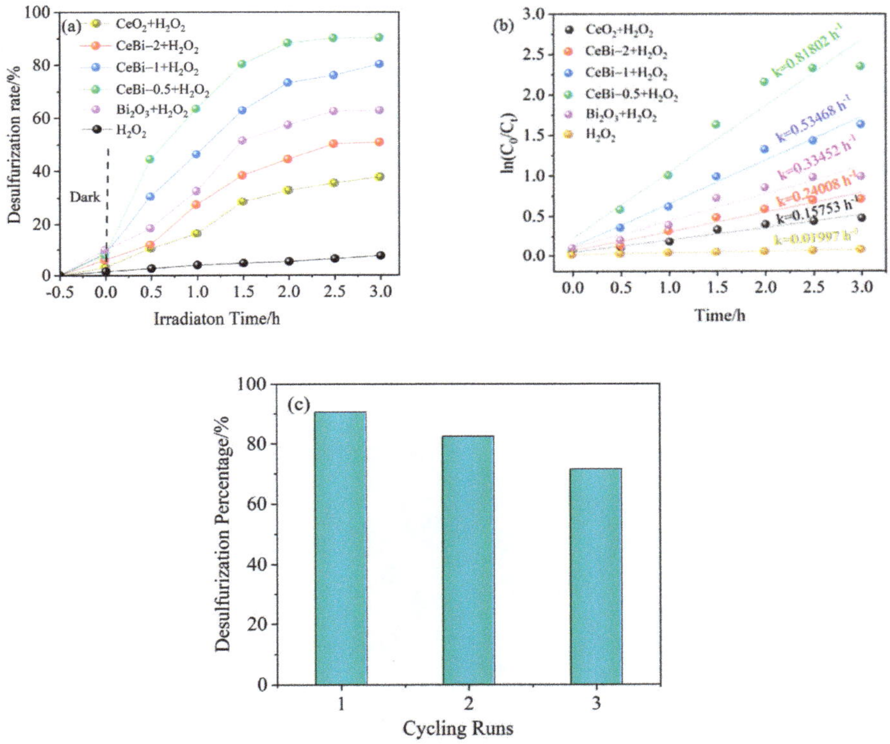

Figure 7. (**a**) Photocatalytic desulfurization rate of different catalysts; (**b**) Kinetic fitting curve; (**c**) Photocatalytic desulfurization cycle over the CeBi-0.5 composite.

2.8. Photoelectrochemical and PL Analysis

In general, the transient photocurrent responses could evaluate the separation and migration efficiency of photogenerated electron-hole pairs in the semiconductor materials The transient photocurrent response curve is shown in Figure 8a. The photocurrent intensity of CeBi-0.5 composite is larger than those of pure CeO_2 and Bi_2O_3, indicating that CeBi-0.5 composite exhibits the higher separation efficiency for electron-hole pairs [46]. Moreover, EIS is used to study the charge transfer resistance of the photogenerated carriers (Figure 8b). Generally speaking, the smaller the arc in the EIS Nyquist curve, the lower the charge transfer resistance on the electrode surface, and the recombination of photo generated charge carriers would be inhibited. The arc radius of CeBi-0.5 composite is significantly smaller than those of pure CeO_2 and Bi_2O_3, indicating that the CeBi-0.5 composite is beneficial for improving the transfer efficiency of photogenerated electron-hole pairs. Fluorescence spectroscopy is also an effective approach to study the separation efficiency of photogenerated electrons and holes. From Figure 8c, it can be found that the PL intensity of CeBi-0.5 composite is lower than that of pure CeO_2 and Bi_2O_3, indicating that the recombination of photogenerated electrons and holes can be greatly inhibited in CeBi-0.5 composite [42]. Therefore, the formation of heterojunction structures between CeO_2 and Bi_2O_3 can effectively promote photo generation and electron separation, thereby improving photocatalytic activity.

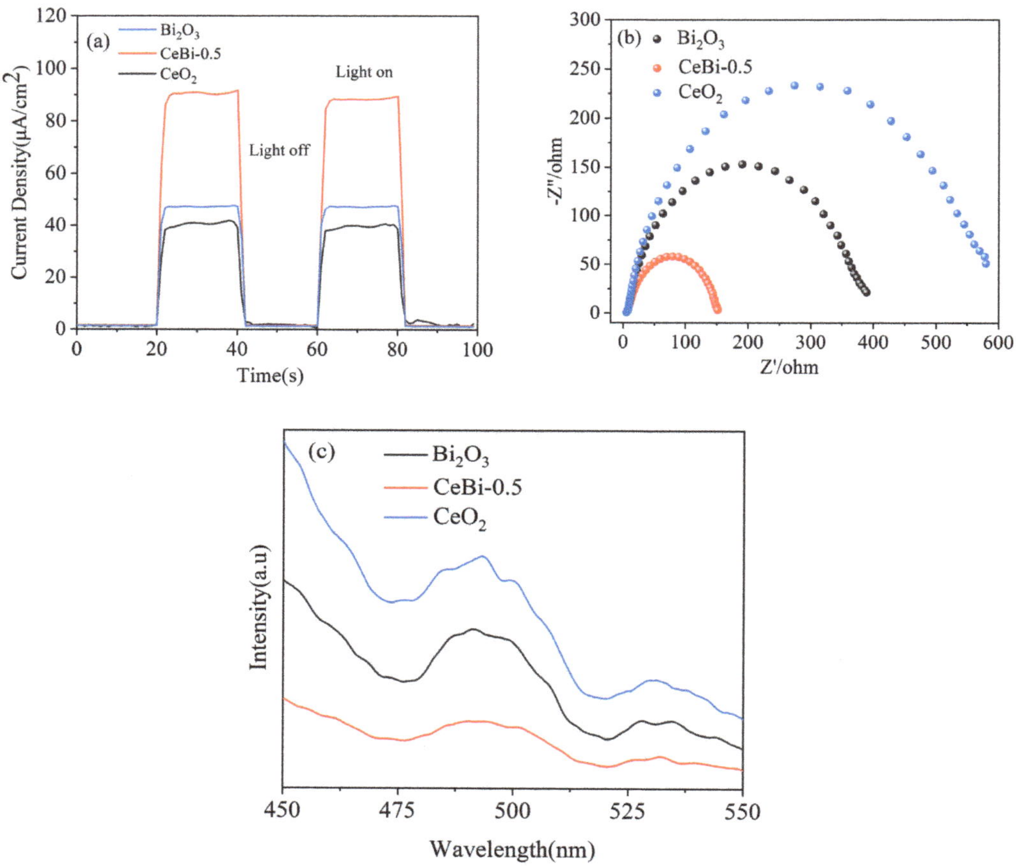

Figure 8. (**a**) Transient photocurrent responses curve, (**b**) Nyquist impedance plots, (**c**) PL spectra of CeO_2, Bi_2O_3 and CeBi-0.5 composite.

2.9. Photocatalysis Mechanism

In general, photoinduced holes (h^+), superoxide radicals ($\cdot O^{2-}$) and hydroxyl radicals ($\cdot OH$) are considered to be the main active species in the photocatalytic degradation of semiconductors [19]. In order to further determine the main active species of CeO_2/Bi_2O_3 composites in photocatalytic reactions, the trapping capture experiment for active species is carried out. Three different kinds of captors, triethanolamine (TEOA; h^+ scavenger), tert-butyl alcohol (TBA; a $\cdot OH$ radical scavenger) and benzoquinone (BQ; a $\cdot O^{2-}$ radical scavenger), are used [47]. As can been seen in Figure 9, the desulfurization rate is slightly decreased using TEOA and BQ as capture agents, indicating that neither h^+ nor $\cdot O^{2-}$ is the main reactive species. However, the addition of TBA causes obvious deactivation of the photocatalyst, reducing the photocatalytic activity for the desulfurization rate from 90.5% to 35%, as shown in the experimental results. This clearly demonstrates that active $\cdot OH$

radicals are the dominant reactive species responsible for the photocatalytic desulfurization system over these photocatalysts.

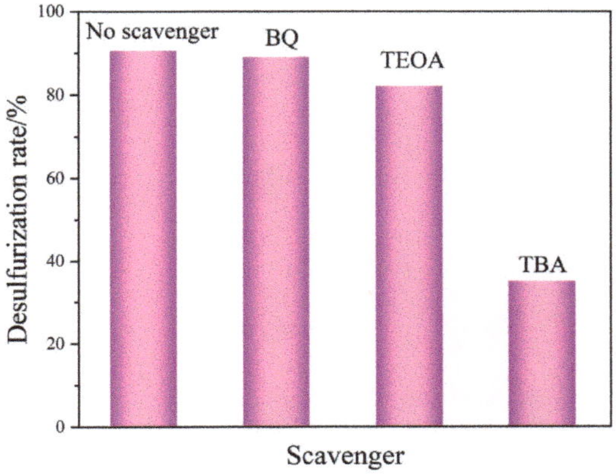

Figure 9. Photocatalytic desulfurization by different scavengers over CeBi-0.5.

To further determine the mechanism of photocatalytic desulfurization, the positions of conductive band (CB) and valence band (VB) need to be determined, according to the following formula [48,49]:

$$E_{CB} = \chi - E_0 - 0.5E_g \quad (3)$$

$$E_{VB} = E_{CB} + E_g \quad (4)$$

where χ is the absolute electronegativity of the semiconductor (the χ values of Bi_2O_3 and CeO_2 are 5.99 eV and 5.56 eV, respectively.), E_0 is the potential energy of the standard hydrogen electrode (4.5 eV), and E_g is the band gap of the semiconductor. Combined with E_g values of Bi_2O_3 and CeO_2 are 2.85 eV and 2.95 eV, and the CB and VB values of Bi_2O_3 are calculated to be 0.07 eV and 2.92 eV, and −0.42 eV and 2.53 eV for Bi_2O_3.

The mechanism of photocatalytic desulfurization is shown in Figure 10. When the composite is irradiated under visible light, both Bi_2O_3 and CeO_2 can be activated and generate electron-hole pairs, due to the CB of CeO_2 being more negative than that of Bi_2O_3. The photogenerated electrons on the CB of CeO_2 tend to migrate to that of Bi_2O_3 through the interface, and then the electrons will react with H_2O_2 to form ·OH, whereas the holes in the VB of Bi_2O_3 are spontaneously transferred to CeO_2, where -OH can be oxidized to ·OH by the holes. Hydroxyl radicals have strong oxidation ability, which can oxidize non-polar DBT molecules adsorbed on the surface of two-dimensional materials into polar dibenzothiophene sulfone ($DBTO_2$). Then $DBTO_2$ can be extracted and removed, due to its strong polarity, and finally the organic sulfur in the model oil is effectively removed [50]. Based on the above discussion, we propose the following reaction equations:

$$CeO_2/Bi_2O_3 + h\nu \rightarrow e^- + h^+ \quad (5)$$

$$e^- + H_2O_2 \rightarrow \bullet OH + OH^- \quad (6)$$

$$h^+ + OH^- \rightarrow \bullet OH \quad (7)$$

$$DBT + \bullet OH \rightarrow DBTO_2 \quad (8)$$

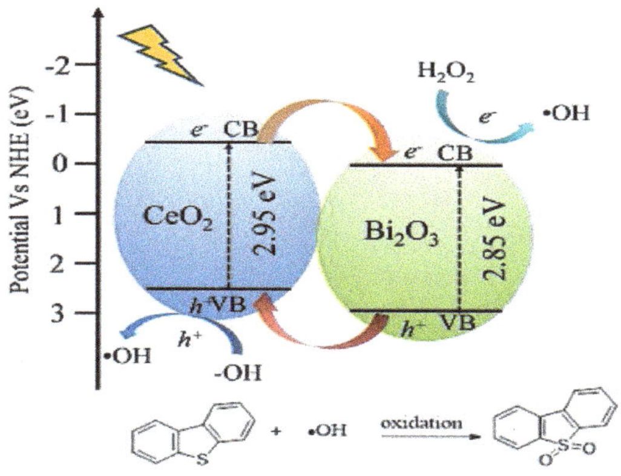

Figure 10. Photocatalytic desulfurization mechanism of CeO_2/Bi_2O_3 nanocomposite.

3. Experimental Section

3.1. Materials

Cerium nitrate hexahydrate ($Ce(NO_3)_3{\cdot}6H_2O$, 99%),Bismuth nitrate pentahydrate ($Bi(NO_3)_3{\cdot}5H_2O$, 99%), Acetonitrile (C_2H_3N, 99%), Anhydrous ethanol (CH_3CH_2OH,) and Tert-butyl alcohol ($(CH_3)_3COH$, TBA) were purchased from China National Medicines Corporation Ltd. Benzo-thiophene ($C_{12}H_8S$, DBT) and n-octane (C_8H_{18}) were purchased from Aladdin Bio-Chem Technology Co., Ltd. (Shanghai, China). Ethylene glycol ($(CH_2OH)_2$), Triethanolamine ($(HOCH_2CH_2)_3N$, TEOA) and Benzoquinone ($C_6H_4O_2$, BQ) were purchased from Shanghai Maclean Biochemical Technology Co. Ltd. All of the reagents in this experiment were analytical grade and directly used without further purification. Distilled water was used throughout this study.

3.2. Preparation of Samples

The typical synthesis process is described as follows: 0.01 mol $Bi(NO_3)_3{\cdot}5H_2O$ was dissolved in a mixed solution of 60 mL ethylene glycol and ethanol (V_{EG}: $V_{Ethanol}$ = 1:2) marked as A, and then a certain stoichiometric ratio of $Ce(NO_3)_3{\cdot}6H_2O$ was dissolved in 10 mL deionized water, marked as B. Subsequently, solution B was slowly dropped into solution A, After the two solutions were fully mixed, the solution was relocated into a 100 mL Teflon liner and then placed in a microwave synthesizer at 160 °C for 30 min, then the precipitate was washed with distilled water and anhydrous ethanol three times, and dried in an oven at 60 °C for 12 h. The composites with molar ratios between CeO_2 and Bi_2O_3 (2:1, 1:1, 1:2) were labeled as CeBi-2, CeBi-1 and CeBi-0.5, respectively. Pure CeO_2 and Bi_2O_3 photocatalysts were also synthesized under similar experimental conditions.

3.3. Characterization

X-ray diffraction (XRD) measurement was performed by a Rigaku, D/max-RB instrument (Tokyo, Japan) between 20 and 80° at a scan rate of 5°/min with Cu Ka radiation; Raman spectra were measured by a Thermo Fisher Scientific DXR spectrometer (Waltham, MA, USA) and the excitation laser wavelength was 532 nm. The morphology was observed by field emission scanning electron microscopy (FESEM, Zeiss Supra55, Jena, Germany) and transition electron microscopy (TEM, JEOL JEM-2100, Tokyo, Japan), equipped with an EDS spectrometer (EDS, Oxford, UK). N_2 adsorption-desorption isotherms were determined using a Micromeritics ASAP 2010 analyzer (Norcross, GA, USA), the surface areas were calculated according to the Brunauer-Emmett-Teller (BET) method, and the pore size distribution was obtained by the Barrett-Joyner-Halenda (BJH) model. X-ray photoelectron spectroscopy (XPS) analysis was carried out by a Thermo Fisher Scientific ESCALAB 250 (Waltham, MA, USA) spectrometer with mono Al Ka radiation (1486.6 eV). Photoluminescence (PL) spectra were collected on a PerkinElmer LS45 fluorescence spectrometer

(Waltham, MA, USA). Ultraviolet-visible diffuse reflectance spectra (UV-Vis DRS) were measured by a Shimadzu UV-2450 spectrophotometer (Kyoto, Japan) equipped with an integrating sphere.

3.4. Photocatalytic Desulfurization Measurement

The model oil with sulfur content of 100 ppm was prepared by dissolving 0.08 g DBT into 200 mL of n-octane. Then, 0.1 g of photocatalysts were added to the above solution with constant stirring; subsequently, an appropriate 30wt% hydrogen peroxide aqueous solution (molar ratio of (H_2O_2/DBT = 4:1) was added. After dark adsorption for 30 min, the suspension was transferred into a photochemical reaction instrument containing a 300 W xenon lamp with an ultraviolet cut-off filter (GHX-2, Yangzhou Science and Technology City Instrument co., Ltd., China). The dispersion was collected every 30 min and extracted with acetonitrile, then the sulfur content was measured using an ultraviolet fluorescence sulfur analyzer (THA2000S, Taizhou Jinhang Analysis Instrument Co., Ltd., Taizhou, China). The desulfurization rate D (%) is obtained according to the following formula:

$$D = (1 - C_t/C_0) \times 100\% \quad (9)$$

where C_0 is the initial sulfur content and C_t is the sulfur content of the solution at reaction time t.

3.5. Photoelectrochemical Measurements

Photoelectrochemical analysis was performed in a standard three electrode system, with Pt foil as the counter electrode, Ag/AgCl electrode as the reference electrode and the prepared sample deposited on FTO glass as the working electrode. Photocurrent was measured on a photoelectrochemical workstation (LK5600, Tianjin Lanlike Chemical Electronics High Technology Co., Ltd., Tianjin, China) with 0.1 M Na_2SO_4 aqueous solution as electrolyte and 300 W xenon lamp as light source. EIS test also used the three-electrode system. The electrolyte solution was 0.5 M KCl containing 0.01 M $K_3Fe(CN)_6/K_4Fe(CN)_6$ (molar ratio 1:1), performed at bias voltages 0.5 V, in the frequency range of 0.1 Hz-100 kHz, with oscillation potential amplitudes of 0.01 V.

4. Conclusions

In summary, a two-dimensional layered CeO_2/Bi_2O_3 photocatalyst was successfully synthesized by microwave solvothermal method and used for photocatalytic desulfurization. The prepared photocatalysis significantly enhances the performance of photocatalytic removal of DBT under visible light irradiation. When the molar ratio of Bi_2O_3 to CeO_2 is 2:1, the prepared sample exhibits the best photocatalytic activity, with removal rate of DBT reaching 90.5% in3 h and the reaction rate constant at 0.81802 h^{-1}, which is approximately 5.19-flod and 2.45-flod of pure CeO_2 and Bi_2O_3, respectively. Free radical scavenging experiments demonstrate that hydroxyl radical is the main active species in the photocatalytic process. The enhanced photocatalytic activity is attributed to the unique two-dimensional structure of the heterojunction formed between cerium oxide and bismuth oxide. This study indicates that the construction of two-dimensional heterostructure photocatalyst is a promising method to improve photocatalytic desulfurization.

Author Contributions: Conceptualization, X.L. and Q.Z.; methodology, X.L. and J.Q.; validation, X.L. and H.H.; formal analysis, X.L., W.C. and H.H.; investigation, X.L. and W.C.; resources, X.L.; data curation, H.H. and W.C.; writing-original draft preparation, X.L. and W.C.; writing-review and editing, H.H. and J.Q.; visualization, X.L. and H.H.; supervision, X.L.; project administration, Q.Z.; funding acquisition, X.L. and Q.Z. All authors have read and agreed to the published version of the manuscript.

Funding: This work is supported by the National Natural Science Foundation of China (No. 12274361), the Natural Science Foundation of Jiangsu Province (BK20211361), College Natural Science Research Project of Jiangsu Province (20KJA430004) and the Funding for school-level research projects of Yancheng Institute of Technology (xjr2019026).

Data Availability Statement: Data available on request from the authors.

Conflicts of Interest: The authors declare no conflict of interest.

References

1. Shindell, D.; Smith, C.J. Climate and air-quality benefits of a realistic phase-out of fossil fuels. *Nature* **2019**, *573*, 408–411. [CrossRef]
2. Shen, N.; Wang, Y.F.; Peng, H.; Hou, Z.P. Renewable energy green innovation, fossil energy consumption, and air pollution-spatial empirical analysis based on China. *Sustainability* **2020**, *12*, 6397. [CrossRef]
3. Tanimu, A.; Alhooshani, K. Advanced Hydrodesulfurization Catalysts: A Review of Design and Synthesis. *Energy Fuels* **2019**, *33*, 2810–2838. [CrossRef]
4. Dedual, G.; MacDonald, M.J.; Alshareef, A.; Wu, Z.; Tsang, D.C.; Yip, A.C. Requirements for effective photocatalytic oxidative desulfurization of a thiophene-containing solution using TiO_2. *J. Environ. Chem. Eng.* **2014**, *2*, 1947–1955. [CrossRef]
5. Bhadra, B.N.; Jhung, S.H. Oxidative desulfurization and denitrogenation of fuels using metal-organic framework-based/-derived catalysts. *Appl. Catal. B* **2019**, *259*, 118021. [CrossRef]
6. Yaseen, M.; Ullah, S.; Ahmad, W.; Subhan, S.; Subhan, F. Fabrication of Zn and Mn loaded activated carbon derived from corn cobs for the adsorptive desulfurization of model and real fuel oils. *Fuel* **2021**, *284*, 119102. [CrossRef]
7. Makoś, P.; Boczkaj, G. Deep eutectic solvents based highly efficient extractive desulfurization of fuels—Eco-friendly approach. *J. Mol. Liq.* **2019**, *296*, 111916. [CrossRef]
8. Manan, N.S.A.; Aldous, L.; Alias, Y.; Murray, P.; Yellowlees, L.J.; Lagunas, M.C.; Hardacre, C. Electrochemistry of Sulfur and Polysulfides in Ionic Liquids. *J. Phys. Chem. B* **2011**, *115*, 13873–13879. [CrossRef] [PubMed]
9. Capecchi, E.; Piccinino, D.; Bizzarri, B.M.; Botta, L.; Crucianelli, M.; Saladino, R. Oxidative Bio-Desulfurization by Nanostructured Peroxidase Mediator System. *Catalysts* **2020**, *10*, 313. [CrossRef]
10. Shafiq, I.; Shafique, S.; Akhter, P.; Ishaq, M.; Yang, W.; Hussain, M. Recent breakthroughs in deep aerobic oxidative desulfurization of petroleum refinery products. *J. Clean. Prod.* **2021**, *294*, 125731. [CrossRef]
11. Zhou, X.; Wang, T.; Liu, H.; Gao, X.; Wang, C.; Wang, G. Desulfurization through Photocatalytic Oxidation: A Critical Review. *ChemSusChem* **2021**, *14*, 492–511. [CrossRef] [PubMed]
12. Shafiq; Shafique, S.; Akhter, P.; Abbas, G.; Qurashi, A.; Hussain, M. Efficient catalyst development for deep aerobic photocatalytic oxidative desulfurization: Recent advances, confines, and outlooks. *Catal. Rev. Sci. Eng.* **2022**, *64*, 789–834. [CrossRef]
13. Gao, X.-M.; Fu, F.; Zhang, L.-P.; Li, W.-H. The preparation of Ag–BiVO4 metal composite oxides and its application in efficient photocatalytic oxidative thiophene. *Phys. B Condens. Matter* **2013**, *419*, 80–85. [CrossRef]
14. Zarrabi, M.; Entezari, M.H.; Goharshadi, E.K. Photocatalytic oxidative desulfurization of dibenzothiophene by C/TiO_2@MCM-41 nanoparticles under visible light and mild conditions. *RSC Adv.* **2015**, *5*, 34652–34662. [CrossRef]
15. Boshagh, F.; Rahmani, M.; Rostami, K.; Yousefifar, M. Key Factors Affecting the Development of Oxidative Desulfurization of Liquid Fuels: A Critical Review. *Energy Fuels* **2021**, *36*, 98–132. [CrossRef]
16. Kalantari, K.; Kalbasi, M.; Sohrabi, M.; Royaee, S.J. Synthesis and characterization of N-doped TiO_2 nanoparticles and their application in photocatalytic oxidation of dibenzothiophene under visible light. *Ceram. Int.* **2016**, *42*, 14834–14842. [CrossRef]
17. Shafiq, I.; Hussain, M.; Shafique, S.; Rashid, R.; Akhter, P.; Ahmed, A.; Jeon, J.-K.; Park, Y.-K. Oxidative desulfurization of refinery diesel pool fractions using $LaVO_4$ photocatalyst. *J. Ind. Eng. Chem.* **2021**, *98*, 283–288. [CrossRef]
18. Shafiq, I.; Hussain, M.; Rashid, R.; Shafique, S.; Akhter, P.; Yang, W.; Nawaz, Z.; Ahmed, A.; Park, Y.-K. Development of hierarchically porous $LaVO_4$ for efficient visible-light-driven photocatalytic desulfurization of diesel. *Chem. Eng. J.* **2021**, *420*, 130529. [CrossRef]
19. Mahboob, I.; Shafiq, I.; Shafique, S.; Akhter, P.; Hussain, M.; Park, Y.-K. Effect of active species scavengers in photocatalytic desulfurization of hydrocracker diesel using mesoporous Ag_3VO_4. *Chem. Eng. J.* **2022**, *441*, 136063. [CrossRef]
20. Shafiq, I.; Hussain, M.; Shafique, S.; Akhter, P.; Ahmed, A.; Ashraf, R.S.; Khan, M.A.; Jeon, B.-H.; Park, Y.-K. Systematic Assessment of Visible-Light-Driven Microspherical V_2O_5 Photocatalyst for the Removal of Hazardous Organosulfur Compounds from Diesel. *Nanomaterials* **2021**, *11*, 2908. [CrossRef]
21. Zhou, X.; Wang, T.; Zhang, L.; Che, S.; Liu, H.; Liu, S.; Wang, C.; Su, D.; Teng, Z. Highly efficient Ag_2O/Na-g-C_3N_4 heterojunction for photocatalytic desulfurization of thiophene in fuel under ambient air conditions. *Appl. Catal. B* **2022**, *316*, 121614. [CrossRef]
22. Belousov, A.S.; Suleimanov, E.V.; Parkhacheva, A.A.; Fukina, D.G.; Koryagin, A.V.; Titaev, D.N.; Lazarev, M.A. Synthesis and Characterization of Bi2WxMo1-xO6 Solid Solutions and Their Application in Photocatalytic Desulfurization under Visible Light. *Processes* **2022**, *10*, 789. [CrossRef]
23. Lu, X.W.; Chen, F.; Qian, J.Q.; Fu, M.; Jiang, Q.; Zhang, Q.F. Facile fabrication of CeF_3/g-C_3N_4 heterojunction photocatalysts with upconversion properties for enhanced photocatalytic desulfurization performance. *J. Rare Earths* **2021**, *39*, 1204–1210. [CrossRef]

24. Tran, D.P.; Pham, M.-T.; Bui, X.-T.; Wang, Y.-F.; You, S.-J. CeO_2 as a photocatalytic material for CO_2 conversion: A review. *Sol. Energy* **2022**, *240*, 443–466. [CrossRef]
25. Mousavi-Kamazani, M.; Ashrafi, S. Single-step sonochemical synthesis of Cu_2O-CeO_2 nanocomposites with enhanced photocatalytic oxidative desulfurization. *Ultrason. Sonochem.* **2020**, *63*, 104948. [CrossRef]
26. Radwan, M.S.; Aboutaleb, W.A.; El Naggar, A.M.; El Sayed, H.A.; Shehata, M.R.; Medany, S.S. Photo-oxidative extractive desulfurization of dibenzothiofene over Fe_2O_3-CeO_2 nanocomposites at visible light irradiation. *J. Photochem. Photobiol. A Chem.* **2022**, *433*, 114137. [CrossRef]
27. Lu, X.W.; Li, X.Z.; Qian, J.C.; Miao, N.M.; Yao, C.; Chen, Z.G. Synthesis and characterization of CeO_2/TiO_2 nanotube arrays and enhanced photocatalytic oxidative desulfurization performance. *J. Alloys Compd.* **2016**, *661*, 363–371. [CrossRef]
28. He, B.W.; Wang, Z.L.; Xiao, P.; Chen, T.; Yu, J.G.; Zhang, L.Y. Cooperative Coupling of H_2O_2 Production and Organic Synthesis over a Floatable Polystyrene-Sphere-Supported TiO_2/Bi_2O_3 S-Scheme Photocatalyst. *Adv. Mater.* **2022**, *34*, 2203225. [CrossRef]
29. Dong, Y.; Ma, A.; Zhang, D.; Gao, Y.; Li, H. Preparation of high-performance α-Bi_2O_3 photocatalysts and their photocatalytic activity. *Surf. Innov.* **2020**, *8*, 295–303. [CrossRef]
30. Gupta, G.; Kaur, M.; Kansal, S.K.; Umar, A.; Ibrahim, A.A. α-Bi_2O_3 nanosheets: An efficient material for sunlight-driven photocatalytic degradation of Rhodamine B. *Ceram. Int.* **2022**, *48*, 29580–29588. [CrossRef]
31. Zhao, Y.; Zhang, S.; Shi, R.; Waterhouse, G.I.; Tang, J.; Zhang, T. Two-dimensional photocatalyst design: A critical review of recent experimental and computational advances. *Mater. Today* **2020**, *34*, 78–91. [CrossRef]
32. Liu, C.; Zhang, Q.; Zou, Z. Recent advances in designing $ZnIn_2S_4$-based heterostructured photocatalysts for hydrogen evolution. *J. Mater. Sci. Technol.* **2023**, *139*, 167–188. [CrossRef]
33. Lu, X.; Li, X.; Chen, F.; Chen, Z.; Qian, J.; Zhang, Q. Biotemplating synthesis of N-doped two-dimensional CeO_2–TiO_2 nanosheets with enhanced visible light photocatalytic desulfurization performance. *J. Alloys Compd.* **2020**, *815*, 152326. [CrossRef]
34. Li, X.Z.; Zhang, Z.S.; Yao, C.; Lu, X.W.; Zhao, X.B.; Ni, C.Y. Attapulgite-CeO_2/MoS_2 ternary nanocomposite for photocatalytic oxidative desulfurization. *Appl. Surf. Sci.* **2016**, *364*, 589–596. [CrossRef]
35. Shen, C.H.; Wen, X.J.; Fei, Z.H.; Liu, Z.T.; Mu, Q.M. Novel Z-scheme $W_{18}O_{49}/CeO_2$ heterojunction for improved photocatalytic hydrogen evolution. *J. Colloid Interface Sci.* **2020**, *579*, 297–306. [CrossRef]
36. Wu, Z.L.; Li, M.J.; Howe, J.; Meyer III, H.M.; Overbury, S.H. Probing defect sites on CeO_2 nanocrystals with well-defined surface planes by Raman spectroscopy and O_2 adsorption. *Langmuir* **2010**, *2*, 16595–16606. [CrossRef]
37. Schmidt, S.; Kubaski, E.T.; Volanti, D.P.; Sequinel, T.; Bezzon, V.D.N.; Beltran, A.; Tebcherani, S.M.; Varela, J.A. Effect of pressure-assisted heat treatment on photoluminescence emission of α Bi_2O_3 needles. *Inorg. Chem.* **2015**, *54*, 10184–10191. [CrossRef]
38. Ravikovitch, P.I.; Neimark, A.V. Characterization of nanoporous materials from adsorption and desorption isotherms. *Colloids Surf. A* **2001**, *187–188*, 11–21. [CrossRef]
39. Anandan, C.; Bera, P. XPS studies on the interaction of CeO_2 with silicon in magnetron sputtered CeO_2 thin films on Si and Si_3N_4 substrates. *Appl. Surf. Sci.* **2013**, *283*, 297–303. [CrossRef]
40. Maheshwaran, S.; Balaji, R.; Chen, S.-M.; Chang, Y.-S.; Tamilalagan, E.; Chandrasekar, N.; Ethiraj, S.; Kumar, M. Ultrasensitive electrochemical detection of furazolidone in biological samples using 1D-2D $BiVO_4$@MoS_2 hierarchical nano-heterojunction composites armed electrodes. *Environ. Res.* **2022**, *205*, 112515. [CrossRef]
41. Shen, C.H.; Chen, Y.; Xu, X.J.; Li, X.Y.; Wen, X.J.; Liu, Z.T.; Xing, R.; Guo, H.; Fei, Z.H. Efficient photocatalytic H_2 evolution and Cr (VI) reduction under visible light using a novel Z-scheme $SnIn_4S_8/CeO_2$ heterojunction photocatalysts. *J. Hazard. Mater.* **2021**, *416*, 126217. [CrossRef] [PubMed]
42. Han, N.; Xu, Q.; Beyene, G.; Zhang, Q.F. Enhanced photocatalytic activity over g-$C_3N_4/(BiO)_2(OH)_xCl_{2-x}$ Z-scheme heterojunction. *Appl. Surf. Sci.* **2020**, *521*, 146464. [CrossRef]
43. Liu, C.; Xiao, W.; Yu, G.Y.; Wang, Q.; Hu, J.W.; Xu, C.B.; Du, X.Y.; Xu, J.G.; Zhang, Q.F.; Zou, Z.G. Interfacial engineering of Ti_3C_2 MXene/CdIn2S4 Schottky heterojunctions for boosting visible-light H_2 evolution and Cr (VI) reduction. *J. Colloid Interface Sci.* **2023**, *640*, 851–863. [CrossRef] [PubMed]
44. Yusuff, A.S.; Popoola, L.T.; Aderibigbe, E.I. Solar photocatalytic degradation of organic pollutants in textile industry wastewater by ZnO/pumice composite photocatalyst. *J. Environ. Chem. Eng.* **2020**, *8*, 103907. [CrossRef]
45. Li, L.; Guo, C.; Ning, J.; Zhong, Y.; Chen, D.; Hu, Y. Oxygen-vacancy-assisted construction of FeOOH/CdS heterostructure as an efficient bifunctional photocatalyst for CO_2 conversion and water oxidation. *Appl. Catal. B Environ.* **2022**, *613*, 764–774. [CrossRef]
46. Banerjee, R.; Pal, A.; Ghosh, D.; Ghosh, A.B.; Nandi, M.; Biswas, P. Improved photocurrent response, photostability and photocatalytic hydrogen generation ability of CdS nanoparticles in presence of mesoporous carbon. *Mater. Res. Bull.* **2021**, *134*, 111085. [CrossRef]
47. Li, X.Z.; Zhu, W.; Lu, X.W.; Zuo, S.X.; Yao, C.; Ni, C.Y. Integrated nanostructures of CeO_2/attapulgite/g-C_3N_4 as efficient catalyst for photocatalytic desulfurization: Mechanism, kinetics and influencing factors. *Chem. Eng. J.* **2017**, *326*, 87–98. [CrossRef]
48. Li, Y.; Wang, J.; Yao, H.; Dang, L.; Li, Z. Chemical etching preparation of $BiOI/Bi_2O_3$ heterostructures with enhanced photocatalytic activities. *Catal. Commun.* **2011**, *12*, 660–664. [CrossRef]

49. Channei, D.; Inceesungvorn, B.; Wetchakun, N.; Ukritnukun, S.; Nattestad, A.; Chen, J.; Phanichphant, S.J.S.R. Photocatalytic Degradation of Methyl Orange by CeO_2 and Fe–doped CeO_2 Films under Visible Light Irradiation. *Sci. Rep.* **2014**, *4*, srep05757. [CrossRef]
50. Wang, C.; Zhu, W.S.; Xu, Y.H.; Xu, H.; Zhang, M.; Chao, Y.H.; Yin, S.; Li, H.M.; Wang, J.G. Preparation of TiO_2/g-C_3N_4 composites and their application in photocatalytic oxidative desulfurization. *Ceram. Int.* **2014**, *40*, 11627–11635. [CrossRef]

catalysts

MDPI

Article

Band Gap Engineering in Quadruple-Layered Sillén–Aurivillius Perovskite Oxychlorides $Bi_7Fe_2Ti_2O_{17}X$ (X = Cl, Br, I) for Enhanced Photocatalytic Performance

Jikun Chen [1,†], Yan Gu [1,†], Shishi Xu [1], Yunxiang Zhang [1,*], Zhe Zhang [1], Lin Shi [1], Zhichao Mu [1], Chenliang Zhou [1], Jiali Zhang [1] and Qinfang Zhang [1,2,*]

1 School of Materials Science and Engineering, Yancheng Institute of Technology, Yancheng 224001, China
2 Jiangsu Provincial Key Laboratory of Eco-Environmental Materials, Yancheng Institute of Technology, Yancheng 224051, China
* Correspondence: chzyx123@163.com (Y.Z.); qfangzhang@gmail.com (Q.Z.)
† These authors contributed equally to this work.

Abstract: Developing efficient photocatalyst for the photoreduction of CO_2 and degradation of organic pollutants is an effective alternative to address increasingly serious energy problems and environmental pollution. Herein, the isostructural Sillén–Aurivillius oxyhalides, $Bi_7Fe_2Ti_2O_{17}X$ (X = Cl, Br, and I; BFTOX), are fabricated for CO_2 reduction and degradation of organic pollutants for the first time. Density functional theory (DFT) calculations show that the valence band maximum (VBM) of BFTOC and BFTOB is contributed by the dispersive $2p$ orbitals of O-atoms, providing the narrow band gap (E_g) and possibly the stability against self-decomposition deactivation. The photocatalytic activities of BFTOX are strongly affected by the halogens (Cl, Br, and I), namely, the BFTOCl sample displays outstanding activity improvement (3.74 $\mu mol \cdot g^{-1} \cdot h^{-1}$) for photocatalytic performance. This is mainly attributed to the high separation of charge carriers, small optical band gap, and extended optical absorption. This work focuses on affording a reference to develop efficient and stable photocatalysts from Sillén-Aurivillius layered oxyhalide materials.

Keywords: Sillén–Aurivillius; $Bi_7Fe_2Ti_2O_{17}X$; photocatalyst; degradation; CO_2 reduction

1. Introduction

With the development of industrialization, environmental problems have become a global issue of general concern in modern society, particularly for the greenhouse effect caused by carbon dioxide (CO_2) and the discharge of various organic pollutants, threatening the survival and development of mankind [1,2]. Photocatalytic technology is an appealing solution for the reduction of CO_2 and the removal of organic pollutants due to its ability to reduce them into high value-added chemicals or completely harmless molecules under sunlight or UV irradiation [3–6]. Moreover, semiconductor photocatalysis has attracted wide attention from researchers with the advantages of low cost, highly efficient, recyclable and does not lead to secondary pollution [7,8]. However, most photocatalytic materials have the problems of wide band gap, serious photogenerated carrier recombination and poor stability under light irradiation [9–12]. Thus, it is necessary to develop new efficient and stable visible light (VL) photocatalysts.

In recent years, bismuth (Bi)-based materials show the potential to be a novel series for VL responsive photocatalysts [13–16]. Consequently, large amounts of Bi-based photocatalysts have been studied, such as Bismuth trioxide (Bi_2O_3) [17,18], BiOX (X = Cl, Br, and I) [19–21], $Bi_4Ti_3O_{12}$ [22,23], and Bi_4MO_8X (M = Nb, Ta; X = Cl, Br) [24,25], etc. Among them, Sillén–Aurivillius layered oxyhalide materials, such as Bi_4MO_8X (M = Nb, Ta; X = Cl, Br), which are composed of $[MO_6]$ and $[(Bi_2O_2)_2X]$ blocks, have attracted considerable interest for photocatalysis with the virtues of the narrower band gap (E_g), ferroelectric polarization, layered crystal structure and stability against self-decomposition deactivation [26–28]. In 2016, Hironori Fujito et al. fabricated Bi_4NbO_8Cl successfully from Bi_2O_3, BiOCl, and Nb_2O_5 powders and the photocatalytic performance for oxygen (O_2) evolution was also evaluated in methanol solution [25]. Simultaneously, their density functional theory (DFT) calculations demonstrated that the valence band maximum (VBM) of Sillén–Aurivillius layered oxyhalide materials is mainly from O-2*p* orbitals [29,30]. This material can avoid the self-decomposition phenomenon of catalysts upon light irradiation because of the high oxidation of stable by the photo-excited holes near VBM [28,31]. Then, Xiaoping Tao et al. proved that Bi_4TaO_8Cl can be used for hydrogen (H_2) evolution under VL irradiation [32]. These results indicate that Sillén–Aurivillius layered oxyhalide materials Bi_4MO_8X (M = Nb, Ta; X = Cl, Br) should be a potential novel visible light photocatalyst material. In recent years, Sillén–Aurivillius materials were further extensively studied, such as the construction of heterojunctions [33,34], surface plasmon nanoparticle modification [26,35], element doping [27,36], etc., which proves that this materials have great development potential for photocatalysis, for example $Bi_7Fe_2Ti_2O_{17}X$ (X = Cl, Br, I) (BFTOX). In 2022, Yan Gu et al. fabricated $Bi_7Fe_2Ti_2O_{17}Cl$ successfully by using a one-step flux route and solid-state reaction, and its degradation efficiency of tetracycline (TC) reaches 90% within 90 min, which confirms that BFTOX should be a potential novel visible light photocatalyst material [37].

In order to further advance the development of Bi-based oxyhalides semiconductor materials, a series of novel Sillén–Aurivillius layered oxyhalide materials, $Bi_7Fe_2Ti_2O_{17}X$ (X = Cl, Br, I) (BFTOX), have been successfully synthesized via a molten-salt method by using the BiOX (X = Cl, Br, and I) precursor powders. Originally, we display the investigation of material characteristics, such as the band structure, effective mass, lattice constant, and E_g of BFTOX influenced by the halogen, Cl, Br, and I, from density functional theory (DFT) calculation. Then, material properties of BFTOX powders are determined by characterization techniques of X-ray diffraction (XRD), scanning electron microscope (SEM), energy dispersive spectroscopy (EDS), and X-ray photoelectron spectroscopy (XPS). Furthermore, the photocatalytic activities of different samples are compared by the reduction of CO_2 and the degradation of methyl orange (MO) and Rhodamine B (RhB), and BFTOX exhibits the best photocatalytic performance. In addition, the possible photocatalysis mechanisms of BFTOX have also been discussed in detail.

2. Results and Discussion

2.1. DFT Calculations

The structure of Sillén-Aurivillius materials is composed of fluorite ($[Bi_2O_2]^{2+}$) [20], like-perovskite ($[A_{n-1}B_nO_{3n+1}]^{3-}$) [21], and halogen ([X]) with the general formula of $[Bi_2O_2]^{2+}[X]^-[Bi_2O_2]^{2+}[A_{n-1}B_nO_{3n+1}]^{3-}$ [22], where n represents the number of perovskite layers (n = 1, 2, 3 …). Meanwhile, with the number of perovskite layers increasing, the difficulty of implementation will also increase, whether in calculation or experiment. After drawing on many previous works, this work put a sight on comparing three quadruple-perovskite layers semiconductor photocatalysts, $Bi_7Fe_2Ti_2O_{17}X$ (X = Cl, Br, I). The layers are constructed in the Cmcm space group (No. 63), aiming to provide the high efficiencies of photocatalysts for the reduction of CO_2 and degradation of organic pollutants. The crystal structure of BFTOX in Figure 1a shows that the novel BFTOX is an n = 4 member of Sillén–Aurivillius type perovskite oxyhalides with alternate stacking of Aurivillius perovskite blocks ($[Bi_3Ti_2Fe_2O_{13}]^{3-}$) and Sillén blocks ($[(Bi_2O_2)_2X]^{3+}$).

In order to further investigate the electronic properties of BFTOX, the energy band structure plots are displayed in Figure 1b–d. We can find that all BFTOX samples belong to the indirect-gap semiconductor. The calculated E_g values of BFTOC, BFTOB, and BFTOI are 2.30 eV, 2.44 eV, and 2.65 eV, respectively. As the halogen atomic number increases from Cl to Br and I, the band gap gradually widens, which may be related to the electronegativity of the halogen atoms. Furthermore, the electronegativity of materials also influences the energy level of the halogen atoms that elements with low electronegativity have high energy level positions, and thus the halogen level (Cl 3*p*, Br 4*p* and I 5*p*) increases as the atomic number increases relative to the O 2*p* orbitals. Quite unexpectedly, the VBM of BFTOC and BFTOB is predominantly composed of O 2*p* orbitals, while that of BFTOI is contributed by I 5p orbitals. For BFTOI, the VBM contribution is similar to that of simple oxyhalides BiOCl and BiOBr. During photocatalytic process, it is inevitable to be oxidized by the holes excited by the light, and the stability will be poor [20]. When it comes to BFTOC and BFTOB, the VBM is mainly composed of the dispersive O 2*p* orbitals. Meanwhile, the calculated effective masses of all samples in Table 1 confirm that the lower values of BFTOC and BFTOB. Hence, we believe that the excited holes in the stable oxygen anion near the VBM of both materials can effectively oxidize water without inactivation due to self-decomposition phenomenon of catalysts upon light irradiation.

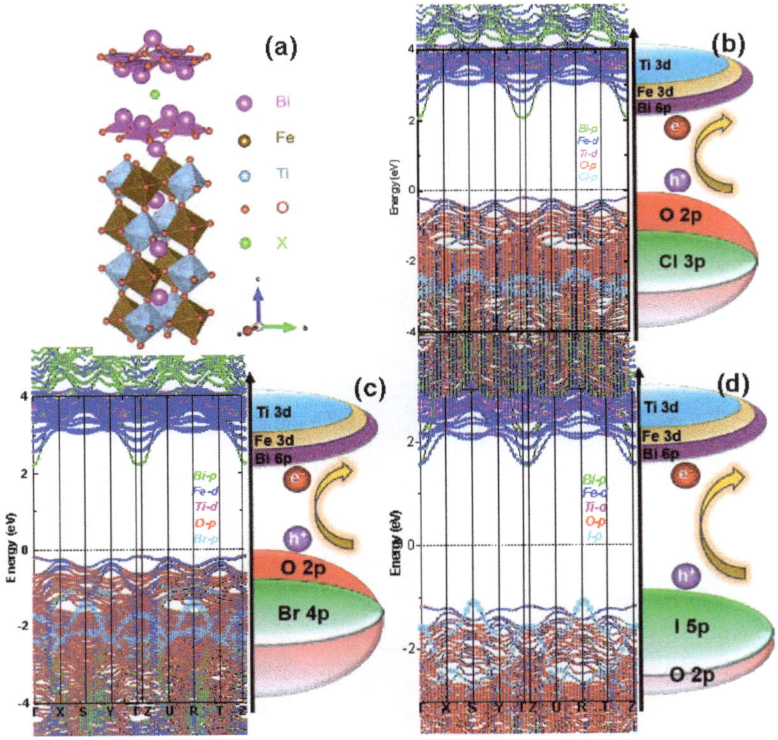

Figure 1. (**a**) Crystal structure of BFTOX. Band structures of (**b**) BFTOC, (**c**) BFTOB and (**d**) BFTOI.

Table 1. Calculated effective masses (unit m_0) of the electrons and holes of BFTOC, BFTOB and BFTOI by different directions.

	BFTOC		BFTOB		BFTOI	
Effective Mass	**Electron**	**Hole**	**Electron**	**Hole**	**Electron**	**Hole**
m_{xx} $(\Gamma - X)$	0.023	0.181	0.013	0.131	0.053	
m_{yy} $(\Gamma - Y)$	0.035	0.099	0.018	0.098	0.054	
m_{zz} $(\Gamma - Z)$	9.879	7.904	19.004	6.228	5.137	
m_{xx} $(S - Y)$						0.035
m_{yy} $(S - X)$						0.031

2.2. Material Properties of BFTOX (X = Cl, Br, and I)

The crystalline phases of as-prepared BFTOX powders were determined by XRD technology, and the corresponding results are displayed in Figure 2a–c. It can be observed that the diffraction peaks at 30.58°, 32.65°, 46.73°, and 56.97° can be indexed to the (116), (020), (220), and (316) crystal planes of BFTOX samples, which are all consistent with the calculated cards from DFT. These results confirm that BFTOX (X = Cl, Br, and I) materials are well-crystallized after calcination at different high temperatures in this experiment. Moreover, from the DFT results in Figure 1a, the structure of BFTOX (X = Cl, Br, and I) is the orthorhombic structure along with the space group of P21cn. In this case, the experimental lattice constants of BFTOX are calculated (BFTOC: a = 5.508 Å, b = 5.481 Å, and c = 26.593 Å; BFTOB: a = 5.501 Å, b = 5.478 Å, and c = 26.592 Å; BFTOI: a = 5.492 Å, b = 5.478 Å, and c = 26.644 Å). These results and the theoretically calculated lattice constants in Figure 2a–c are extremely similar, further indicating the successful preparation of BFTOX.

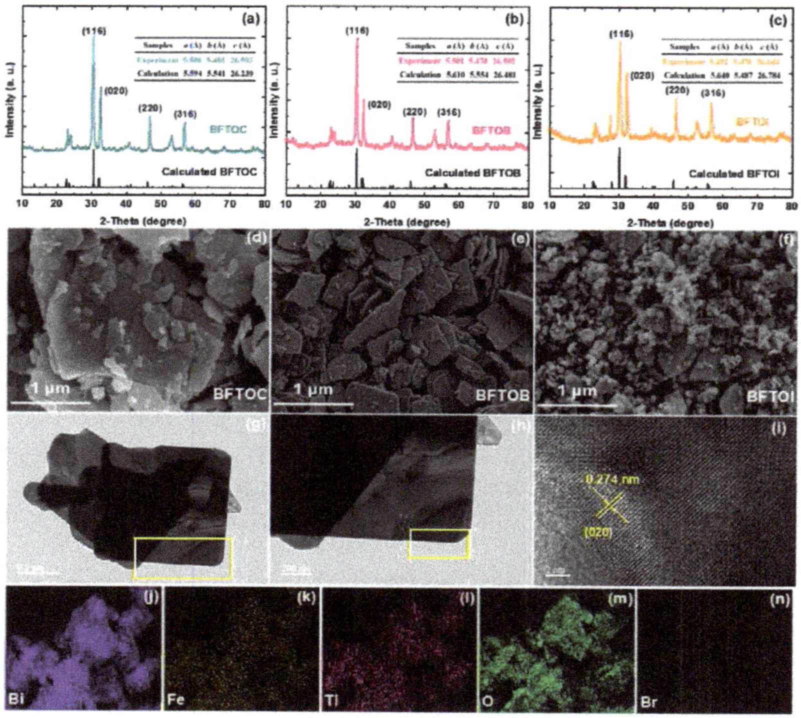

Figure 2. XRD patterns and their corresponding calculated cards of (**a**) BFTOC, (**b**) BFTOB, and (**c**) BFTOI. SEM images of (**d**) BFTOC, (**e**) BFTOB, and (**f**) BFTO-I. (**g**) TEM and the corresponding HRTEM of (**h**,**i**) BFTOC nanoplates in yellow squares. The corresponding SEM-EDS elemental mappings of (**j**) Bi, (**k**) Fe, (**l**) Ti, (**m**) O, and (**n**) Br for BFTOB sample.

Afterwards, the microstructure of the as-prepared BFTOX samples is determined by using SEM. As depicted in Figure 2d–f, all samples possess a nanoplate-like structure, which may be caused by the weak Van der Waals' force between the halogen layer [X] and the fluorite layer [Bi_2O_2]. Meanwhile, the size of nanoplates of BFTOX samples gradually decreased from the halogen variation in Sillén–Aurivillius material, particularly for BFTOI. One possible reason for this phenomenon is the stability of the material structure. Because the iodine element (I) is easy to sublimate during the high temperature calcination process, it is easy to destroy the crystal structure, so the crystallinity is poor and it is easy to form nano-agglomerates. In addition, the SEM-EDS mappings of BFTOB nanoplates in Figure 2g–k confirm the existence of Bi, Fe, Ti, O, and X and they are evenly distributed on the whole nanoplates. A similar phenomenon is also observed in BFTOC and BFTOI samples, seen the supporting information in Figure S1. These further demonstrate the successful synthesis of the BFTOX materials.

The elements of Bi, Fe, Ti, O and Br are found in the survey XPS spectra of BFTOBr (Figure 3a), which is in accordance with the results of SEM-EDS. The peaks at about 159.10 and 164.42 eV in Figure 3b, obtained from BFTOB, are ascribed to the Bi $4f_{7/2}$ and Bi $4f_{5/2}$ of Bi^{3+} [38]. In the Fe 2*p* region of BFTOB (Figure 3c), there were two peaks at 711.09 and 725.36 eV, referring to Fe $2p_{1/2}$ and Fe $2p_{3/2}$, respectively, manifesting the existence of Fe^{3+} ions [39]. There were two fitted Ti 2p peaks centered at 458.12 and 465.73 eV in Figure S2, which were attributed to Ti $2p_{3/2}$ and Ti $2p_{1/2}$ to the oxidation state of Ti^{4+} [40,41]. The O 1s peaks at binding energies of 529.50, 530.17 and 531.99 eV, which arises from the Bi-O, Fe-O and Ti-O, respectively (Figure S3). In addition, the binding energy peaks of Br $3d_{5/2}$ and Br $3d_{3/2}$ are located at 68.46 and 79.45 eV, respectively (Figure 2d) [42,43].

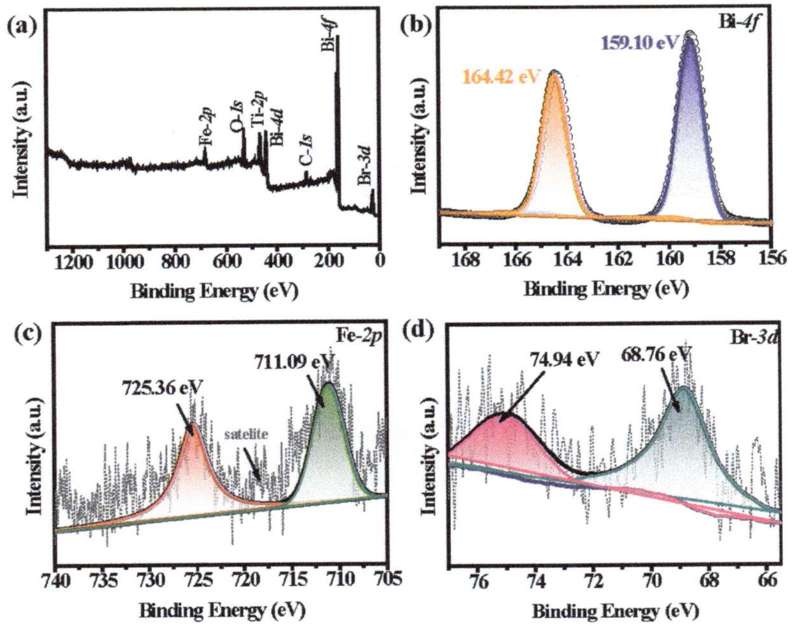

Figure 3. (**a**) XPS survey spectra of BFTOB corresponding to Bi, Fe, Ti, O and Br. High resolution XPS spectra of (**b**) Bi 4*f*, (**c**) Fe 2*p*, and (**d**) Br 3*d* of BFTOB, respectively.

2.3. Photocatalytic CO_2 Reduction Performance for BFTOX Materials

In order to evaluate the photocatalytic activities of BFTOX affected by the halogens, Cl, Br, and I, we conducted CO_2 reduction under the light irradiation. Under the simulated sunlight illumination, all BFTOX (X = Cl, Br, and I) samples can reduce the CO_2 to carbon monoxide (CO) in a sealed gas-solid phase reaction vessel without any organic sacrificial reagent. Figure 4a–c displays the relatively steady CO yields as a function of the irradiation time generated by BFTOC, BFTOB, and BFTOI. The yield values are about 14 μmol, 12 μmol, and 10 μmol for 210 min, respectively. To compare the performance clearly, the average yields of CO in 3.5 h are calculated in Figure 4d. It can be observed that the production rate of CO for BFTOC and BFTOB are 3.74 μmol g^{-1} h^{-1} and 3.6 μmol g^{-1} h^{-1}, which are approximately 1.4 times higher than that of BFTOI (2.66 μmol g^{-1} h^{-1}). This result means that the trend of photocatalytic activities of BFTOX is BFTOC > BFTOB > BFTOI. The decrease in the CO production rate for samples may be related to the absorption of solar photons.

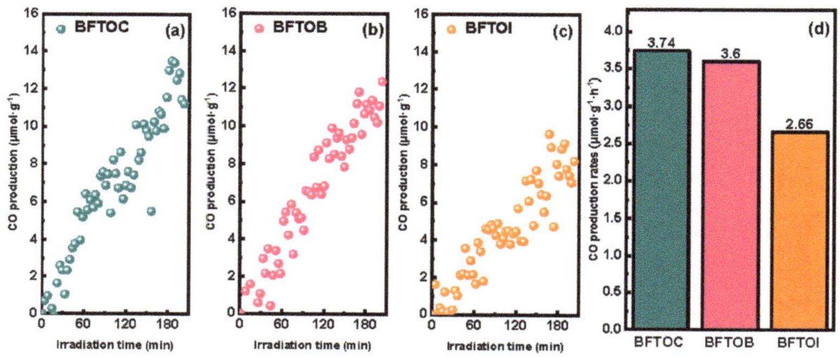

Figure 4. CO production of (**a**) BFTOC, (**b**) BFTOB, and (**c**) BFTOI, respectively. (**d**) The average CO production rates of BFTOX (X = Cl, Br, and I).

Subsequently, the UV–Vis diffuse absorbance spectrum (*DRS*) is used to evaluate the light absorption of BFTOX samples. It can be seen that all samples exhibit a strong photoresponse in the range of 390–550 nm in Figure 5a, which indicates that the BFTOX (X= Cl, Br, and I) materials are visible light catalysts. It is observed also that, the order of light response intensities is BFTOC > BFTOB > BFTOI, which can be ascribed to the sub-band gap absorption of powder materials. As we know, the significant sub-band gap absorption of films is indicative of the presence of the secondary phases or the band tailing [44–47]. This observation suggests the BFTOC powder has the lowest defect concentrations and photogenerated carrier recombination compared with the two other samples, which should be the reason for the highest performance of CO_2 reduction for BFTOC sample, as is consistent with the SEM results.

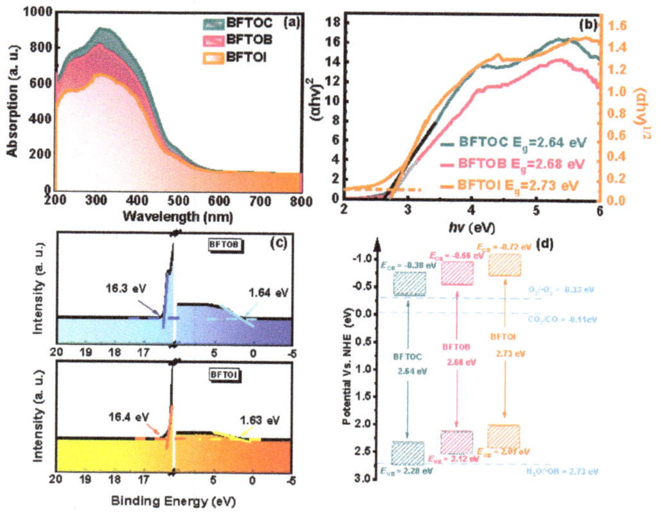

Figure 5. (**a**) UV-vis diffuse reflectance spectrum (*DRS*) of BFTOX (X = Cl, Br, and I) samples. (**b**) The calculated E_g of BFTO-X from Tauc's band-gap plots. (**c**) UPS valence band spectra of BFTOBr and BFTOI; (**d**) The energy band structure of BFTOX (X = Cl, Br, and I).

In addition, the BFTOC powder has the lowest defect concentrations compared with the two other samples. This should be the reason for the highest performance of CO_2 reduction for BFTOC sample, as is consistent with the SEM results. Furthermore, as the atomic number of the halogen increases in BFTOX (X = Cl, Br, and I), the absorption edge decreases, thus widening the E_g of samples. For a clearer comparison, the E_g values are determined by applying the Kabelka-Munk function [48–50]:

$$(\alpha hv)^{1/m} = A(hv - E_g) \tag{1}$$

where A, α, and hv are constant, absorption coefficient, and photon energy, respectively. The constant m is connected with the type of materials: the direct-gap semiconductor or the indirect-gap semiconductor [49]. From DFT calculation results, BFTOX is an indirect-gap semiconductor. Therefore, the value of m is 2. By the linear intersection in Figure 5b, the corresponding E_g values of BFTOX (X = Cl, Br, and I) are 2.64 eV, 2.68 eV, and 2.74 eV, respectively. This result is consistent with the trend of DFT calculation, as depicted in Figure 1.

In general, the positions of the valence band maximum (VBM) and conduction band minimum (CBM) of photocatalysts determines the photocatalytic activities of CO_2 reduction under light irradiation. Here, ultraviolet photoelectron spectroscopy (UPS) is employed to clarify the VBM and CBM positions of BFTOX samples, and the results of BFTOB and BFTOI are shown in Figure 5c. The result of BFTOC can be found in our previous work [44]. It can be calculated that the values of secondary cutoff energy ($E_{\text{cut-off}}$) of BFTOX are 16.30 eV, 16.30 eV, and 16.40 eV, respectively. And the position energy of corresponding Fermi levels (E_F) can be deduced from the equation [49]:

$$\varphi = hv - E_{\text{cut-off}} \tag{2}$$

where hv is about 21.22 eV for the He I electron energy. Therefore, the E_F values of BFTOC, BFTOB, and BFTOI are 4.92 eV, 4.92 eV, and 4.82 eV, respectively. In addition, the on-set energy ($E_{\text{on-set}}$) of samples can be applied to determining the gap between valence band and E_F. From Figure 5c, the $E_{\text{on-set}}$ values of BFTOC, BFTOB, and BFTOI are 1.80 eV, 1.64 eV, and 1.63 eV. Therefore, the valence band positions of these samples are 6.72 eV, 6.56 eV, and 6.45 eV, respectively. As the E_g values are depicted in Figure 5b, the conduction band positions are 4.06 eV, 3.88 eV, and 3.72 eV from the equation that $E_{CB} = E_{VB} - E_g$. According to previous works, the relationship between the vacuum energy (E_{abs}) and the normal electrode potential ($E^{\ominus}$) can be described as follows [51]:

$$E_{abs} = -E^{\ominus} - 4.44 \tag{3}$$

Correspondingly, the valence band positions of BFTOC, BFTOB, and BFTOI (vs. NHE) are 2.28 eV, 2.12 eV, and 2.01 eV, while the valence band positions of samples are −0.38 eV, −0.56 eV, and −0.72 eV, respectively. Based on the above discussions, the energy band diagrams of the CBM and VBM potentials for BFTOX samples are exhibited in Figure 5d. We can find that as the halogen in BFTOX is substituted from Cl to I, both the conduction and valence bands of samples are shifted upward. Meanwhiles, the E_g values are increased from 2.64 eV to 2.74, this result means that the widening of the band gap is mainly due to the shifting of the conduction band. Since the CBM of BFTOX (−0.38 eV, −0.56 eV, and −0.72 eV) are more negative than that of the CO_2/CO (−0.11 eV) redox potential, it confirms that the valence bands of BFTOX (X = Cl, Br, and I) are capable of CO_2 reduction, as the results in Figure 4.

In general, the size of photocatalyst nanoplates can directly affect the photocatalytic activity. When the nanoplate size is smaller, the number of nanoplates per unit mass is larger, and the specific surface area is larger. According to SEM results in Figure 2, the size of nanoplates of BFTOX samples gradually decreased from the halogen variation in Sillén–Aurivillius material, particularly for BFTOI. These results indicate that the BFTOI sample has the largest specific surface area, and it should have the maximum photocatalytic efficiency. However, this result is the opposite, which may be related to the generation and recombination of photogenerated charge carriers. As we know, the photocatalytic performance of BFTOX samples is affected by many factors, such as the separation and transfer of photogenerated electrons and holes. Here, photoelectrochemical characterizations, transient photocurrent densities (TPC) and electrochemical impedance spectra (EIS), are employed to investigate the charge transfer capability of the catalysts. Figure 6a shows the photocurrent response curves of BFTOX under light irradiation. It can be found that the trend of the photocurrent density of BFTOC is BFTOC > BFTOB > BFTOI. This result reveals the higher efficient separation of photo-generated electrons and holes for the BFTOC sample. Meanwhiles, a similar conclusion is also observed in the results of EIS in Figure 6b. The diameter of the arc radius of BFTOC on the EIS is much smaller that these of BFTOB and BFTOI, indicating that BFTOC sample possesses a faster transfer of the photo-generated electrons and holes. This excellent performance of separation and transfer of photogenerated carriers for BFTOC can be related to the highest quality of powder material along with the lowest defect concentrations, as exhibited in Figure 5a. In addition, the calculated effective masses of the electrons and holes in Table 1 shows that the values of BFTOB is smallest than other two samples, it suggests that the electron-hole pair recombination of BFTOB should be lowest and the BFTOB sample possesses the highest photocatalytic efficiency. However, the sample with the highest photocatalytic efficiency is BFTOC in experiment. The reason for this result is that the BFTOC sample has higher crystallization compared with the other two samples.

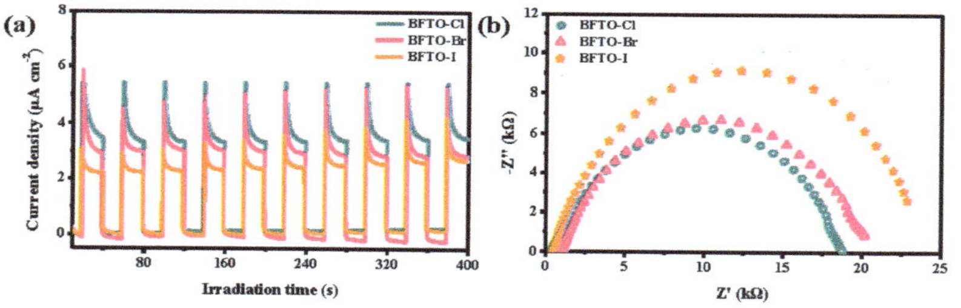

Figure 6. (**a**) Transient photocurrent densities and (**b**) EIS Nyquist plots of BFTOX (X = Cl, Br, and I).

2.4. *Photocatalytic Activity Enhancement Mechanism for the Photodegradation of Organic Pollutants*

The photocatalytic activities of BFTOX samples are also measured by using MO and RhB (10 mg/L) as the simulated pollutant under light irradiation, the results are displayed in Figure 7. To get a clear contrast, the degradation efficiencies of MO and RhB by BFTOX at different concentrations are calculated by the following equation [36]:

$$\eta = (C_0 - C) \;/\; C_0 \times 100\% \tag{4}$$

where C and C_0 are the sampling concentration of the reaction solution at different times and the initial concentration of MO and RhB solutions before light irradiation, respectively. From Figure 7a, we can find that the trend of degradation efficiencies for MO is BFTOC > BFTOB > BFTOI, and its result is similar for that of RhB solutions. Moreover, according to the fitting of pseudo-first-order: $\ln(C_0/C) = kt$, the degradation constants (k) are evaluated, as depicted in Figure 7b.

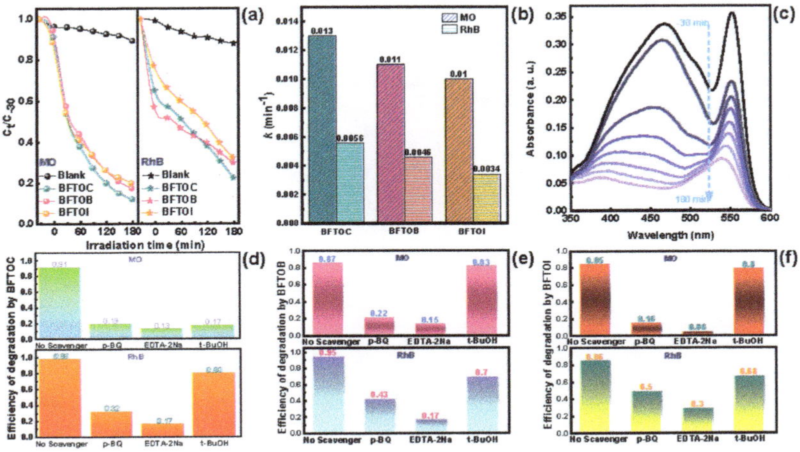

Figure 7. (**a**) The photocatalytic MO and RhB degradation activities over BFTOX. (**b**) The corresponding k values from kinetic linear fitting curves. (**c**) Degradation of the mixture of MO and RhB as a function of irradiation time. Degradation rate of MO/RhB over (**d**) BFTOC, (**e**) BFTOB, and (**f**) BFTOI with the addition of p-BQ, EDTA-2Na and TBA.

For MO, the BFTOC sample exhibits the highest apparent rate constant of about 0.011 min^{-1}, which is around 1.18 and 1.30 times of bare BFTOB and BFTOI samples, respectively. In addition, it can be observed that the degradation efficiencies of MO over BFTOX are higher than that of RhB, which indicates that that photocatalytic degradation of BFTOC powder is best, particularly for MO. Besides, BFTOC sample was also used to degrade the MO and RhB mixed pollutants, and its result is shown in Figure 7c. Obviously, the degradation efficiencies of MO at the wavelength of approximately 460 nm is higher compared with that of RhB at 550 nm. This further confirms the above conclusions. More importantly, the degradation efficiency of the mixture of MO and RhB is higher than that of single MO, which might be ascribed to the electrostatic force promoted the degradation of MO [52]. Moreover, the degradation efficiency of the mixture of MO and RhB is lower than that of single RhB due to the active sites of BFTOX being encircled intermediates, decelerating the production of electro-hole pairs and restraining the formation of radicals. In addition, the degradation efficiencies of MO are also higher than that of RhB in the mixture of MO and RhB. Namely, the degradation rates of MO over the BFTOX nanoplates are always higher than those of RhB, it may be mainly attributed to the molecular structure of oxychlorides, which needs further research in the future.

Afterwards, the active species trapping experiments of BFTOX (X = Cl, Br, and I) were carried out under light irradiation to evaluated the exact photocatalytic degradation mechanism. Radical scavenging materials, such as BQ, EDTA-2Na and TBA, are employed as the scavengers of $\cdot OH$, h^+, and $\cdot O_2^-$, respectively. The results in Figure 7d exhibit that the degradation efficiencies of MO and RhB over BFTOC are significantly inhibited under BQ and EDTA-2Na, while TBA only slightly reduces the degradation. In addition, similar capture trends are also found for BFTOB (Figure 7e) and BFTOI (Figure 7f). Therefore, h^+ and $\cdot O_2^-$ should be the major active species for MO and RhB degradation by BFTOC.

In fact, the stability and durability of photocatalysts is very important in practical applications. Figure 8a exhibits that the RhB degradation efficiency over BFTOC nanoplates with the same photocatalyst for five cycles. It can be found that the excellent photocatalysis performance of BFTOC is still significant after five cycles use, which confirms that the good recycling ability of BFTOC nanoplates. In addition, the stability of sample was also ascertained from XRD pattern in Figure 8b. The crystallinity of the as-prepared sample is similar to that of the used samples, it indicates that BFTOC nanoplates have the potential for the practical applications.

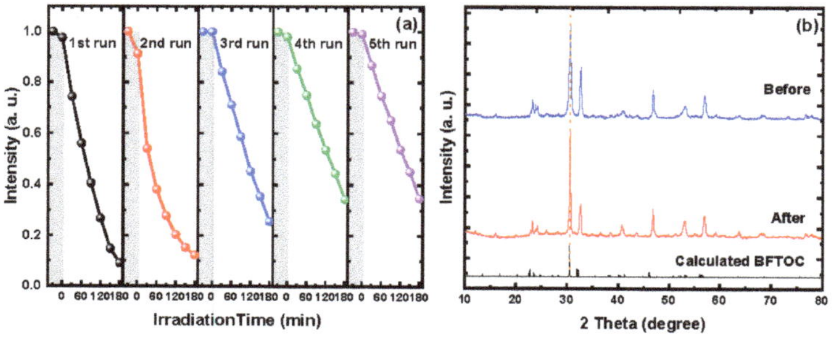

Figure 8. Time profiles of RhB degradation for five successive cycles with BFTOC in the photocatalytic process (**a**) and its XRD patterns (**b**).

Based on the aforementioned discussions, the probable photocatalytic degradation mechanism of organic pollutants over BFTOX materials is proposed in Figure 9. Under the light illumination, the e^- and h^+ will be produced on the CBM and VBM of BFTOX. The accumulated e^- on the CBM of BFTOX readily can reduce oxygen to $\cdot O_2^-$, which has more negative potentials (−0.38 eV, −0.56 eV, and −0.72 eV vs. NHE) than the O_2/O_2^- redox potential (−0.33 eV vs. NHE). The reason why ·OH ions do not contribute to the photocatalytic degradation process of organic pollutants is that in the acidic reaction solution, the E_{VB}s of BFTOX (BFTOC 2.28 eV; BFTOB 2.12 eV; BFTOI 2.01 eV) are more negative than the redox potential of $H_2O/\cdot OH$ (2.73 eV), and photogenerated holes cannot induce H_2O to ·OH, as shown in Figure 5d. Finally, the free radicals $\cdot O_2^-$ and e^-, which migrate to the surface of the powder and have good oxidation ability, directly degrade MO and RhB molecules into oxidation products. The possible photocatalytic mechanism of BFTOX powders can be described as follows:

$$Bi_7Fe_2Ti_2O_{17}X\ (X = Cl,\ Br,\ and\ I) + hv + (ultrasound) \rightarrow e^- + h^+ \tag{5}$$

$$O_2 + e^- \rightarrow \cdot O_2^- \tag{6}$$

$$\cdot O_2^- + h^+ + RhB/MO \rightarrow Intermediates + CO_2 + H_2O \tag{7}$$

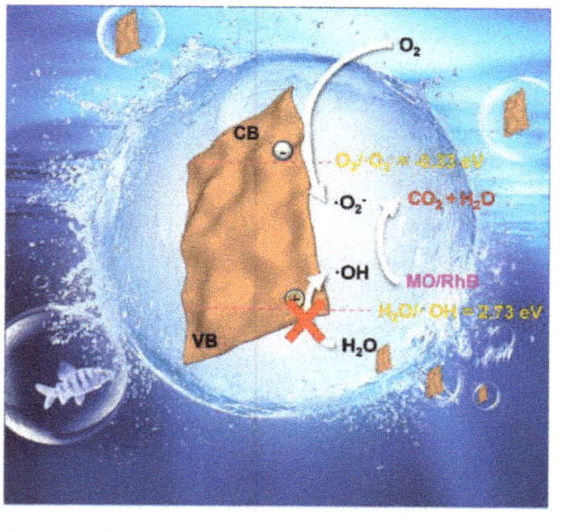

Figure 9. Schematic illustration of the proposed photocatalytic mechanism in the BFTOX powders.

3. Materials and Methods

3.1. Materials

Bismuth trioxide (Bi_2O_3) was purchased from Shanghai Aladdin Biochemical Technology CO., Ltd. Ferric oxide (Fe_2O_3) was purchased from Shanghai Shanpu Chemical Co., Ltd., Shanghai, China. Titanium dioxide (TiO_2), polyethylene glycol (EG), potassium chloride (KCl), potassium bromide (KBr), potassium iodide (KI), sodium chloride (NaCl), sodium bromide (NaBr), and sodium iodide (KI) were brought from Sinopharm Chemical Reagent Co., Ltd., Shanghai, China. All chemical reagents were analytical grade and directly used without any further treatment.

3.2. Preparation of Photocatalysts

3.2.1. Preparation of BiOX (X = Cl, Br, and I)

In this experiment, BiOX (X = Cl, Br, I) was prepared by a facile soft liquid deposition method, as reported in our previous works [27,49–51]. 24 mmol of $Bi(NO_3)_3{\cdot}5H_2O$ was dissolved into 240 mL of ethylene glycol solution (EG) to obtain solution A at room temperature. Then, 24 mmol of KX (X = Cl, Br, and I) was dissolved into 120 mL of deionized water to obtain solutions B, which was dropped slowly into the previous solution A. The mixed solution was magnetically stirred for 1 h at room temperature. Subsequently, the products were obtained by washing them with deionized water several times and dried at 60 °C for 12 h.

3.2.2. Preparation of $Bi_7Fe_2Ti_2O_{17}X$ (X = Cl, Br, and I)

BFTOX powders were synthesized by a flux method using Bi_2O_3, BiOX, Fe_2O_3 and TiO_2 as the precursor materials. KCl/NaCl, KBr/NaBr, and KI/NaI with a ratio of 1:1 were served as the molten salt. The molten salt was mixed with the solute of Bi_2O_3, BiOX, Fe_2O_3 and TiO_2 at the stoichiometric molar ratio for BFTOX (3:1:1:2). The solute concentration is about 6.25 mol% (BFTOX/(BFTOX + flux)). 10 mol% excess of BiOX over the stoichiometric ratio was added to prevent volatilization. Afterwards, three different mixtures were transferred to alumina crucibles for the high-temperature calcination. The $Bi_7Fe_2Ti_2O_{17}Cl$ (BFTOC), $Bi_7Fe_2Ti_2O_{17}Br$ (BFTOB), and $Bi_7Fe_2Ti_2O_{17}I$ (BFTOI) were obtained at the calcination temperature of 750 °C, 750 °C, and 700 °C for 6 h, respectively. After natural cooling, the obtained products were collected by filtration and thoroughly washed with the deionized water for several times. Finally, the products were dried at 60 °C for 12 h.

3.3. Characterization

X-ray diffraction (XRD) (X' Pert3 Powder, PANalytical, Almelo, The Netherlands, λ = 0.15406 nm) was used to analyze the crystal structure of BFTOX samples with Cu Kα radiation. The sample morphologies were examined by field emission scanning electron microscope (FE-SEM) (FEI, Hillsboro, OR, USA). X-ray photoelectron spectroscopy (XPS) was carried out to investigate the chemical nature of the samples via a Thermo Scientific Escalab 250 Xi spectrometer (Thermo Fisher Scientific, Carlsbad, CA, USA). The UV-vis diffuse reflectance spectroscopy (UV-vis DRS) was attained by using a UV-2550 (Shimadzu, Kyoto, Japan). The UV-vis absorption spectra of the samples were performed by UV-2450 (Shimadzu, Kyoto, Japan).

3.4. Photocatalytic Experiments

Photocatalytic experiments of the samples were discussed with a 300 W Xenon lamp (CEL-HXF300 Beijing China Education Au-light Co., Ltd., Beijing, China) as the light source. 100 mg of photocatalysts were separately added into 100 mL of 10 mg/L MO and RhB, and then stirred in the dark for 30 min to reach adsorption-desorption equilibrium. 4 mL solution was withdrawn every 30 min. The above suspension was centrifuged and the residual concentration of dye was detected by UV-vis spectrophotometer. During the degradation process of MO and RhB over BFTOX, $\cdot OH$, h^+, and $\cdot O_2^-$ were captured by a certain proportion of *tert*-butyl alcohol (TBA, 1 mM), EDTA (1 mM), and benzoquinone (BQ, 0.25 mM), respectively.

The experiments on the reduction of CO_2 were performed in a gas closed-circulation system. Photocatalyst powders of 0.01 g were dispersed in the absolute ethyl alcohol. Then, the photocatalysts solution was coated on a glass slide and placed in a reaction chamber (1 L). Deionized water entered the reaction chamber in the form of steam, and 1000 ppm CO_2 was introduced. After the whole system was stabilized, a 300 W Xe-arc lamp was applied to simulate sunlight.

3.5. Electrochemical Measurement

Photocurrent density versus time curves and electrochemical impedance spectra (EIS) were performed by a CHI660D electrochemical analyzer (Chenhua Instruments Co., Ltd., Shanghai, China) with a standard three-electrode system, including a working electrode, a reference electrode (saturated Ag/AgCl), and a counter electrode (Pt wire). The working electrode was obtained: 5 mg of as-prepared photocatalysts were dispersed into 40 μL of naphthol to make the slurry, and the slurry was deposited on the fluorine-tin oxide (FTO)glass. Afterwards, the substrate was dried at 80 °C for 4 h and calcined in a muffle furnace at 120 °C for 1 h.

3.6. Theoretical Calculations

In this work, the calculation of DFT with the projector-augmented wave (PAW) was carried out by VASP code [53]. The generalized gradient approximation (GGA) for the electron exchange and correlation corrections used Perdew-Burke-Ernzerhof (PBE) [54] to optimize the geometric structure and the next procedures were performed within the Heyd-Scuseria-Ernzerhof screened hybrid functional (HSE) [55], including band structures. In order to obtain accurate results, the cut-off energy was set to 520 eV and the irreducible Brillouin zone grid was used to $5 \times 5 \times 1$ mesh. The energy convergence criterion was 1×10^{-4} eV as well as -1×10^{-2} eV for atomic force convergence. In addition, the magnetic moment of the Fe atom was set to 4 μB. Due to the unpredictability of the synthesized samples in the experiment, Fe and Ti alternate arrangement is used to simulate the disorder situation in this DFT calculated process.

4. Conclusions

In summary, a series of novel Sillén–Aurivillius oxygen halides, $Bi_7Fe_2Ti_2O_{17}X$ (X = Cl, Br, and I), were successfully synthesized by molten salt method for the efficient CO_2 reduction and degradation of organic pollutants. The highly dispersive O 2p band of BFTOC and BFTOB samples contributing to the VBM not only renders the VBM essentially negative, but also explains its stability against self-decomposition deactivation from theory and experiment. For BFTOI, self-oxidation deactivation of I 5p orbitals leads to inferior CO_2 reduction in BFTOI. BFTOC and BFTOB samples possess outstanding activity for CO_2 reduction to CO, and the production rates of CO are 3.74 μmol g^{-1} h^{-1} and 3.6 μmol g^{-1} h^{-1}, respectively. More importantly, the BFTOX with unique nanosheet morphology displays a superb photocatalytic performance on MO and RhB mixture degradation. The removal efficiencies of the MO and RhB for BFTOC and BFTOB are 90% and 85% higher than BFTOI, which is a result of the improved separation of photogenerated charge carriers and optical absorption. To be more specific, the radical trapping tests and possible mechanism of BFTOX for the photocatalytic degradation of MO and RhB indicated that the h^+ and $\cdot O_2^-$ were the major active species to boost photocatalytic reactions. As an attractive photocatalyst, BFTOX has promising potential for CO_2 reduction and the degradation of organic pollutants.

Supplementary Materials: The following supporting information can be downloaded at: https://www.mdpi.com/article/10.3390/catal13040751/s1, Figure S1: The SEM-EDS elemental mappings of (a) Bi, (b) Fe, (c) Ti, (d) O, and (e) Br for BFTOC sample. The SEM-EDS elemental mappings of (f) Bi, (g) Fe, (h) Ti, (i) O, and (j) Br for BFTOI sample; Figure S2: Ti 2p XPS spectrum of the BFTO-Br sample; Figure S3: O 1*s* XPS spectrum of the BFTO-Br sample; Figure S4: Degradation rate of MO/RhB over (a) BFTOC and (b) BFTOC with the addition of p-BQ, EDTA-2Na and TBA; Table S1: The average CO production rates of BFTOX at different times.

Author Contributions: J.C. and Y.G.: conceptualization, methodology, investigation, testing, editing original draft and theoretical calculation; S.X., Z.Z., Z.M., C.Z. and J.Z.: conceptualization, methodology, investigation, testing; L.S.: theoretical calculation; Y.Z.: writing—review & editing, supervision, data curation. Q.Z.: project administration, funding acquisition, writing—review & editing. All authors have read and agreed to the published version of the manuscript.

Funding: This work is supported by the National Natural Science Foundation of China (No. 12274361), the Natural Science Foundation of Jiangsu Province (BK20211361), and College Natural Science Research Project of Jiangsu Province (20KJA430004), the Postgraduate Research & Practice Innovation Program of Jiangsu Province (KYCX21_3145) and the school-level research projects of Yancheng Institute of Technology (xjr2019028, xjr2019059).

Data Availability Statement: The data presented in this study are available on request from the corresponding author.

Conflicts of Interest: The authors declare no conflict of interest.

References

1. Yan, J.; Wang, C.; Ma, H.; Li, Y.; Liu, Y.; Suzuki, N.; Terashima, C.; Fujishima, A.; Zhang, X. Photothermal synergic enhancement of direct Z-scheme behavior of $Bi_4TaO_8Cl/W_{18}O_{49}$ heterostructure for CO_2 reduction. *Appl. Catal. B Environ.* **2020**, *268*, 118401. [CrossRef]
2. Zhang, Y.; Lin, S.; Cheng, S.; He, Z.; Hu, Z.; Zhou, Z.; Liu, W.; Sun, Y. Boosting Cu(In,Ga)Se_2 Thin Film Growth in Low-Temperature Rapid-Deposition Processes: An Improved Design for the Single-Heating Knudsen Effusion Cell. *Engineering* **2021**, *7*, 534–541. [CrossRef]
3. Katsumata, H.; Islam Molla, M.A.; Islam, J.B.; Tateishi, I.; Furukawa, M.; Kaneco, S. Dual Z-scheme heterojunction g-C_3N_4/Ag_3PO_4/AgBr photocatalyst with enhanced visible-light photocatalytic activity. *Ceram. Int.* **2022**, *48*, 21898–21905. [CrossRef]
4. Lu, X.; Quan, L.; Hou, H.; Qian, J.; Liu, Z.; Zhang, Q. Fabrication of 1D/2D Y-doped $CeO_2/ZnIn_2S_4$ S-scheme photocatalyst for enhanced photocatalytic H_2 evolution. *J. Alloy. Compd.* **2022**, *925*, 166552. [CrossRef]
5. Melchionna, M.; Fornasiero, P. Updates on the Roadmap for Photocatalysis. *ACS Catal.* **2020**, *10*, 5493–5501. [CrossRef]
6. Kumar, A.; Kumar, A.; Krishnan, V. Perovskite Oxide Based Materials for Energy and Environment-Oriented Photocatalysis. *ACS Catal.* **2020**, *10*, 10253–10315. [CrossRef]
7. Kumar, A.; Pathania, D.; Gupta, N.; Raj, P.; Sharma, A. Photo-degradation of noxious pollutants from water system using Cornulaca monacantha stem supported $ZnFe_2O_4$ magnetic bio-nanocomposite. *Sustain. Chem. Pharm.* **2020**, *18*, 100290. [CrossRef]
8. Arush, S.; Deepak, P.; Ajay, K. Bio-Polymer Based Tragacanth Gum (TG) Loaded Fe_3O_4 Nanocomposite for the Sequestration of Tenacious Congo Red Dye from Waste Water. *J. Mater. Sci. Technol. Res.* **2020**, *7*, 92–100. [CrossRef]
9. Sun, C.; Yang, J.; Xu, M.; Cui, Y.; Ren, W.; Zhang, J.; Zhao, H.; Liang, B. Recent intensification strategies of SnO_2-based photocatalysts: A review. *Chem. Eng. J.* **2022**, *427*, 131564. [CrossRef]
10. Xu, K.; Wang, L.; Xu, X.; Dou, S.X.; Hao, W.; Du, Y. Two dimensional bismuth-based layered materials for energy-related applications. *Energy Storage Mater.* **2019**, *19*, 446–463. [CrossRef]
11. Lu, X.; Liu, Z.; Zhao, X.; Xu, W.; Hou, H.; Qian, J. CdS Nanoparticles Decorated 1D CeO_2 Nanorods for Enhanced Photocatalytic Desulfurization Performance. *Catalysts* **2022**, *12*, 1478. [CrossRef]
12. Chandel, M.; Thakur, M.; Sharma, A.; Pathania, D.; Kumar, A.; Singh, L. Chlorophyll sensitized $(BiO)_2CO_3$/ $CdWO_4$/rGO nano-hybrid assembly for solar assisted photo-degradation of chlorzoxazone. *Chemosphere* **2022**, *305*, 135472. [CrossRef] [PubMed]
13. Ren, X.; Gao, M.; Zhang, Y.; Zhang, Z.; Cao, X.; Wang, B.; Wang, X. Photocatalytic reduction of CO_2 on BiOX: Effect of halogen element type and surface oxygen vacancy mediated mechanism. *Appl. Catal. B: Environ.* **2020**, *274*, 119063. [CrossRef]
14. Zhang, L.; Li, Y.; Li, Q.; Fan, J.; Carabineiro, S.A.C.; Lv, K. Recent advances on Bismuth-based Photocatalysts: Strategies and mechanisms. *Chem. Eng. J.* **2021**, *419*, 129484. [CrossRef]
15. He, R.; Cao, S.; Zhou, P.; Yu, J. Recent advances in visible light Bi-based photocatalysts. *Chin. J. Catal.* **2014**, *35*, 989–1007. [CrossRef]

16. Miao, Z.; Wang, Q.; Zhang, Y.; Meng, L.; Wang, X. In situ construction of S-scheme AgBr/BiOBr heterojunction with surface oxygen vacancy for boosting photocatalytic CO_2 reduction with H_2O. *Appl. Catal. B Environ.* **2022**, *301*, 120802. [CrossRef]
17. Meng, X.; Zhang, Z. Bismuth-based photocatalytic semiconductors: Introduction, challenges and possible approaches. *J. Mol. Catal. A Chem.* **2016**, *423*, 533–549. [CrossRef]
18. Divya, J.; Shivaramu, N.J.; Roos, W.D.; Purcell, W.; Swart, H.C. Synthesis, surface and photoluminescence properties of Sm^{3+} doped α-Bi_2O_3. *J. Alloy Compd.* **2021**, *854*, 157221. [CrossRef]
19. Zhao, Q.; Liu, X.; Xing, Y.; Liu, Z.; Du, C. Synthesizing Bi_2O_3/BiOCl heterojunctions by partial conversion of BiOCl. *J. Mater. Sci.* **2016**, *52*, 2117–2130. [CrossRef]
20. Shang, J.; Chen, T.; Wang, X.; Sun, L.; Su, Q. Facile fabrication and enhanced photocatalytic performance: From BiOCl to element-doped BiOCl. *Chem. Phys. Lett.* **2018**, *706*, 483–487. [CrossRef]
21. Han, N.; Zhang, Q. Stoichiometry-dependent photocatalytic performance of bismuth-based oxychlorides BiOyCl. *Appl. Surf. Sci.* **2021**, *562*, 150215. [CrossRef]
22. He, H.; Yin, J.; Li, Y.; Zhang, Y.; Qiu, H.; Xu, J.; Xu, T.; Wang, C. Size controllable synthesis of single-crystal ferroelectric $Bi_4Ti_3O_{12}$ nanosheet dominated with {0 0 1} facets toward enhanced visible-light-driven photocatalytic activities. *Appl. Catal. B Environ.* **2014**, *156–157*, 35–43. [CrossRef]
23. Xie, Z.; Tang, X.; Shi, J.; Wang, Y.; Yuan, G.; Liu, J.-M. Excellent piezo-photocatalytic performance of Bi4Ti3O12 nanoplates synthesized by molten-salt method. *Nano Energy* **2022**, *98*, 107247. [CrossRef]
24. Bhat, S.S.M.; Sundaram, N.G. Photocatalysis of Bi4NbO8Cl hierarchical nanostructure for degradation of dye under solar/UV irradiation. *New J. Chem.* **2015**, *39*, 3956–3963. [CrossRef]
25. Fujito, H.; Kunioku, H.; Kato, D.; Suzuki, H.; Higashi, M.; Kageyama, H.; Abe, R. Layered Perovskite Oxychloride Bi4NbO8Cl: A Stable Visible Light Responsive Photocatalyst for Water Splitting. *J. Am. Chem. Soc.* **2016**, *138*, 2082–2085. [CrossRef] [PubMed]
26. Li, S.; Wang, C.; Li, D.; Xing, Y.; Zhang, X.; Liu, Y. Bi_4TaO_8Cl/Bi heterojunction enables high-selectivity photothermal catalytic conversion of CO_2-H_2O flow to liquid alcohol. *Chem. Eng. J.* **2022**, *435*, 135133. [CrossRef]
27. Sun, J.; Han, N.; Gu, Y.; Lu, X.; Si, L.; Zhang, Q. Hole Doping to Enhance the Photocatalytic Activity of Bi4NbO8Cl. *Catalysts* **2020**, *10*, 1425. [CrossRef]
28. Zhong, C.; Kato, D.; Takeiri, F.; Fujii, K.; Yashima, M.; Nishiwaki, E.; Fujii, Y.; Koreeda, A.; Tassel, C.; Abe, R.; et al. Single Crystal Growth of Sillén–Aurivillius Perovskite Oxyhalides Bi4NbO8X (X = Cl, Br). *Inorganics* **2018**, *6*, 41. [CrossRef]
29. Kunioku, H.; Higashi, M.; Tomita, O.; Yabuuchi, M.; Kato, D.; Fujito, H.; Kageyama, H.; Abe, R. Strong hybridization between Bi-6s and O-2p orbitals in Sillén–Aurivillius perovskite Bi4MO8X (M = Nb, Ta; X = Cl, Br), visible light photocatalysts enabling stable water oxidation. *J. Mater. Chem. A* **2018**, *6*, 3100–3107. [CrossRef]
30. Sharma, A.; Kanth, S.K.; Xu, S.; Han, N.; Zhu, L.; Fan, L.; Liu, C.; Zhang, Q. Visible light driven g-C3N4/Bi4NbO8X (X Cl, Br) heterojunction photocatalyst for the degradation of organic pollutants. *J. Alloy. Compd.* **2022**, *895*, 162576. [CrossRef]
31. Kato, D.; Hongo, K.; Maezono, R.; Higashi, M.; Kunioku, H.; Yabuuchi, M.; Suzuki, H.; Okajima, H.; Zhong, C.; Nakano, K.; et al. Valence Band Engineering of Layered Bismuth Oxyhalides toward Stable Visible-Light Water Splitting: Madelung Site Potential Analysis. *J. Am. Chem. Soc.* **2017**, *139*, 18725–18731. [CrossRef]
32. Tao, X.; Zhao, Y.; Mu, L.; Wang, S.; Li, R.; Li, C. Bismuth Tantalum Oxyhalogen: A Promising Candidate Photocatalyst for Solar Water Splitting. *Adv. Energy Mater.* **2018**, *8*, 1701392. [CrossRef]
33. Majumdar, A.; Ghosh, U.; Pal, A. Novel 2D/2D g-C(3)N(4)/Bi(4)NbO(8)Cl nano-composite for enhanced photocatalytic degradation of oxytetracycline under visible LED light irradiation. *J. Colloid. Interface Sci.* **2021**, *584*, 320–331. [CrossRef] [PubMed]
34. Qu, X.; Liu, M.; Zhang, W.; Sun, Z.; Meng, W.; Shi, L.; Du, F. A facile route to construct $NiTiO_3$/Bi_4NbO_8Cl heterostructures for enhanced photocatalytic water purification. *J. Mater. Sci.* **2020**, *55*, 9330–9342. [CrossRef]
35. Shi, L.; Wu, X.; Xu, C.; Bai, Q.; Nie, Z.; Ji, C.; Zhang, Y.; Yin, Z.; Zhang, S.; Qu, X.; et al. Defect-engineering of Pt/Bi_4NbO_8Br heterostructures for synergetic promotional photocatalytic removal of versatile organic contaminants. *J. Mater. Chem. C* **2021**, *9*, 2784–2792. [CrossRef]
36. Feng, Z.; Li, G.; Yuan, X.; Zhang, Z.; Tang, M.; Zhang, R. Improved photocatalytic activity of Bi_4TaO_8Cl by Gd^{3+} doping. *J. Am. Ceram. Soc.* **2018**, *102*, 2390–2397. [CrossRef]
37. Gu, Y.; Yu, F.; Chen, J.; Zhang, Q. Facile Synthesis of Sillén-Aurivillius Layered Oxide $Bi_7Fe_2Ti_2O_{17}Cl$ with Efficient Photocatalytic Performance for Degradation of Tetracycline. *Catalysts* **2022**, *12*, 221. [CrossRef]
38. Wang, Y.; Sun, X.; Xian, T.; Liu, G.; Yang, H. Photocatalytic purification of simulated dye wastewater in different pH environments by using $BaTiO_3$/Bi_2WO_6 heterojunction photocatalysts. *Opt. Mater.* **2021**, *113*, 110853. [CrossRef]
39. Shao, L.; Xia, X.; Wei, G.; Qin, J.; Liu, Y. A dramatic enhancement of antibiotic photodegradation catalyzed by red mud-derived $Bi_5FeTi_3O_{15}$. *Sep. Purif. Technol.* **2021**, *275*, 119244. [CrossRef]
40. Yin, X.; Li, X.; Liu, H.; Gu, W.; Zou, W.; Zhu, L.; Fu, Z.; Lu, Y. Realizing selective water splitting hydrogen/oxygen evolution on ferroelectric Bi_3TiNbO_9 nanosheets. *Nano Energy* **2018**, *49*, 489–497. [CrossRef]
41. Wang, Z.; Li, C.; Domen, K. Recent developments in heterogeneous photocatalysts for solar-driven overall water splitting. *Chem. Soc. Rev.* **2019**, *48*, 2109–2125. [CrossRef] [PubMed]
42. Li, K.-L.; Lee, W.W.; Lu, C.-S.; Dai, Y.-M.; Chou, S.-Y.; Chen, H.-L.; Lin, H.-P.; Chen, C.-C. Synthesis of BiOBr, Bi_3O_4Br, and $Bi_{12}O_{17}Br_2$ by controlled hydrothermal method and their photocatalytic properties. *J. Taiwan Inst. Chem. Eng.* **2014**, *45*, 2688–2697. [CrossRef]

43. Lin, H.-P.; Lee, W.W.; Huang, S.-T.; Chen, L.-W.; Yeh, T.-W.; Fu, J.-Y.; Chen, C.-C. Controlled hydrothermal synthesis of $PbBiO_2Br/BiOBr$ heterojunction with enhanced visible-driven-light photocatalytic activities. *J. Mol. Catal. A Chem.* **2016**, *417*, 168–183. [CrossRef]
44. Zhang, Y.; Hu, Z.; Lin, S.; Cheng, S.; He, Z.; Wang, C.; Zhou, Z.; Sun, Y.; Liu, W. Facile Silver-Incorporated Method of Tuning the Back Gradient of $Cu(In,Ga)Se_2$ Films. *ACS Appl. Energy Mater.* **2020**, *3*, 9963–9971. [CrossRef]
45. Zhang, Y.; Hu, Z.; Lin, S.; Wang, C.; Cheng, S.; He, Z.; Zhou, Z.; Sun, Y.; Liu, W. Silver Surface Treatment of $Cu(In,Ga)Se_2$ (CIGS) Thin Film: A New Passivation Process for the CdS/CIGS Heterojunction Interface. *Sol. RRL* **2020**, *4*, 2000290. [CrossRef]
46. Zhang, Y.; Gao, Q.; Ao, J.; Zhang, Y.; Bi, J.; Guo, J.; Han, Y.; Sun, G.; Zhang, Y.; Liu, W.; et al. Enhancing Surface Properties for Electrodeposited $Cu(In,Ga)Se_2$ Films by $(NH_4)_2S$ Solution at Room Temperature. *ACS Appl. Energy Mater.* **2021**, *4*, 3822–3831. [CrossRef]
47. Qing, G.; Ao, J.; Zhang, Y.; Zhang, Y.; Guo, J.; Sun, G.; Liu, W.; Liu, F.; Zhang, Y. Pulse Selenization in $Cu(In,Ga)Se_2$ Solar Cells: A Promising Approach to Achieve High Efficiency by Electrodeposition. *ACS Appl. Energy Mater.* **2021**, *4*, 8322–8329. [CrossRef]
48. Makula, P.; Pacia, M.; Macyk, W. How To Correctly Determine the Band Gap Energy of Modified Semiconductor Photocatalysts Based on UV-Vis Spectra. *J. Phys. Chem. Lett.* **2018**, *9*, 6814–6817. [CrossRef]
49. Zhang, Y.; Xu, S.; Mu, Z.; Liu, K.; Chen, J.; Zhou, C.; Yao, Y.; Chen, X.; Shi, L.; Wang, Z.; et al. Simulation analysis of Cd-free $Cu(In,Ga)Se_2$ solar cells with novel BiOX (X=Cl, Br) buffer layers. *Vacuum* **2022**, *206*, 111569. [CrossRef]
50. Xu, S.; Zhang, Y.; Wang, Y.; Chen, J.; Zhou, C.; Mu, Z.; Zhang, Z.; Zhang, J.; Wang, J.; Zhang, Q. Coupled Ferroelectric Polarization in Novelty Sillén–Aurivillius $BaBi_4TiNbO_{11}Cl$ Material for Photocatalysis. *J. Alloy Compd.* **2023**, *954*, 169932. [CrossRef]
51. Han, N.; Xu, S.; Zhang, Q. Flux synthesis of Bi_2MO_4Cl (M = Gd and Nd) nanosheets for high-efficiency photocatalytic oxygen evolution under visible light. *J. Mater. Sci.* **2022**, *57*, 2870–2882. [CrossRef]
52. Wang, S.; Teng, F.; Zhao, Y. Effect of the molecular structure and surface charge of a bismuth catalyst on the adsorption and photocatalytic degradation of dye mixtures. *RSC Adv.* **2015**, *5*, 76588–76598. [CrossRef]
53. Kresse, G.; Hafner, J. Ab initio molecular dynamics for liquid metals. *Phys. Rev. B Condens. Matter.* **1993**, *47*, 558–561. [CrossRef]
54. Perdew, J.P.; Burke, K.; Ernzerhof, M. Generalized gradient approximation made simple. *Phys. Rev. Lett.* **1996**, *77*, 3865. [CrossRef] [PubMed]
55. Heyd, J.; Scuseria, G.E.; Ernzerhof, M.J. Hybrid functionals based on a screened Coulomb potential. *J. Chem. Phys.* **2003**, *118*, 8207–8215. [CrossRef]

catalysts

MDPI

Article

$BiPO_4$/Ov-BiOBr High-Low Junctions for Efficient Visible Light Photocatalytic Performance for Tetracycline Degradation and H_2O_2 Production

Minghui Tang [1], Xibao Li [1,2,*], Fang Deng [2], Lu Han [3], Yu Xie [2], Juntong Huang [1], Zhi Chen [1], Zhijun Feng [1] and Yingtang Zhou [4,*]

1 School of Materials Science and Engineering, Nanchang Hangkong University, Nanchang 330063, China
2 National-Local Joint Engineering Research Center of Heavy Metal Pollutants Control and Resource Utilization, Nanchang Hangkong University, Nanchang 330063, China
3 School of Materials and Metallurgy, University of Science and Technology Liaoning, Anshan 114051, China
4 National Engineering Research Center for Marine Aquaculture, Marine Science and Technology College, Zhejiang Ocean University, Zhoushan 316004, China
* Correspondence: lixibao@nchu.edu.cn (X.L.); zhouyingtang@zjou.edu.cn (Y.Z.)

Abstract: Through a two-step solvothermal method, different molar ratios of $BiPO_4$ were grown in situ on the surface of oxygen-vacancy-rich BiOBr (Ov-BiOBr), successfully constructing a $BiPO_4$/Ov-BiOBr heterojunction composite material. By constructing a novel type I high-low junction between the semiconductor $BiPO_4$ and Ov-BiOBr, stronger oxidative holes or reductive electrons were retained, thereby improving the redox performance of the photocatalyst. The composite catalyst with a 10% molar content of $BiPO_4$ demonstrated the highest degradation rate of tetracycline (TC), degrading over 95% within 90 min, with a rate constant of 0.02534 min^{-1}, which is 2.3 times that of Ov-BiOBr and 22 times that of $BiPO_4$. The 10% $BiPO_4$/Ov-BiOBr sample displayed the best photocatalytic activity, producing 139 $\mu mol \cdot L^{-1}$ H_2O_2 in 120 min, which is 3.6 times the efficiency of Ov-BiOBr and 19 times that of $BiPO_4$. This was due to the appropriate bandgap matching between $BiPO_4$ and Ov-BiOBr, the photo-generated electron transfer channel via Bi-bridge, and efficient charge separation. It was inferred that the free radical species $\cdot OH$ and $\cdot O_2^-$ played the dominant role in the photocatalytic process. Based on experimental and theoretical results, a possible photocatalytic mechanism was proposed.

Keywords: $BiPO_4$; BiOBr; oxygen vacancy; high-low junction; photocatalysis

1. Introduction

Since the acceleration of urbanization and rapid industrial development in modern times, environmental pollution has created increasingly serious problems, among which organic pollution has gradually become one of the major issues faced by human society in the 21st century. Tetracycline (TC), a widely used antibiotic, is present in numerous wastewater streams and environmental samples. Its excessive use and discharge have led to environmental pollution issues, such as the spread of antibiotic resistance genes. Therefore, studying the photocatalytic degradation of tetracycline is of significant environmental importance. As a pollutant with representative antibiotic structural features, tetracycline's photocatalytic degradation helps in better understanding and evaluating the treatment effects of different photocatalysts on similar contaminants. The high stability and low natural degradation of tetracycline in water present challenges for the design and optimization of photocatalysts. The aim is to discover novel and efficient photocatalysts through the study of tetracycline's photocatalytic degradation process. Various methods have been employed for tetracycline degradation, including biodegradation, photochemical degradation, advanced oxidation processes (AOPs), membrane filtration technology and adsorption. However, these methods have drawbacks such as slow degradation rates, high energy consumption, susceptibility to environmental conditions, generation of secondary pollutants, stringent operating conditions, only achieving separation and concentration rather than true degradation of pollutants, reaching adsorption saturation easily, and the potential for membrane fouling. The photocatalytic method is chosen for tetracycline degradation in this research due to its environmental friendliness, low energy consumption, and lack of secondary pollution. By designing and optimizing photocatalysts, the goal is to improve degradation efficiency and reduce energy consumption. In terms of environmental purification, semiconductor catalysis technology has become a promising green energy utilization technology, which has significant advantages over traditional organic treatment methods, such as no secondary pollution, low cost, and strong oxidation, and has received widespread attention. The research focus of photocatalytic semiconductor catalysis technology is the selection, synthesis, and performance improvement of semiconductors [1–4].

The photocatalytic performance of heterojunctions mainly depends on the generation of electron-hole pairs in the conduction and valence bands after absorbing light energy and maintaining spatial separation to participate in redox reactions, thus achieving photocatalytic effects [5–7]. However, due to the structural characteristics of the semiconductor material itself, the recombination rate of photo-generated charge carriers is usually greater than the migration rate. In addition, the bandgap width of the semiconductor material itself is too large, and the wavelength range of light absorption is too narrow, resulting in a decrease in the absorption and utilization efficiency of visible light, which leads to the excited electrons and holes not being able to participate in redox reactions in a timely manner; instead, they are lost in the form of thermal energy or light energy [8]. Therefore, the recombination of electron-hole pairs and the wavelength range of light absorption by semiconductor materials are the two main factors affecting the photocatalytic performance of composite photocatalysts [9,10].

BiOBr is a ternary bismuth-based layered semiconductor photocatalyst. The staggered arrangement of its layered structure leads to an inherent electric field between the Br- layer and the $[Bi_2O_2]^{2+}$ layer, facilitating better separation of electrons and holes [11–14]. Moreover, BiOBr has oxygen vacancy defect sites, which can capture excited-state electrons and create spatial separation between different photogenerated charge carriers, preventing recombination and improving the efficiency of redox reactions. The layered structure of BiOBr can effectively reduce the probability of recombination of photogenerated charge carriers; moreover, BiOBr has good electrical, optical, and magnetic properties. Density functional theory calculations show that under illumination, electrons in Br 4p of BiOBr are excited and pass through the K layer to reach the conduction band, making it difficult for holes and electrons to recombine. Wu et al. prepared a new photocatalyst, p-n heterojunction $BiOBr/La_2Ti_2O_7$, and used in situ FT-IR dynamic monitoring of the adsorption-reaction process of visible light oxidation of NO [15]. They found that the built-in electric field could accelerate the separation of e^--h^+ pairs, thus improving the photocatalytic efficiency. In addition, Ye et al. anchored Fe^{3+} sites on oxygen-deficient BiOBr nanosheets, effectively regulating charge/energy transfer and adsorption of reactants, thereby enhancing the dual functions of photocatalytic reaction and highly selective activation of O_2 to $^1O_2$ [16]. $BiPO_4$ is an n-type semiconductor with three crystal structures: low-temperature monoclinic phase, high-temperature monoclinic phase, and hexagonal phase. The low-temperature monoclinic phase has the highest photocatalytic efficiency, especially when combined with a staggered heterojunction formed with BiOBr, which promotes the separation of electron-hole pairs. Despite its small bandgap width of only 3.8–4.2 eV, $BiPO_4$ has a narrow absorption range and can only absorb UV light [17,18]. However, $BiPO_4$ contains the non-metallic oxide ${PO_4}^{3-}$, which induces the separation of photoinduced electron-hole pairs and the abundance of negatively charged ${PO_4}^{3-}$ facilitates this separation, thereby enhancing photocatalytic activity [19]. $BiPO_4$ nanoparticles are actually aggregates of $BiPO_4$ quantum dots with an average diameter of about 5 nm [20]. Therefore, ultrathin

Bi nanosheets can increase edge sites and promote the conversion of CO_2 to carboxylates, increasing the redox reaction sites and improving photocatalytic efficiency. Currently, research mainly focuses on adjusting the composite structure of $BiPO_4$ to improve its solar energy conversion efficiency and quantum yield [21]. Lu et al. used ion exchange to form in-situ Bi_2S_3 nanoparticles on the surface of $BiPO_4$ rods, achieving an 80% degradation rate of methylene blue under visible light irradiation within 3 h [22]. Zhang et al. prepared $Ag/BiPO_4$ composite photocatalysts using a combination of hydrothermal synthesis and impregnation methods, which showed much higher efficiency for the degradation of methylene blue under UV light irradiation compared to $BiPO_4$ alone [23]. Phosphates also provide clear directions and strategies for designing new photocatalysts. Oxygen vacancy defects are one of the important factors in improving photocatalytic efficiency [24–26]. In BiOBr, oxygen defect sites can capture photoexcited electrons, thereby reducing the recombination of electron-hole pairs and achieving spatial separation of photo-generated charge carriers. Introducing high-concentration oxygen vacancy defects can extend the range of the light response wavelength and increase photocatalytic activity. Research by Ren et al. showed that BiOX with oxygen vacancy defects can photocatalytically reduce CO_2 under simulated solar light irradiation, and oxygen vacancy defects can promote the adsorption and activation of CO_2 [27]. Chen et al. introduced oxygen vacancy (Ov) defects on the exposed surface of BiOBr nanosheets, modifying the band structure of BiOBr, enhancing its adsorption capacity, and lengthening the N≡N bond of adsorbed N_2 molecules [28].

This study aimed to prepare $BiPO_4$/Ov-BiOBr Type I heterojunction composites using a two-step solvothermal method to achieve exceptional visible-light photocatalytic performance for TC degradation and H_2O_2 generation [29–31]. The preparation parameters were adjusted to obtain optimal morphology, structure, and optical properties. The resulting composites were comprehensively characterized using various techniques. The photocatalytic degradation performance of the composites under visible-light irradiation was evaluated and compared with pure-phase $BiPO_4$ and Ov-BiOBr materials, revealing the advantages of heterojunction structures. The photocatalytic H_2O_2 generation performance of the composites under visible-light irradiation was investigated, and efforts were made to optimize photocatalytic efficiency by adjusting the composition and structure of the materials. The mechanisms of photocatalytic degradation and H_2O_2 generation were studied, with a focus on the generation, separation, migration, and consumption of photogenerated electron-hole pairs. Finally, the stability and reusability of the composites in photocatalytic processes were assessed through continuous cycling experiments, verifying their practicality and potential applications.

2. Results and Discussion

2.1. Crystal Structure Analysis

Figure 1 presents the X-ray diffraction (XRD) patterns of $BiPO_4$, Ov-BiOBr, and $BiPO_4$/Ov-BiOBr. The diffraction peaks of pure Ov-BiOBr and $BiPO_4$ in the XRD patterns match well with the standard data of BiOBr and $BiPO_4$ (BiOBr: PDF#73-2061; $BiPO_4$: PDF#15-0767). For $BiPO_4$/Ov-BiOBr composite, the diffraction peaks appearing at 2θ values of 21.4°, 25.3°, 29.1°, 31.2°, 38.6°, 41.6°, and 46.3° correspond to the (101), (111), (120), (012), (022), (031), and (310) crystal planes of $BiPO_4$, respectively, indicating the presence of $BiPO_4$ in the composite material [32,33]. The diffraction peaks appearing at 2θ values of 25.3°, 27.1°, 32.3°, 38.6°, and 46.3° correspond to the (111), (200), (110), (022), and (310) crystal planes of BiOBr, respectively, indicating the presence of BiOBr in the composite material [16,34]. Due to the effects of high temperature and ethylene glycol, the Bi-O bond on the $[Bi_2O_2]^{2+}$ layer breaks leading to the precipitation of some Bi atoms from the BiOBr lattice [34]. The diffraction peak at 2θ of 27.1° corresponds to the (012) crystal plane of Bi phase (PDF#85-1329). Furthermore, the significant decrease in the intensity of the (120) diffraction peak of $BiPO_4$ in the $BiPO_4$/Ov-BiOBr composite is likely due to the formation of a heterojunction structure between the (110) plane of $BiPO_4$ and BiOBr, thereby reducing its diffraction peak intensity. In Figure 1, it can be observed that the diffraction peaks of the composite materials are significantly shifted to the right, indicating the strong interaction between the two components in the composite.

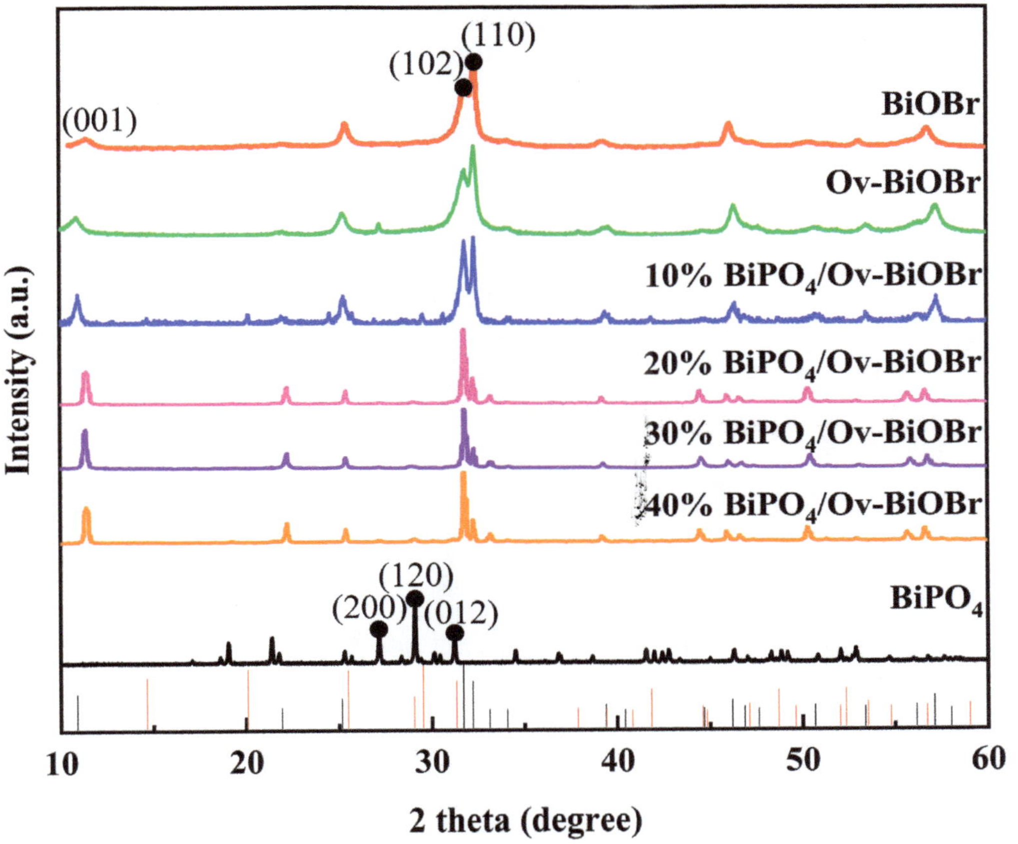

Figure 1. XRD patterns of Ov-BiOBr, $BiPO_4$ and $BiPO_4$/Ov-BiOBr composites.

2.2. TEM Analysis

Figure 2a shows the TEM image of the high-low heterojunction of $BiPO_4$/Ov-BiOBr, which reveals the nano-rod structure of $BiPO_4$ which is deposited on the spherical Ov-BiOBr. To confirm the nanorods as $BiPO_4$, the HRTEM images were analyzed, as shown in Figure 2b, which displays clear lattice fringes of $BiPO_4$ and good contact at the interface between $BiPO_4$ and Ov-BiOBr. The interplanar spacing of the (110) crystal plane in Ov-BiOBr was found to be 0.227 nm, while that of the (200) crystal plane in $BiPO_4$ was 0.327 nm [14,17]. At the interface of $BiPO_4$ and Ov-BiOBr, the main crystal plane of Bi was found to be (110), corresponding to interplanar spacing of 0.227 nm [13,34]. It is likely that the heterojunction was formed due to the in-situ deposition of $BiPO_4$ on the Bi sites in Ov-BiOBr [35–38]. The elemental mapping of Bi, Br, O, and P, as shown in Figure 2c–f, demonstrated that the O, Bi, Br, and P elements were evenly distributed on the $BiPO_4$/Ov-BiOBr composite photocatalyst [39]. These results indicate that the two-step hydrothermal method can effectively promote the in-situ deposition of $BiPO_4$ on Ov-BiOBr nanospheres, eventually forming the $BiPO_4$/Ov-BiOBr heterojunction.

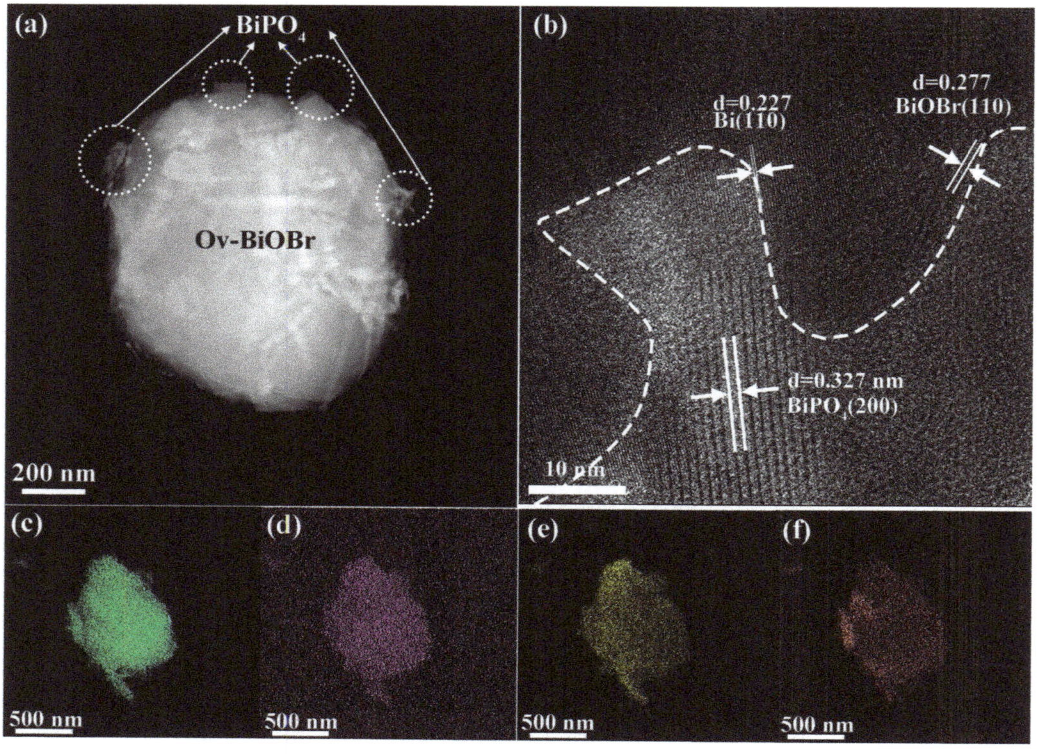

Figure 2. TEM image (**a**), HRTEM image (**b**), and elemental mapping of Bi (**c**), Br (**d**), O (**e**), and P (**f**) of $BiPO_4$/Ov-BiOBr.

2.3. XPS Analysis

The change in binding energy of elements can directly reflect the transfer of electrons and charge transfer processes in heterogeneous catalysts [40,41]. Therefore, monitoring the changes in binding energy can be used to study the direction of electron transfer in heterojunction photocatalysts [42]. In this study, in situ irradiation XPS analysis was performed on the $BiPO_4$/Ov-BiOBr composite before and after irradiation with a 300 W xenon lamp (420 nm < λ < 780 nm), and changes in the binding energies of Bi 4f, Br 3d, O 1s, and P 2p were observed. In the Bi 4f spectrum, the binding energy of Bi $4f_{5/2}$ decreased by 0.12 eV after irradiation, possibly due to electron transfer from PO_4^{3-} in $BiPO_4$ to BiOBr (Figure 3a) [17,43]. In the Br 3d spectrum, the binding energy of Br $3d_{5/2}$ decreased by 0.1 eV after irradiation, indicating a decrease in the electron density of BiOBr. These results indicate an increase in the electron density on BiOBr in the irradiated composite (Figure 3b). In the O 1s spectrum, the characteristic peak binding energies of lattice oxygen and surface hydroxyls at 533 eV and 529.9 eV increased by 0.1 eV after irradiation, possibly due to a decrease in the binding energy of OVs in the composite [44,45]. These changes indicate that the O in PO_4^{3-} provides electrons and transfers to BiOBr, and the O in the surface hydroxyls of $BiPO_4$ also provides electrons and participates in electron transfer (Figure 3c). In the P 2p spectrum, the binding energy of P 2p increased by 0.1 eV after irradiation, indicating an increase in the electron density of $BiPO_4$. This further confirms that electrons transfer from PO_4^{3-} in $BiPO_4$ to BiOBr (Figure 3d) [46]. Taken together, these results demonstrate the existence of the $BiPO_4$/Ov-BiOBr heterojunction and electron transfer between the heterojunctions.

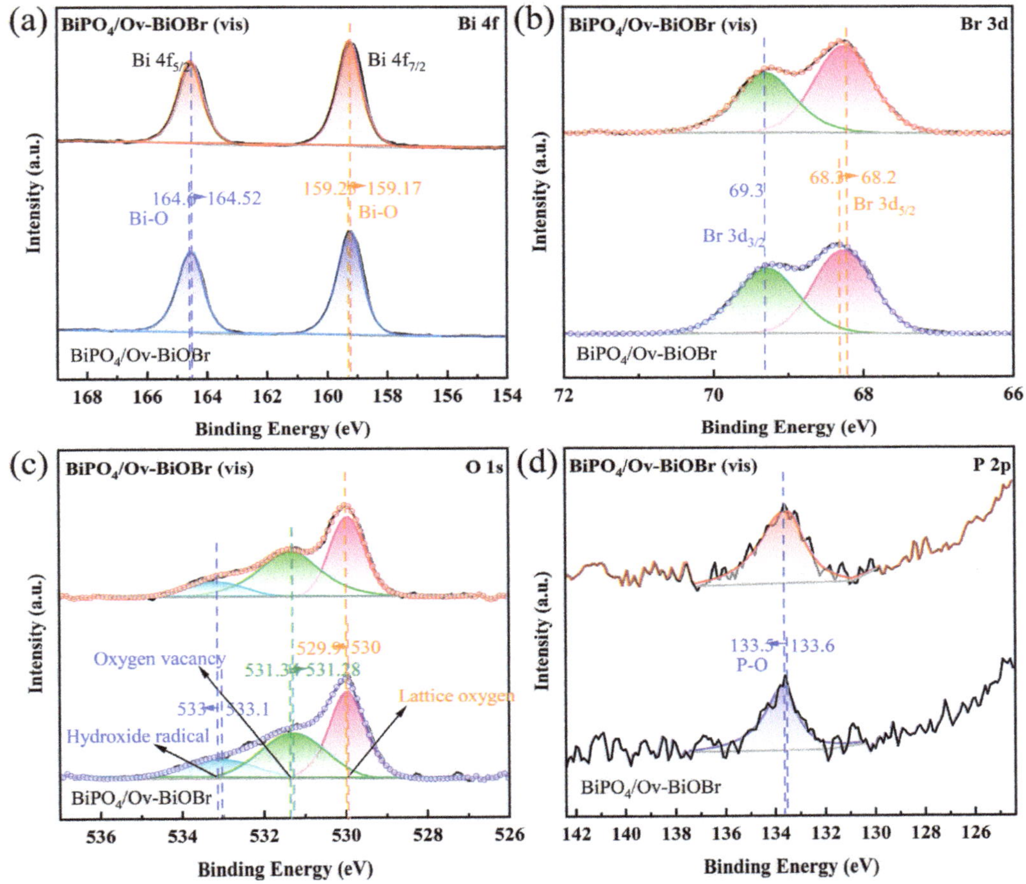

Figure 3. In situ irradiation XPS spectra of $BiPO_4$/Ov-BiOBr (**a**); high-resolution XPS maps of Bi 4f (**b**), Br 3d (**c**), O 1s (**d**) and P 2p (**e**).

Based on the experimental results, under visible light irradiation, the photogenerated electrons in the $BiPO_4$/Ov-BiOBr sample show a transfer direction from $BiPO_4$ to Ov-BiOBr. This finding provides strong evidence for the transfer direction of photogenerated electrons in $BiPO_4$/Ov-BiOBr heterojunction.

2.4. Raman Analysis

To verify the presence of precipitated metallic elemental Bi in Ov-BiOBr, Raman spectra of Ov-BiOBr and $BiPO_4$/Ov-BiOBr samples were analyzed (Figure 4). In the Raman spectrum of BiOBr, three distinct bands can be observed at 56 cm^{-1}, 112 cm^{-1}, and 374 cm^{-1}, respectively. The 56 cm^{-1} and 112 cm^{-1} Raman peaks correspond to the A1g external and Eg internal Bi-O stretching modes of BiOBr, while the relatively weaker peak at 374 cm^{-1} can be attributed to Eg and B1g bands caused by oxygen atom vibration [47]. For the $BiPO_4$/Ov-BiOBr composite, the band at 56 cm^{-1} can be attributed to the first-order scattering A1g mode of elemental Bi, the 112 cm^{-1} band corresponds to the A1g internal Bi-O stretching mode, and the 397 cm^{-1} band corresponds to Bi-O stretching [48,49]. These results further confirm the presence of precipitated metallic elemental Bi in Ov-BiOBr and the distinct Raman spectral features of the $BiPO_4$/Ov-BiOBr composite as compared to those of BiOBr.

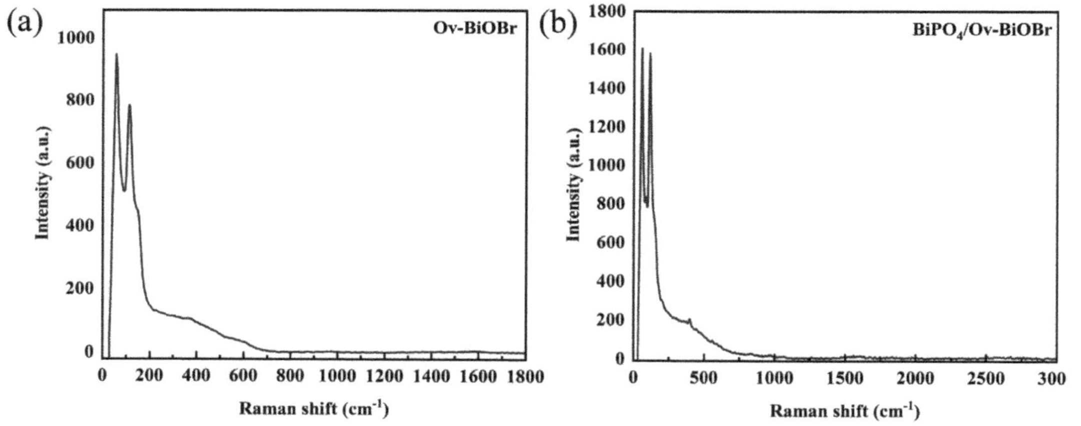

Figure 4. Raman spectra of Ov-BiOBr (**a**) and $BiPO_4$/Ov-BiOBr (**b**).

2.5. UV–Vis Diffuse Reflectance Spectra

To investigate the photoresponsive properties of $BiPO_4$/Ov-BiOBr photocatalyst, the band structure analysis is essential. The band structure of a photocatalyst is closely related to its photoresponsive properties. UV-Vis diffuse reflectance spectroscopy was used to analyze $BiPO_4$, Ov-BiOBr, and $BiPO_4$/Ov-BiOBr composite materials [50]. As shown in Figure 5a, the light absorption edge of $BiPO_4$ is between 300–400 nm, indicating that it is mainly responsive to UV light [51,52]. The light absorption edge of Ov-BiOBr is between 400–800 nm, indicating it has good visible light responsiveness. $BiPO_4$/Ov-BiOBr composite material has an absorption edge larger than 500 nm and strong visible light absorption characteristics. By converting the diffuse reflectance spectrum data, the band gap width diagrams of $BiPO_4$ and Ov-BiOBr can be obtained, as shown in Figure 5b. Extending the linear part of the curve to the intersection with the abscissa axis gives a band gap width of 4.2 eV and 2.7 eV for $BiPO_4$ and Ov-BiOBr, respectively. According to the formula Eg = E(VB) − E(CB), the valence band of $BiPO_4$ is 3.6 eV and the valence band of Ov-BiOBr is 2.22 eV. From the perspective of the band structure, n-type semiconductors with staggered band structures such as $BiPO_4$ and Ov-BiOBr can form effective heterojunctions.

2.6. Photoelectrochemical Properties

Band structure and light responsiveness are critical factors influencing the performance of photocatalysts. Therefore, in the photoelectrochemical experiments, we also detected and analyzed the Mott–Schottky (MS) curves of $BiPO_4$ and Ov-BiOBr [53]. Figure 6a,b show the MS curves at different frequencies, and the Erof values were calculated to be approximately −0.60 V and −0.38 V (vs. NHE, pH = 7) for $BiPO_4$ and Ov-BiOBr, respectively. In addition, we observed that the MS curves of $BiPO_4$ and Ov-BiOBr had a positive slope in the linear part, indicating that both materials have n-type semiconductor properties. According to the rule that the conduction band bottom potential of n-type semiconductors is approximately equal to the flat band potential, the conduction band potential (CB) of $BiPO_4$ was calculated to be −0.60 eV and that of Ov-BiOBr was −0.38 eV.

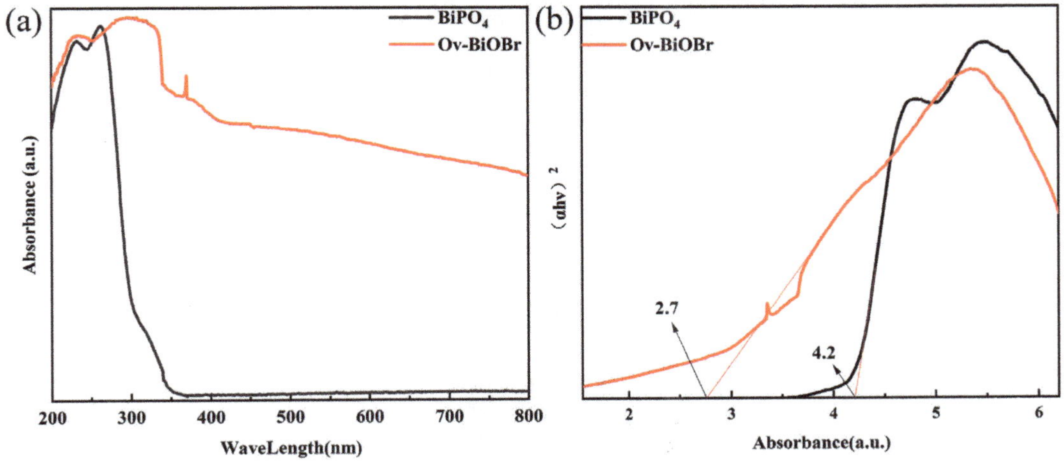

Figure 5. UV-Vis DRS (**a**) and forbidden band width (**b**) plots of the prepared samples.

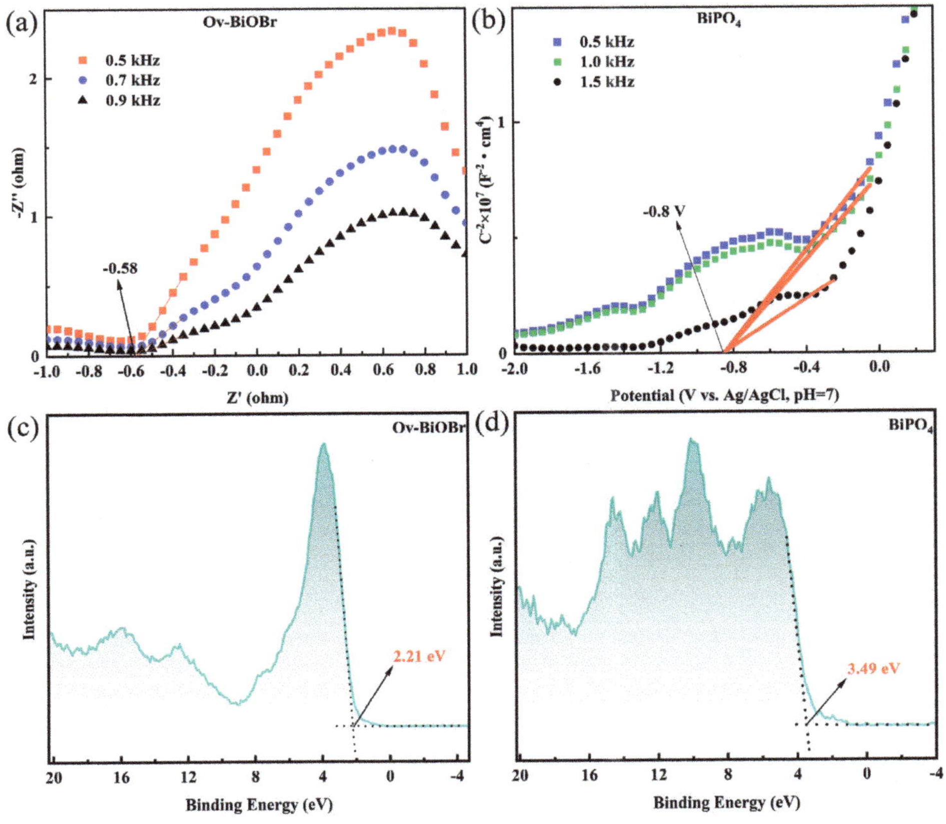

Figure 6. M-S plots (**a**,**b**) and VB-XPS (**c**,**d**) of Ov-BiOBr, $BiPO_4$.

Figure 6c,d show the VB-XPS spectra of Ov-BiOBr and $BiPO_4$. By reading the position of the intersection of the image, E_{vb} values of 2.21 and 3.49 eV for Ov-BiOBr and $BiPO_4$ were obtained, respectively. Considering the possibility of a potential difference between the sample and the instrument, we calculated the VB potential of Ov-BiOBr and $BiPO_4$ using the formula $E_{VB\text{-}NHE} = \varphi + E_{VB\text{-}XPS} - 4.44$, where φ is the instrument work function of the XPS analyzer (4.55 eV). Therefore, the EVB-NHE values of Ov-BiOBr and $BiPO_4$ are 2.32 and 3.6 eV, respectively. This is consistent with the results of the MS measurement of the valence band position. In conclusion, we have successfully determined the band structure of the $BiPO_4$/Ov-BiOBr heterojunction and demonstrated the successful formation of the $BiPO_4$/Ov-BiOBr composite with a type-I band alignment.

The photoelectrochemical measurements on $BiPO_4$, Ov-BiOBr, and 10% $BiPO_4$/Ov-BiOBr composite photocatalysts were performed. The photocurrent (Figure 7a) of pure $BiPO_4$ and pure Ov-BiOBr under visible light irradiation is 2 μA and 12 μA, respectively. Among all the composite catalysts, 10% $BiPO_4$/Ov-BiOBr displays the best photocurrent response (39 μA), which is three times that of Ov-BiOBr. In addition, the electrochemical impedance spectroscopy (EIS) diagram (Figure 7b) shows that 10% $BiPO_4$/Ov-BiOBr has smaller charge transfer resistance and larger transient photocurrent response. This indicates that the excellent photoelectrochemical performance of the 10% $BiPO_4$/Ov-BiOBr heterojunction can be attributed to its strong interface interaction, resulting in smaller photo-generated electron transfer resistance.

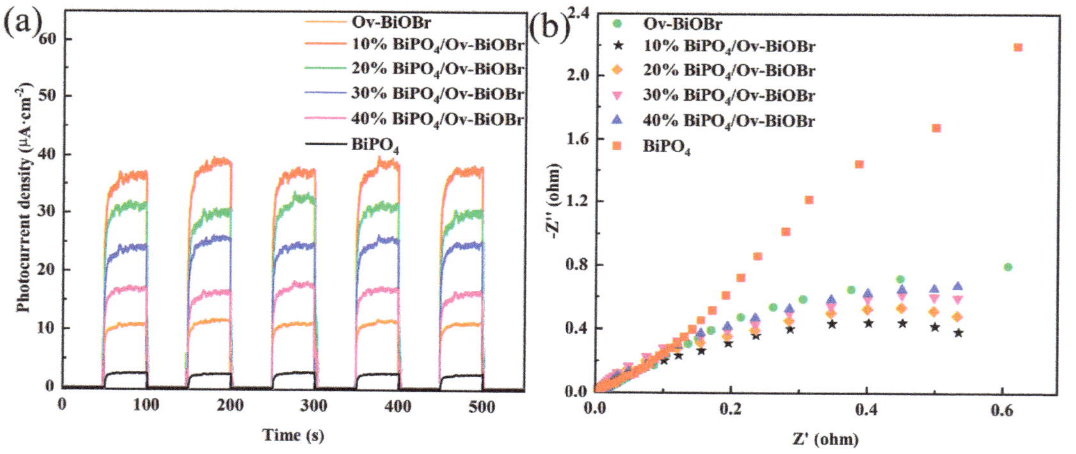

Figure 7. Photocurrent profiles (**a**) and impedance profiles (**b**) of the samples.

2.7. Photocatalytic Activity

The photocatalytic performance of samples was evaluated by the photocatalytic degradation experiment of tetracycline (TC). Compared to pure Ov-BiOBr, the composite material $BiPO_4$/Ov-BiOBr shows a significant enhancement in photocatalytic activity, with the 10% mol $BiPO_4$/Ov-BiOBr sample exhibiting the best degradation efficiency, degrading over 80% of TC under visible light irradiation in 30 min, far higher than other samples (Figure 8a). The photocatalytic degradation of TC by $BiPO_4$/Ov-BiOBr follows first-order reaction kinetics. As observed from Figure 8b, the reaction rate constants (K) of pure Ov-BiOBr, 10% $BiPO_4$/Ov-BiOBr, 20% $BiPO_4$/Ov-BiOBr, 30% $BiPO_4$/Ov-BiOBr, 40% $BiPO_4$/Ov-BiOBr, and pure $BiPO_4$ are 0.00944, 0.02534, 0.01376, 0.01043, 0.00991, and 0.00114 min^{-1}, respectively. The rate constant of 10% $BiPO_4$/Ov-BiOBr is 22, 2.68, 2.56, 2.42, and 1.84 times that of pure $BiPO_4$, pure Ov-BiOBr, 40% $BiPO_4$/Ov-BiOBr, 30% $BiPO_4$/Ov-BiOBr, and 20% $BiPO_4$/Ov-BiOBr, respectively. To detect the stability of $BiPO_4$/Ov-BiOBr, cyclic tests of TC photocatalytic degradation and XRD detection before and after illumination were carried out (Figure 8c,d). The experimental results show that after four cycles of photocatalytic degradation, the degradation rate of TC by 10% $BiPO_4$/Ov-BiOBr still reached 90%. Subsequently, 39 more cycles were conducted, and after 3.25 days, the degradation efficiency decreased to 45%. The experiments indicate that the synthesized catalyst has relatively high stability. Meanwhile, the XRD pattern remained unchanged, indicating the relatively stable photocatalytic activity and structural stability of 10% $BiPO_4$/Ov-BiOBr during the photocatalytic process. The comparison of the degradation performance of the samples prepared in this study with other reported Bi-based photocatalysts is shown in Table 1.

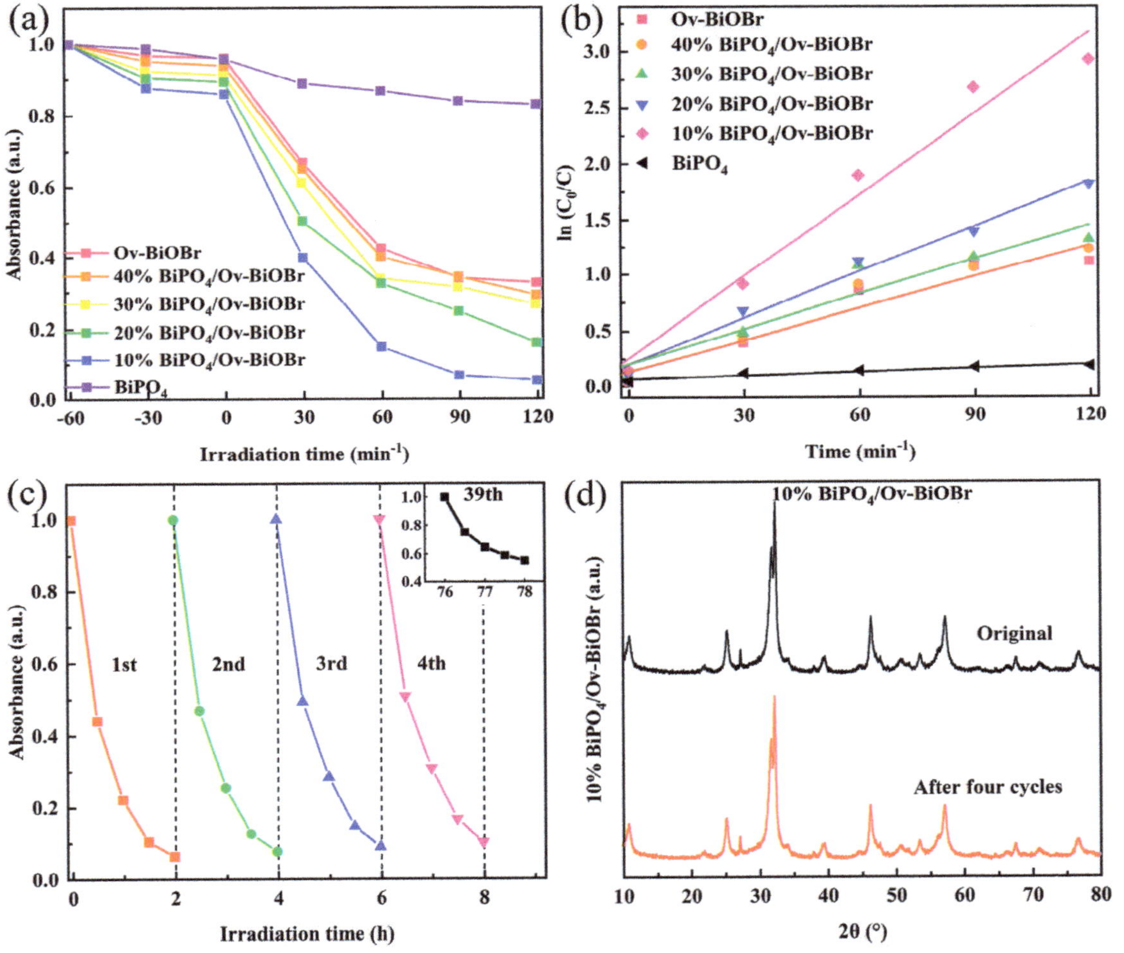

Figure 8. Photodegradation curve (**a**) and photodegradation kinetic curve (**b**) of TC; photodegradation cycle test curve (**c**) and XRD detection before and after illumination of $BiPO_4$/Ov-BiOBr (**d**).

Table 1. Recent work of BiOBr-based heterojunctions in degradation of TC.

Catalysts	Light Source (λ > 420 nm)	Concentration ($g \cdot L^{-1}$)	Pollutant ($mg \cdot L^{-1}$)	k (min^{-1})	References
$BiPO_4$/Ov-BiOBr	300 W Xe lamp	1	50(TC)	0.02534	This work
$W_{18}O_{49}$/BiOBr	300 W Xe lamp	1	20(TC)	0.014	[54]
UiO-66-NH_2/Bi/BiOBr	350 W Xe lamp	1	30(TC)	0.00838	[55]
CN@BOB-CQDs	LED light irradiation	1	20(TC)	0.221	[56]
BiOBr/Bi_2WO_6	500 W Xe lamp	1	10(TC)	0.0229	[57]
BiOBr/Bi_3O_4Br	500 W Xe lamp	1	10(TC)	0.02013	[58]

The study also investigated the photocatalytic activity of samples prepared under visible light irradiation (λ > 420 nm) for the in-situ production of H_2O_2 [59,60]. By substituting the absorbance peak values at different time points into the equation, the change in H_2O_2 concentration was calculated and a curve was plotted as shown in Figure 9. The results show that 10%$BiPO_4$/Ov-BiOBr has the best photocatalytic activity for producing H_2O_2, and after 120 min of visible light irradiation, the H_2O_2 concentration in the reactor reaches 139 $\mu mol \cdot L^{-1}$, which is 3.6 times that of BiOBr and 19 times that of $BiPO_4$. The free radical reaction process of this photocatalytic reaction includes $O_2 + 2e^- + 2H^+ \rightarrow H_2O_2$ (0.69 V vs. NHE) and $h^+ + OH^- \rightarrow \cdot OH + \cdot OH \rightarrow H_2O_2$ (2.29 V vs. NHE). Therefore, 10% $BiPO_4$/Ov-BiOBr obtained by solvothermal treatment may generate more photogenerated electrons and transfer them to the molecular surface for reacting with dissolved O_2. The successful construction of $BiPO_4$/Ov-BiOBr type I high-low junction retains the h^+ on the deep valence band of $BiPO_4$, thereby oxidizing OH^- to produce H_2O_2, which generates more H_2O_2 during the photocatalytic reaction. These results indicate that the nanoheterojunction photocatalyst with a more compact interface formed by solvothermal treatment is beneficial for improving the photocatalytic efficiency.

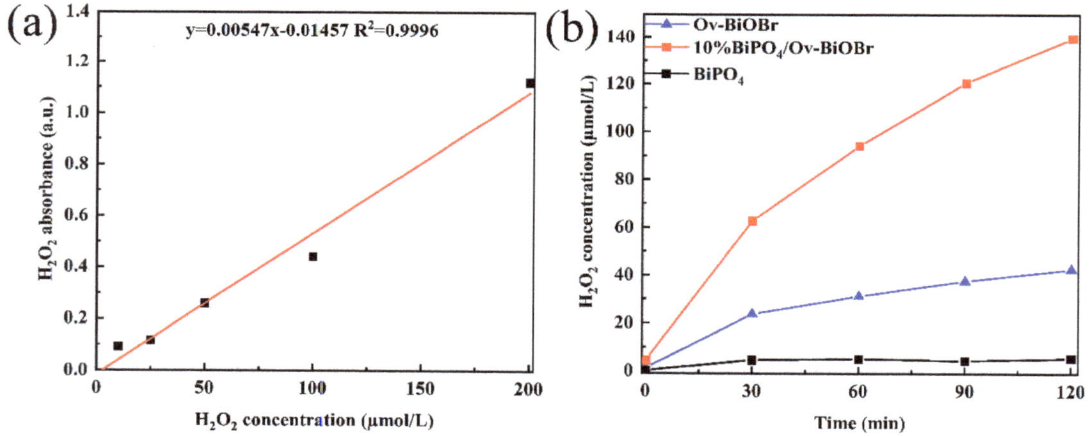

Figure 9. Concentration-absorbance curve of H_2O_2 (**a**); time-concentration curve of H_2O_2 produced under visible light irradiation (**b**).

2.8. Free Radical Capture

To further determine the main active species of $BiPO_4$/Ov-BiOBr in photocatalytic reactions, ESR and free radical capture experiments were conducted. The results are shown in Figure 10. IPA, BQ, and AO were used as free radical capture agents for $\cdot OH$, $\cdot O_2^-$, and h^+, respectively [18]. The experimental results show that $\cdot OH$ and $\cdot O_2^-$ are the main active species in the TC photocatalytic degradation process. In addition, Figure 10b,c show the characteristic peaks of DMPO-$\cdot OH$ and DMPO-$\cdot O_2^-$ observed under visible light irradiation, while no signal was detected in the dark state. This indicates that $\cdot OH$ and $\cdot O_2^-$

play a crucial role in the photocatalytic process of $BiPO_4$/Ov-BiOBr. The ESR free radical detection experiment further confirms the results of the free radical capture experiment, thereby determining that $\cdot OH$ and $\cdot O_2^-$ are the main active species in the photocatalytic degradation reaction of $BiPO_4$/Ov-BiOBr.

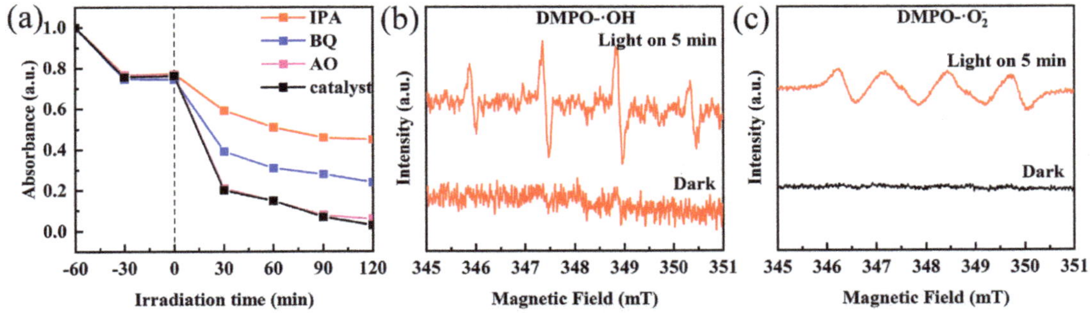

Figure 10. Experiments of radical trapping (**a**) ESR detection for 10% $BiPO_4$/Ov-BiOBr under dark and visible light irradiation: DMPO-$\cdot OH$ (**b**) and DMPO-$\cdot O_2^-$ (**c**).

2.9. Photocatalytic Mechanism

The difference in work function (Φ) or Fermi level (E_f) of semiconductor photocatalysts in heterojunction nanocomposites is considered an intrinsic driving force based on the VB-XPS measurement of the work function of Ov-BiOBr, $BiPO_4$, and $BiPO_4$/Ov-BiOBr; it can be observed that the material with a higher work function attracts electrons while the material with a lower work function loses electrons until the Fermi levels of both materials reach equilibrium. Consequently, an internal electric field (IEF) is formed at the interface, with the material with the higher work function being negatively charged and the material with the lower work function being positively charged [61,62]. Using the formula $\Phi = \Delta V + \varphi$ (where ΔV is the contact potential difference between the sample and the instrument sample holder, φ is the work function of the XPS analyzer, taken as 4.55 eV, and Φ is the work function of the sample) [30,63], the work functions of Ov-BiOBr, $BiPO_4$, and $BiPO_4$/Ov-BiOBr are calculated to be 6.18, 6.96, and 5.99 eV, respectively, and the distances between their two IP points are 1.63, 2.41, and 1.44 eV, respectively (Figure 11a–c). Since the Fermi level (E_{f1}) of $BiPO_4$ is lower than that of Ov-BiOBr (E_{f2}) (Figure 11d), when $BiPO_4$ comes into close contact with Ov-BiOBr, electrons will spontaneously diffuse from the material with the lower work function (Ov-BiOBr) to the material with the higher work function ($BiPO_4$) until the Fermi levels reach a new equilibrium (E_{f3}). After reaching equilibrium, an electron accumulation layer will form on the surface of $BiPO_4$, and an electron depletion layer will form on the surface of Ov-BiOBr, resulting in $BiPO_4$ being negatively charged and Ov-BiOBr being positively charged, thereby creating an IEF pointing from Ov-BiOBr to $BiPO_4$ (Figure 11e). This IEF provides a driving force for the directed movement of photogenerated charge carriers between different semiconductors. After photoexcitation, photogenerated electrons transfer from the conduction band of $BiPO_4$ to the conduction band of Ov-BiOBr, consistent with the results of in situ XPS measurement of electron transfer. Upon contact between the two semiconductors, the energy band of $BiPO_4$ and Ov-BiOBr will bend, leading to the migration of electrons from the conduction band of $BiPO_4$ to that of Ov-BiOBr. On the other hand, the photogenerated holes on the valence band of $BiPO_4$ will stay in place due to the band bending effect and Coulomb repulsion (Figure 11f). In summary, by matching the energy band structure, IEF, and energy band bending, the charge carriers with strong redox ability can be preserved, which creates conditions for improving photocatalytic activity.

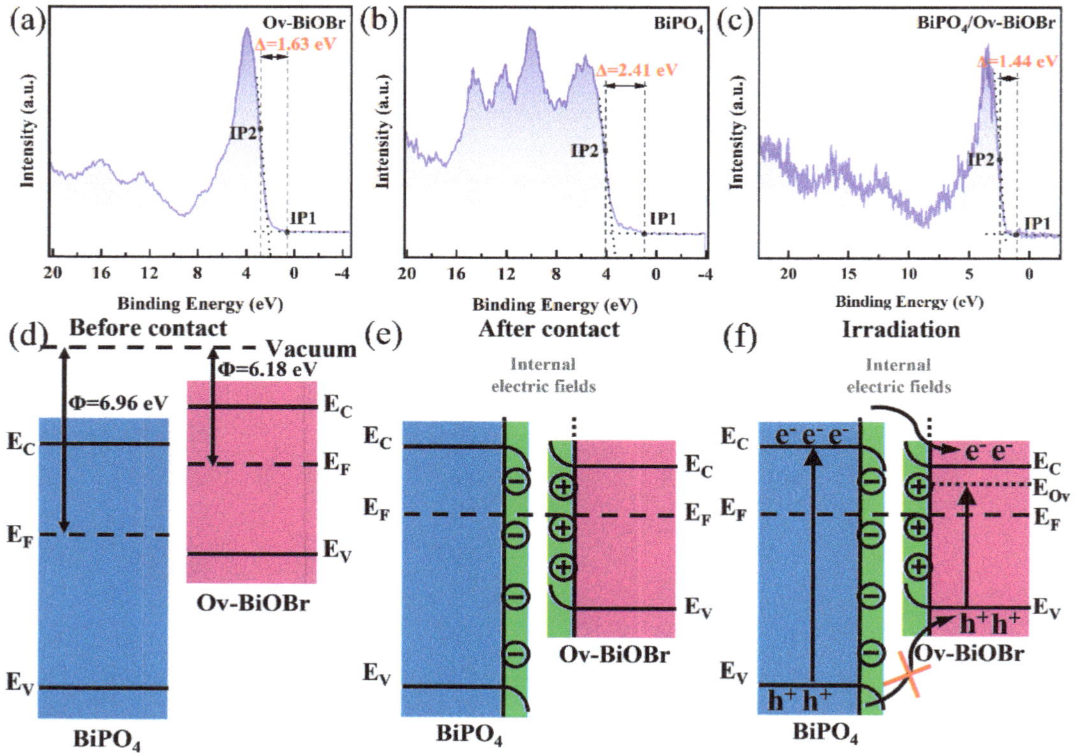

Figure 11. Work function of Ov-BiOBr, $BiPO_4$ and $BiPO_4$/Ov-BiOBr (**a–c**), carrier transfer mechanism of $BiPO_4$/Ov-BiOBr (**d–f**).

According to the experimental results, a possible reaction mechanism of the $BiPO_4$/Ov-BiOBr heterojunction photocatalyst is proposed (Figure 12). After illumination, the photogenerated electrons in $BiPO_4$ flow to Ov-BiOBr via the metallic Bi bridge, while the defect energy levels between the conduction and valence bands in Ov-BiOBr are excited [48,64–67]. Due to the effect of IEF and band bending, the holes in $BiPO_4$ cannot transfer to Ov-BiOBr. Therefore, the strongly oxidizing holes in $BiPO_4$ are retained. Since the conduction band potential of Ov-BiOBr is lower than that of dissolved oxygen, the photogenerated electrons on Ov-BiOBr can react with dissolved oxygen to produce $\cdot O_2^-$. In addition, the holes in $BiPO_4$ and Ov-BiOBr can effectively oxidize H_2O to generate ·OH. Since the ·OH has strong oxidizing properties, it can participate in the degradation of pollutants. Therefore, $\cdot O_2^-$ and ·OH are the main active species in the process of pollutant degradation. Based on the analysis of the energy band structure, electron transfer, work function, and active species detection results of the $BiPO_4$/Ov-BiOBr heterojunction, the possible charge transfer mechanism and reaction mechanism of the $BiPO_4$/Ov-BiOBr type I high-low junctions are proposed.

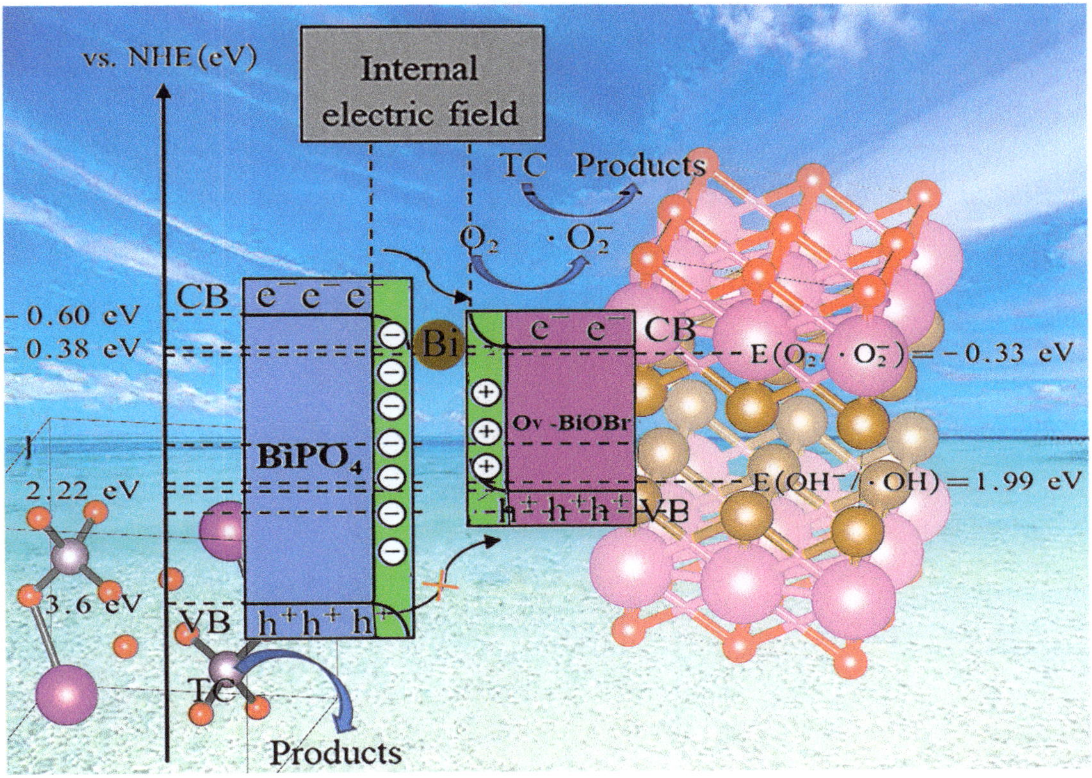

Figure 12. Photocatalytic mechanism of $BiPO_4$/Ov-BiOBr high-low junction.

3. Experimental Section

3.1. Chemicals

Bismuth nitrate pentahydrate ($Bi(NO_3)_3{\cdot}5H_2O$, 99%), potassium bromide (KBr, 99%), sodium dihydrogen phosphate (NaH_2PO_4, 99%), isopropanol (C_3H_8O), were purchased from Shanghai Tikem Industrial Co. Ethylene glycol ($(CH_2OH)_2$) (Shanghai, China) was purchased from Xilong Chemical Co. Tetracycline ($C_{22}H_{24}N_2O_8$) (Guangdong, China) was purchased from Shanghai Maclean Biochemical Technology Co. Anhydrous ethanol (CH_3CH_2OH) (Shanghai, China) was purchased from Xilong Science Co.

3.2. The Preparation of Ov-BiOBr

1 mmol $Bi(NO_3)_3{\cdot}5H_2O$ was added into 15 mL of ethylene glycol and stirred thoroughly to dissolve $Bi(NO_3)_3{\cdot}5H_2O$ completely in ethylene glycol, obtaining solution A [26]. Next, 1 mmol KBr was dissolved in 15 mL of ethylene glycol using the same method as for solution A, obtaining solution B. Solutions A and B were placed on a magnetic stirrer and stirred for 30 min at room temperature. Then, solution B was added dropwise into solution A and stirring was continued for 1 h. The mixed solution was transferred to a Teflon-lined autoclave and heated at 180 °C for 24 h. After cooling, the mixture was centrifuged at 9000 r/min for 5 min to ensure complete separation of the supernatant and the precipitate. The precipitate was repeatedly washed with ethanol and distilled water, and dried at 80 °C for 12 h. Finally, the sample was collected and ground as the precursor for the synthesis of the composite material [28,34,37].

3.3. The Preparation of $BiPO_4$ and $BiPO_4$/Ov-BiOBr

To prepare the $BiPO_4$/Ov-BiOBr heterojunction, a suitable amount of $Bi(NO_3)_3{\cdot}5H_2O$ was dispersed in 20 mL of distilled water. In another container, 50 mmol of Ov-BiOBr was dispersed in 20 mL of distilled water and sonicated for 30 min before being mixed with the $Bi(NO_3)_3{\cdot}5H_2O$ solution. Then, under vigorous stirring, an equimolar amount of NaH_2PO_4 compared to $Bi(NO_3)_3{\cdot}5H_2O$ was added dropwise into the mixed solution. The resulting mixture was transferred to a Teflon-lined high-pressure reaction vessel and heated at 170 °C for 12 h. After natural cooling, the sample was centrifuged to remove the supernatant and repeatedly washed with ethanol and distilled water. Finally, the sample was dried at 60 °C for 12 h. When preparing $BiPO_4$/Ov-BiOBr photocatalysts, different proportions of $BiPO_4$ (10%, 20%, 30%, and 40%) can be used to obtain various ratios. If Ov-BiOBr is not added, the resulting sample will be pure $BiPO_4$. The specific preparation process is shown in Figure 13.

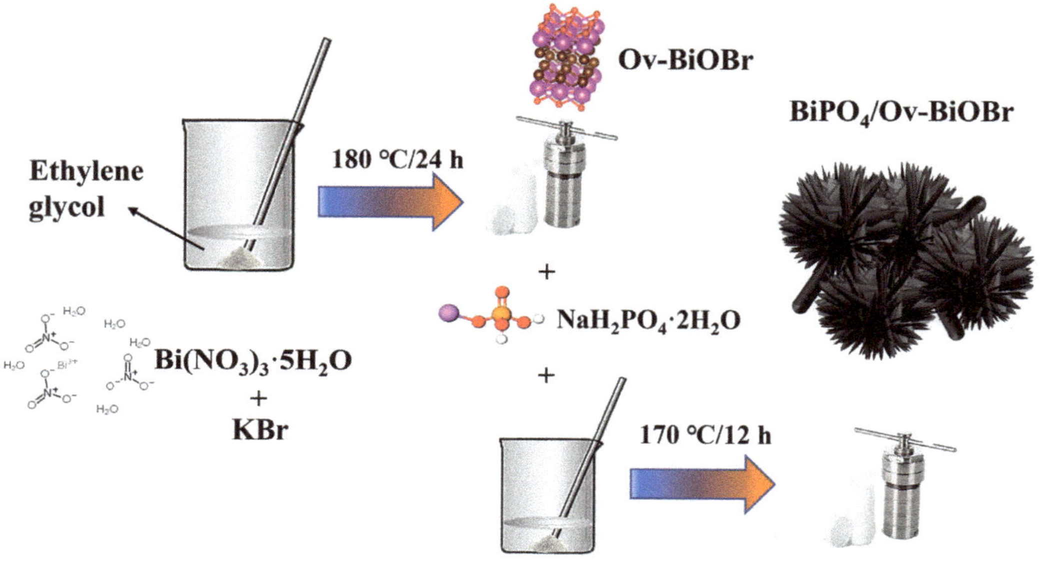

Figure 13. Preparation process of $BiPO_4$/Ov-BiOBr composite photocatalyst.

3.4. Chracterization

The crystal structures were analyzed using a D8 ADVANCE X-ray diffractometer (XRD, Bruker, Billerica, MA, USA). The morphology of $BiPO_4$/Ov-BiOBr was observed using a FEI Talos F200X transmission electron microscope (TEM, Waltham, MA, USA). Elemental analysis was performed using an Axis Ultra DLD X-ray photoelectron spectrometer (XPS, Kratos analytical, Manchester, UK). The Bi-O bond in Ov-BiOBr and $BiPO_4$/Ov-BiOBr was detected using Portman polarized Raman spectroscopy (e.g., Hengyuan, Shanghai, China). The UV-visible diffuse reflectance spectra (DRS) of Ov-BiOBr, $BiPO_4$, and $BiPO_4$/Ov-BiOBr were obtained using a UV-9000 H spectrophotometer (Metash, Shanghai, China) equipped with an integrating sphere. The photoelectrochemical properties of Ov-BiOBr, $BiPO_4$, and $BiPO_4$/Ov-BiOBr were analyzed using a CHI660E electrochemical analyzer (Chenhua, Shanghai, China) equipped with a three-electrode system. The degradation pathway of TC was analyzed using an LC-20A liquid chromatograph (Shimadzu, Kyoto, Japan). The detection of free radicals and characterization of defects in $BiPO_4$/Ov-BiOBr were tested using a JES-FA200 ESR spectrometer (JEOL, Tokyo, Japan).

3.5. Assessment of Photocatalytic Activity

In this experiment, the photocatalytic degradation performance of $BiPO_4$/Ov-BiOBr catalyst under visible light irradiation was investigated. Firstly, a 300 W xenon arc lamp (PLS-SXE 300, Beijing Perfectlight Technology Co., Ltd., Beijing, China) was fixed approximately 15 cm above a double-walled cylindrical quartz reactor. Then, 100 mg of photocatalyst powder with different ratios was mixed with 100 mL of tetracycline solution at a concentration of 50 $mg \cdot L^{-1}$. With the light source turned off or blocked, the suspension underwent 30 min of dark adsorption and was stirred in the dark with a magnetic stirrer to establish a dynamic equilibrium between adsorption and desorption prior to illumination. To verify whether tetracycline would decompose under light, a control experiment was conducted. In this experiment, no catalyst was added, and the degradation of tetracycline (TC) was directly analyzed using a UV-visible spectrophotometer. During the experiment, approximately 4 mL of the suspension was withdrawn with a pipette every 30 min and transferred to a test tube. The suspension was centrifuged twice at 9000 r/min for 5 min each time to ensure complete separation of the supernatant and precipitate. Then, the supernatant was transferred with a plastic dropper to a colorimetric dish for absorbance and concentration measurements of degradation products. The qualitative detection of the target compound concentration was performed by detecting the specific absorption spectrum generated by the sample molecule upon excitation by UV or visible light. Quantitative analysis was carried out by determining the specific concentration of the target compound through the posiive correlation between the concentration and absorbance intensity of the same substance at the maximum absorption peak, as described by the Lambert–Beer law.

3.6. Active Species Trapping Experiments

Under the conditions of isopropyl alcohol (IPA, 10 mmol/L) and ammonium oxalate (AO, 10 mmol/L), the $BiPO_4$/Ov-BiOBr heterojunction was used for photocatalytic degradation of TC, while detecting the generation of hydroxyl radicals ($\cdot OH$) and holes (h^+). To determine the presence of superoxide radicals ($\cdot O_2^-$), an appropriate amount of photocatalyst was added to benzenequinone (BQ, 0.1 mmol/L), and the same experimental procedure for evaluating photocatalytic activity was performed. Finally, according to the Lambert–Beer law, the concentration of TC solution was analyzed at 423 nm using an UV-visible spectrophotometer equipped with an integrating sphere. ESR testing was conducted to further verify free radicals.

4. Conclusions

In summary, this study successfully synthesized a $BiPO_4$/Ov-BiOBr type-I high-low junction photocatalyst using a two-step solvothermal method, which exhibited higher photocatalytic activity compared to single $BiPO_4$ and Ov-BiOBr. The experimental results demonstrated that the $BiPO_4$ content had a significant impact on the catalyst performance, with the optimal addition amount being 10 mol%. The prepared sample displayed the best catalytic activity, with a reaction rate constant of 0.02534 min^{-1}, which was 2.3 times that of pure Ov-BiOBr. This was attributed to the synergistic effect of the bandgap matching between $BiPO_4$ and Ov-BiOBr, including factors such as charge transfer channels and efficient charge separation. Through free radical scavenging experiments, it was found that $\cdot OH$ and $\cdot O_2^-$ free radicals were the main active species in the photocatalytic process. This study provides a new approach for the development of efficient visible-light type-I high-low junction photocatalysts.

Author Contributions: Conceptualization, X.L. and Y.Z.; methodology, J.H.; software, Z.F.; validation, X.L., Y.Z. and J.H.; formal analysis, F.D.; investigation, Z.C.; resources, Z.F.; data curation, Z.C.; writing—original draft preparation, M.T.; writing—review and editing, X.L.; visualization, L.H.; supervision, F.D.; project administration, Y.X.; funding acquisition, X.L. All authors have read and agreed to the published version of the manuscript.

Funding: The research was funded by the National Natural Science Foundation of China (Grant No. 22262024, 51962023, 51862024, 51978324).

Data Availability Statement: The data generated during and/or analysed during the study are available from the corresponding author on reasonable request.

Acknowledgments: The authors acknowledge the financial support from the National Natural Science Foundation of China (Grant No. 22262024, 51962023, 51862024, 51978324), Zhejiang Province Key Research and Development Project (2023 C01191), the Double Thousand Talent Plan of Jiangxi Province, the Jiangxi Academic and Technical Leader of Major Disciplines (20213BCJL22053), the Key Laboratory of Jiangxi Province for Persistent Pollutants Control and Resources Recycling (Nanchang Hangkong University) (Grant No. ES202002077), Postgraduate Innovation Special Fund of NCHU (Nanchang Hangkong University), China (YC2021-003).

Conflicts of Interest: The authors declare no conflict of interest.

References

1. Chen, J.; Abazari, R.; Adegoke, K.A.; Maxakato, N.W.; Bello, O.S.; Tahir, M.; Tasleem, S.; Sanati, S.; Kirillov, A.M.; Zhou, Y. Metal–Organic Frameworks and Derived Materials as Photocatalysts for Water Splitting and Carbon Dioxide Reduction. *Coord. Chem. Rev.* **2022**, *469*, 214664. [CrossRef]
2. Sun, L.; Han, L.; Huang, J.; Luo, X.; Li, X. Single-Atom Catalysts for Photocatalytic Hydrogen Evolution: A Review. *Int. J. Hydrog. Energy* **2022**, *47*, 17583–17599. [CrossRef]
3. Zhao, C.; Zhang, Y.; Jiang, H.; Chen, J.; Liu, Y.; Liang, Q.; Zhou, M.; Li, Z.; Zhou, Y. Combined Effects of Octahedron NH_2 -UiO-66 and Flowerlike $ZnIn_2S_4$Microspheres for Photocatalytic Dye Degradation and Hydrogen Evolution under Visible Light. *J. Phys. Chem. C* **2019**, *123*, 18037–18049. [CrossRef]
4. Sun, L.-J.; Su, H.-W.; Xu, D.-F.; Wang, L.-L.; Tang, H.; Liu, Q.-Q. Carbon Hollow Spheres as Cocatalyst of Cu-Doped TiO_2 Nanoparticles for Improved Photocatalytic H_2 Generation. *Rare Met.* **2022**, *41*, 2063–2073. [CrossRef]
5. Lu, X.; Chen, F.; Qian, J.; Fu, M.; Jiang, Q.; Zhang, Q. Facile Fabrication of CeF_3/g-C_3N_4 Heterojunction Photocatalysts with Upconversion Properties for Enhanced Photocatalytic Desulfurization Performance. *J. Rare Earths* **2021**, *39*, 1204–1210. [CrossRef]
6. Li, X.; Liu, J.; Huang, J.; He, C.; Feng, Z.; Chen, Z.; Wan, L.; Deng, F. All Organic S-Scheme Heterojunction PDI-Ala/S-C_3N_4 Photocatalyst with Enhanced Photocatalytic Performance. *Acta Phys.-Chim. Sin.* **2021**, *37*, 2010030. [CrossRef]
7. Zhang, J.; Li, Z.; Meng, T.; Zang, S.; Yang, C.; Luo, X.; Wang, H.; Chen, J.; Jing, F.; Wang, C.; et al. Monolithic All-Weather Solar-Thermal Interfacial Membrane Evaporator. *Chem. Eng. J.* **2022**, *450*, 137893. [CrossRef]
8. Dong, S.; Zhao, Y.; Yang, J.; Liu, X.; Li, W.; Zhang, L.; Wu, Y.; Sun, J.; Feng, J.; Zhu, Y. Visible-Light Responsive PDI/RGO Composite Film for the Photothermal Catalytic Degradation of Antibiotic Wastewater and Interfacial Water Evaporation. *Appl. Catal. B Environ.* **2021**, *291*, 120127. [CrossRef]
9. Li, S.; Cai, M.; Liu, Y.; Wang, C.; Yan, R.; Chen, X. Constructing $Cd_{0.5}Zn_{0.5}S/Bi_2WO_6$ S-Scheme Heterojunction for Boosted Photocatalytic Antibiotic Oxidation and Cr(VI) Reduction. *Adv. Powder Mater.* **2023**, *2*, 100073. [CrossRef]
10. Guo, Y.; Yan, B.; Deng, F.; Shao, P.; Zou, J.; Luo, X.; Zhang, S.; Li, X. Lattice Expansion Boosting Photocatalytic Degradation Performance of $CuCo_2S_4$ with an Inherent Dipole Moment. *Chin. Chem. Lett.* **2023**, *34*, 107468. [CrossRef]
11. Li, S.; Cai, M.; Wang, C.; Liu, Y.; Li, N.; Zhang, P.; Li, X. Rationally designed Ta_3N_5/BiOCl S-scheme heterojunction with oxygen vacancies for elimination of tetracycline antibiotic and Cr(VI): Performance, toxicity evaluation and mechanism insight. *J. Mater. Sci. Technol.* **2022**, *123*, 177–190. [CrossRef]
12. Binas, V.D.; Sambani, K.; Maggos, T.; Katsanaki, A.; Kiriakidis, G. Synthesis and Photocatalytic Activity of Mn-Doped TiO_2 Nanostructured Powders under UV and Visible Light. *Appl. Catal. B-Environ.* **2012**, *113*, 79–86. [CrossRef]
13. Li, X.; Kang, B.; Dong, F.; Deng, F.; Han, L.; Gao, X.; Xu, J.; Hou, X.; Feng, Z.; Chen, Z.; et al. BiOBr with Oxygen Vacancies Capture 0D Black Phosphorus Quantum Dots for High Efficient Photocatalytic Ofloxacin Degradation. *Appl. Surf. Sci.* **2022**, *593*, 153422. [CrossRef]
14. Gao, M.; Zhang, D.; Pu, X.; Ma, H.; Su, C.; Gao, X.; Dou, J. Surface Decoration of BiOBr with $BiPO_4$ Nanoparticles to Build Heterostructure Photocatalysts with Enhanced Visible-Light Photocatalytic Activity. *Sep. Purif. Technol.* **2016**, *170*, 183–189. [CrossRef]
15. Wu, H.; Chen, R.; Wang, H.; Cui, W.; Li, J.; Wang, J.; Yuan, C.; Zhuo, L.; Zhang, Y.; Dong, F. An Atomic Insight into BiOBr/$La_2Ti_2O_7$ p-n Heterojunctions: Interfacial Charge Transfer Pathway and Photocatalysis Mechanism. *Catal. Sci. Technol.* **2020**, *10*, 826–834. [CrossRef]
16. Ye, X.; Li, Y.; Luo, P.; He, B.; Cao, X.; Lu, T. Iron Sites on Defective BiOBr Nanosheets: Tailoring the Molecular Oxygen Activation for Enhanced Photocatalytic Organic Synthesis. *Nano Res.* **2022**, *15*, 1509–1516. [CrossRef]
17. Liu, Z.S.; Wu, B.T.; Niu, J.N.; Feng, P.Z.; Zhu, Y.B. $BiPO_4$/BiOBr p–n Junction Photocatalysts: One-Pot Synthesis and Dramatic Visible Light Photocatalytic Activity. *Mater. Res. Bull.* **2015**, *63*, 187–193. [CrossRef]

18. Shi, J.; Meng, X.; Hao, M.; Cao, Z.; He, W.; Gao, Y.; Liu, J. Enhanced Photoactivity of $BiPO_4$/(001) Facet-Dominated Square BiOBr Flakes by Combining Heterojunctions with Facet Engineering Effects. *J. Phys. Chem. Solids* **2018**, *113*, 142–150. [CrossRef]
19. An, W.; Cui, W.; Liang, Y.; Hu, J.; Liu, L. Surface Decoration of $BiPO_4$ with BiOBr Nanoflakes to Build Heterostructure Photocatalysts with Enhanced Photocatalytic Activity. *Appl. Surf. Sci.* **2015**, *351*, 1131–1139. [CrossRef]
20. Pan, C.; Li, D.; Ma, X.; Chen, Y.; Zhu, Y. Effects of Distortion of PO_4 Tetrahedron on the Photocatalytic Performances of $BiPO_4$. *Catal. Sci. Technol.* **2011**, *1*, 1399–1405. [CrossRef]
21. Xu, H.; Xu, Y.; Li, H.; Xia, J.; Xiong, J.; Yin, S.; Huang, C.; Wan, H. Synthesis, Characterization and Photocatalytic Property of AgBr/$BiPO_4$ Heterojunction Photocatalyst. *Dalton Trans.* **2012**, *41*, 3387–3394. [CrossRef]
22. Lu, M.; Yuan, G.; Wang, Z.; Wang, Y.; Guo, J. Synthesis of $BiPO_4$/Bi_2S_3 Heterojunction with Enhanced Photocatalytic Activity under Visible-Light Irradiation. *Nanoscale Res. Lett.* **2015**, *10*, 385. [CrossRef]
23. Zhang, Y.; Fan, H.; Li, M.; Tian, H. Ag/$BiPO_4$ Heterostructures: Synthesis, Characterization and Their Enhanced Photocatalytic Properties. *Dalton Trans.* **2013**, *42*, 13172–13178. [CrossRef]
24. Zhu, C.; Zhu, M.; Sun, Y.; Zhou, Y.; Gao, J.; Huang, H.; Liu, Y.; Kang, Z. Carbon-Supported Oxygen Vacancy-Rich Co_3O_4 for Robust Photocatalytic H_2O_2 Production via Coupled Water Oxidation and Oxygen Reduction Reaction. *ACS Appl. Energy Mater.* **2019**, *2*, 8737–8746. [CrossRef]
25. Li, Y.; Liu, T.; Cheng, Z.; Peng, Y.; Yang, S.; Zhang, Y. Facile Synthesis of High Crystallinity and Oxygen Vacancies Rich Bismuth Oxybromide Upconversion Nanosheets by Air-Annealing for UV-Vis-NIR Broad Spectrum Driven Bisphenol A Degradation. *Chem. Eng. J.* **2021**, *421*, 127868. [CrossRef]
26. Sun, Z.; Yang, X.; Yu, X.-F.; Xia, L.; Peng, Y.; Li, Z.; Zhang, Y.; Cheng, J.; Zhang, K.; Yu, J. Surface Oxygen Vacancies of Pd/Bi_2MoO_{6-x} Acts as "Electron Bridge" to Promote Photocatalytic Selective Oxidation of Alcohol. *Appl. Catal. B Environ.* **2021**, *285*, 119790. [CrossRef]
27. Ren, X.; Gao, M.; Zhang, Y.; Zhang, Z.; Cao, X.; Wang, B.; Wang, X. Photocatalytic Reduction of CO_2 on BiOX: Effect of Halogen Element Type and Surface Oxygen Vacancy Mediated Mechanism. *Appl. Catal. B Environ.* **2020**, *274*, 119063. [CrossRef]
28. Chen, X.; Zhang, X.; Li, Y.-H.; Qi, M.-Y.; Li, J.-Y.; Tang, Z.-R.; Zhou, Z.; Xu, Y.-J. Transition Metal Doping BiOBr Nanosheets with Oxygen Vacancy and Exposed {102} Facets for Visible Light Nitrogen Fixation. *Appl. Catal. B Environ.* **2021**, *281*, 119516. [CrossRef]
29. Hu, Y.; Li, X.; Wang, W.; Deng, F.; Han, L.; Gao, X.; Feng, Z.; Chen, Z.; Huang, J.; Zeng, F.; et al. Bi and S Co-doping g-C_3N_4 to Enhance Internal Electric Field for Robust Photocatalytic Degradation and H_2 Production. *Chin. J. Struct. Chem.* **2022**, *41*, 2206069–2206078. [CrossRef]
30. Xiong, J.; Li, X.; Huang, J.; Gao, X.; Chen, Z.; Liu, J.; Li, H.; Kang, B.; Yao, W.; Zhu, Y. CN/RGO@BPQDs High-Low Junctions with Stretching Spatial Charge Separation Ability for Photocatalytic Degradation and H_2O_2 Production. *Appl. Catal. B Environ.* **2020**, *266*, 118602. [CrossRef]
31. Yan, B.; Peng, J.; Deng, F.; Liu, L.; Li, X.; Shao, P.; Zou, J.; Zhang, S.; Wang, J.; Luo, X. Novel $ZnFe_2O_4$/Bi_2S_3 High-Low Junctions for Boosting Tetracycline Degradation and Cr(VI) Reduction. *Chem. Eng. J.* **2023**, *452*, 139353. [CrossRef]
32. Zhang, Z.; Liu, H.; Xu, J.; Zeng, H. CuTCPP/$BiPO_4$ Composite with Enhanced Visible Light Absorption and Charge Separation. *J. Photochem. Photobiol. A Chem.* **2017**, *336*, 25–31. [CrossRef]
33. Pan, C.; Xu, J.; Wang, Y.; Li, D.; Zhu, Y. Dramatic Activity of C_3N_4/$BiPO_4$ Photocatalyst with Core/Shell Structure Formed by Self-Assembly. *Adv. Funct. Mater.* **2012**, *22*, 1518–1524. [CrossRef]
34. Li, X.; Liu, Q.; Deng, F.; Huang, J.; Han, L.; He, C.; Chen, Z.; Luo, Y.; Zhu, Y. Double-Defect-Induced Polarization Enhanced O_V-BiOBr/$Cu_{2-X}S$ High-Low Junction for Boosted Photoelectrochemical Hydrogen Evolution. *Appl. Catal. B-Environ.* **2022**, *314*, 121502. [CrossRef]
35. Zhu, K.; Yao, W.; Zhu, Y. Preparation of Bismuth Phosphate Photocatalyst with High Dispersion by Refluxing Method. *Acta Phys. -Chim. Sin.* **2016**, *32*, 1519–1526. [CrossRef]
36. Jin, M.; Lu, S.; Ma, L.; Gan, M. One-Step Synthesis of in Situ Reduced Metal Bi Decorated Bismuth Molybdate Hollow Microspheres with Enhancing Photocatalytic Activity. *Appl. Surf. Sci.* **2017**, *396*, 438–443. [CrossRef]
37. Ma, H.; Zhao, M.; Xing, H.; Fu, Y.; Zhang, X.; Dong, X. Synthesis and Enhanced Photoreactivity of Metallic Bi-Decorated BiOBr Composites with Abundant Oxygen Vacancies. *J. Mater. Sci.-Mater. Electron.* **2015**, *26*, 10002–10011. [CrossRef]
38. Liu, J.; Wang, T.; Nie, Q.; Hu, L.; Cui, Y.; Tan, Z.; Yu, H. One-Step Synthesis of Metallic Bi Deposited Bi_2WO_6 Nanoclusters for Enhanced Photocatalytic Performance: An Experimental and DFT Study. *Appl. Surf. Sci.* **2021**, *559*, 149970. [CrossRef]
39. Zhu, J.; Li, Y.; Wang, X.; Zhao, J.; Wu, Y.; Li, F. Simultaneous Phosphorylation and Bi Modification of BiOBr for Promoting Photocatalytic CO_2 Reduction. *ACS Sustain. Chem. Eng.* **2019**, *7*, 14953–14961. [CrossRef]
40. Tougaard, S. Practical Guide to the Use of Backgrounds in Quantitative XPS. *J. Vac. Sci. Technol. A* **2021**, *39*, 011201. [CrossRef]
41. Catalin, N.; Cernea, M. Characterization of $BaTi_4O_9$ Ceramics by Raman Spectroscopy and XPS after Ion Etching. *J. Optoelectron. Adv. Mater.* **2006**, *8*, 1879–1883.
42. Xia, P.; Cao, S.; Zhu, B.; Liu, M.; Shi, M.; Yu, J.; Zhang, Y. Designing a 0D/2D S-Scheme Heterojunction over Polymeric Carbon Nitride for Visible-Light Photocatalytic Inactivation of Bacteria. *Angew. Chem. Int. Ed.* **2020**, *59*, 5218–5225. [CrossRef]

43. Zou, X.; Dong, Y.; Zhang, X.; Cui, Y.; Ou, X.; Qi, X. The Highly Enhanced Visible Light Photocatalytic Degradation of Gaseous o-Dichlorobenzene through Fabricating like-Flowers $BiPO_4$/BiOBr p-n Heterojunction Composites. *Appl. Surf. Sci.* **2017**, *391*, 525–534. [CrossRef]
44. Xu, L.; Jiang, D.; Zhao, Y.; Yan, P.; Dong, J.; Qian, J.; Ao, H.; Li, J.; Yan, C.; Li, H. Integrated $BiPO_4$ Nanocrystal/BiOBr Heterojunction for Sensitive Photoelectrochemical Sensing of 4-Chlorophenol. *Dalton Trans.* **2018**, *47*, 13353. [CrossRef] [PubMed]
45. Zhang, H.; Zhao, L.; Wang, L.; Hao, J.; Meng, X. Fabrication of Oxygen-Vacancy-Rich Black-BiOBr/BiOBr Heterojunction with Enhanced Photocatalytic Activity. *J. Mater. Sci.* **2020**, *55*, 10785–10795. [CrossRef]
46. Hu, X.; Li, C.; Song, J.; Zheng, S.; Sun, Z. Multidimensional Assembly of Oxygen Vacancy-Rich Amorphous TiO_2-BiOBr-Sepiolite Composite for Rapid Elimination of Formaldehyde and Oxytetracycline under Visible Light. *J. Colloid Interface Sci.* **2020**, *574*, 61–73. [CrossRef]
47. Sultana, R.; Gurjar, G.; Gahtori, B.; Patnaik, S.; Awana, V.P.S. Flux Free Single Crystal Growth and Detailed Physical Property Characterization of $Bi_{1-X}Sb_X$ (X = 0.05, 0.1 and 0.15) Topological Insulator. *Mater. Res. Express* **2019**, *6*, 106102. [CrossRef]
48. Deng, F.; Zhang, Q.; Yang, L.; Luo, X.; Wang, A.; Luo, S.; Dionysiou, D.D. Visible-Light-Responsive Graphene-Functionalized Bi-Bridge Z-Scheme Black BiOCl/Bi_2O_3 Heterojunction with Oxygen Vacancy and Multiple Charge Transfer Channels for Efficient Photocatalytic Degradation of 2-Nitrophenol and Industrial Wastewater Treatment. *Appl. Catal. B Environ.* **2018**, *238*, 61–69. [CrossRef]
49. Zhu, B.; Hong, X.; Tang, L.; Liu, Q.; Tang, H. Enhanced Photocatalytic CO_2 Reduction over 2D/1D $BiOBr_{0.5}Cl_{0.5}$/WO_3 S-Scheme Heterostructure. *Acta Phys. -Chim. Sin.* **2022**, *38*, 2111008. [CrossRef]
50. Dong, S.; Cui, L.; Tian, Y.; Xia, L.; Wu, Y.; Yu, J.; Bagley, D.M.; Sun, J.; Fan, M. A Novel and High-Performance Double Z-Scheme Photocatalyst ZnO-SnO_2-Zn_2SnO_4 for Effective Removal of the Biological Toxicity of Antibiotics. *J. Hazard. Mater.* **2020**, *399*, 123017. [CrossRef]
51. Kumar, A.; Sharma, S.K.; Sharma, G.; Al-Muhtaseb, A.H.; Naushad, M.; Ghfar, A.A.; Stadler, F.J. Wide Spectral Degradation of Norfloxacin by Ag@$BiPO_4$/BiOBr/$BiFeO_3$ Nano-Assembly: Elucidating the Photocatalytic Mechanism under Different Light Sources. *J. Hazard. Mater.* **2019**, *364*, 429–440. [CrossRef] [PubMed]
52. Chu, L.; Han, X.; Zhang, J.; Jiang, S.; Xue, R.; Dong, S.; Sun, J. $BiPO_4$ Coupled Bi^{3+} Doping SnO_2 QDs with Improved Performance towards Various Organic Pollutants Clearance. *Colloids Surf. A Physicochem. Eng. Asp.* **2021**, *612*, 125915. [CrossRef]
53. Li, X.; Xiong, J.; Gao, X.; Ma, J.; Chen, Z.; Kang, B.; Liu, J.; Li, H.; Feng, Z.; Huang, J. Novel BP/BiOBr S-Scheme Nano-Heterojunction for Enhanced Visible-Light Photocatalytic Tetracycline Removal and Oxygen Evolution Activity. *J. Hazard. Mater.* **2020**, *387*, 121690. [CrossRef] [PubMed]
54. Ji, S.; Dong, J.; Ji, M.; Zou, W.; Yin, S.; Chen, Z.; Xia, J. Rapid Dual-Channel Electrons Transfer via Synergistic Effect of LSPR Effect and Build-in Electric Field in Z-Scheme $W_{18}O_{49}$/BiOBr Heterojunction for Organic Pollutants Degradation. *Inorg. Chem. Commun.* **2022**, *138*, 109283. [CrossRef]
55. Zhang, X.; Cao, W.; Chen, C.; Jiang, C.; Wang, Y. Photocatalytic Activity and Degradation Mechanism of In-Situ Reduction Bi-Doped Self-Assembled 3D Flower-Sphere UiO-66-NH_2/BiOBr. *Opt. Mater. X* **2021**, *12*, 100090. [CrossRef]
56. Huang, L.; Liu, J.; Li, P.; Li, Y.; Wang, C.; Shu, S.; Song, Y. CQDs Modulating Z-Scheme g-C_3N_4/BiOBr Heterostructure for Photocatalytic Removing RhB, BPA and TC and E. *Coli by LED Light. J. Alloys Compd.* **2022**, *895*, 162637. [CrossRef]
57. Liu, Y.; He, J.; Qi, Y.; Wang, Y.; Long, F.; Wang, M. Preparation of Flower-like BiOBr/Bi_2WO_6 Z-Scheme Heterojunction through an Ion Exchange Process with Enhanced Photocatalytic Activity. *Mater. Sci. Semicond. Process.* **2022**, *137*, 106195. [CrossRef]
58. Zhang, F.; Xiao, X.; Xiao, Y. In Situ Fabrication of Type II 3D Hierarchical Flower-like BiOBr/Bi_3O_4Br Heterojunction with Improved Photocatalytic Activity. *J. Alloys Compd.* **2022**, *923*, 166417. [CrossRef]
59. Wang, W.; Li, X.; Deng, F.; Liu, J.; Gao, X.; Huang, J.; Xu, J.; Feng, Z.; Chen, Z.; Han, L. Novel Organic/Inorganic PDI-Urea/BiOBr S-Scheme Heterojunction for Improved Photocatalytic Antibiotic Degradation and H_2O_2 Production. *Chin. Chem. Lett.* **2022**, *33*, 5200–5207. [CrossRef]
60. Zan, Z.; Li, X.; Gao, X.; Huang, J.; Luo, Y.; Han, L. 0D/2D Carbon Nitride Quantum Dots (CNQDs)/BiOBr S-Scheme Heterojunction for Robust Photocatalytic Degradation and H_2O_2 Production. *Acta Phys.-Chim. Sin.* **2023**, *39*, 2209016. [CrossRef]
61. Yuan, J.; Liu, X.; Tang, Y.; Zeng, Y.; Wang, L.; Zhang, S.; Cai, T.; Liu, Y.; Luo, S.; Pei, Y.; et al. Positioning Cyanamide Defects in G-C_3N_4: Engineering Energy Levels and Active Sites for Superior Photocatalytic Hydrogen Evolution. *Appl. Catal. B Environ.* **2018**, *237*, 24–31. [CrossRef]
62. Dong, S.; Xia, L.; Chen, X.; Cui, L.; Zhu, W.; Lu, Z.; Sun, J.; Fan, M. Interfacial and Electronic Band Structure Optimization for the Adsorption and Visible-Light Photocatalytic Activity of Macroscopic $ZnSnO_3$/Graphene Aerogel. *Compos. Part B Eng.* **2021**, *215*, 108765. [CrossRef]
63. Li, X.; Kang, B.; Dong, F.; Zhang, Z.; Luo, X.; Han, L.; Huang, J.; Feng, Z.; Chen, Z.; Xu, J.; et al. Enhanced Photocatalytic Degradation and H_2/H_2O_2 Production Performance of S-PCN/$WO_{2.72}$ S-Scheme Heterojunction with Appropriate Surface Oxygen Vacancies. *Nano Energy* **2021**, *81*, 105671. [CrossRef]
64. Chung, D.Y.; Uher, C.; Kanatzidis, M.G. Sb and Se Substitution in $CsBi_4Te_6$: The Semiconductors CsM_4Q_6 (M = Bi, Sb; Q = Te, Se), $Cs_2Bi_{10}Q_{15}$, and $CsBi_5Q_8$. *Chem. Mater.* **2012**, *24*, 1854–1863. [CrossRef]

65. Jiang, X.; Yu, F.; Wu, D.; Tian, L.; Zheng, L.; Chen, L.; Chen, P.; Zhang, L.; Zeng, H.; Chen, Y.; et al. Isotypic heterojunction based on Fe-doped and terephthalaldehyde-modified carbon nitride for improving photocatalytic degradation with simultaneous hydrogen production. *Chin. Chem. Lett.* **2021**, *32*, 2782–2786. [CrossRef]
66. Bai, S.; Qiu, H.; Song, M.; He, G.; Wang, F.; Liu, Y.; Guo, L. Porous fixed-bed photoreactor for boosting C–C coupling in photocatalytic CO2 reduction. *eScience* **2022**, *2*, 428–437. [CrossRef]
67. Hao, W.; Teng, F.; Liu, Z.; Yang, Z.; Liu, Z.; Gu, W. Synergistic Effect of Mott-Schottky Junction with Oxygen Defect and Dramatically Improved Charge Transfer and Separation of Bi@Bi-Doped TiO_2. *Sol. Energy Mater. Sol. Cells* **2019**, *203*, 110198. [CrossRef]

MDPI AG
Grosspeteranlage 5
4052 Basel
Switzerland
Tel.: +41 61 683 77 34

Catalysts Editorial Office
E-mail: catalysts@mdpi.com
www.mdpi.com/journal/catalysts

www.ingramcontent.com/pod-product-compliance
Lightning Source LLC
Chambersburg PA
CBHW061558210926
51850CB00009B/2148